Recent Developments in Asymmetric Organocatalysis

RSC Catalysis Series

Series Editors:
Professor James J Spivey, *Louisiana State University, Baton Rouge, USA*

Advisor to the Board:
Krijn P de Jong, *University of Utrecht, The Netherlands*, James A Dumesic, *University of Wisconsin-Madison, USA*, Chris Hardacre, *Queen's University Belfast, Northern Ireland*, Enrique Iglesia, *University of California at Berkeley, USA*, Zinfer Ismagilov, *Boreskov Institute of Catalysis, Novosibirsk, Russia*, Johannes Lercher, *TU München, Germany*, Umit Ozkan, *Ohio State University, USA*, Chunshan Song, *Penn State University, USA*

Titles in the Series:
1: Carbons and Carbon Supported Catalysis in Hydroprocessing
2: Chiral Sulfur Ligands: Asymmetric Catalysis
3: Recent Developments in Asymmetric Organocatalysis

How to obtain future titles on publication:
A standing order plan is available for this series. A standing order will bring delivery of each new volume immediately on publication.

For further information please contact:
Book Sales Department, Royal Society of Chemistry, Thomas Graham House, Science Park, Milton Road, Cambridge, CB4 0WF, UK
Telephone: +44 (0)1223 420066, Fax: +44 (0)1223 420247, Email: books@rsc.org
Visit our website at http://www.rsc.org/Shop/Books/

Recent Developments in Asymmetric Organocatalysis

Hélène Pellissier
CNRS and Paul Cézanne University Aix-Marseille III, Marseille, France

RSCPublishing

RSC Catalysis Series No. 3

ISBN: 978-1-84973-054-9
ISSN: 1757-6725

A catalogue record for this book is available from the British Library

Published by The Royal Society of Chemistry,
Thomas Graham House, Science Park, Milton Road,
Cambridge CB4 0WF, UK

Registered Charity Number 207890

For further information see our web site at www.rsc.org

Preface

The enantioselective production of compounds is a central theme in current research. The broad utility of synthetic chiral molecules as single-enantiomer pharmaceuticals, in electronic and optical devices, as components in polymers with novel properties and as probes of biological function, has made asymmetric catalysis a prominent area of investigation. Until a few years ago, it was generally accepted that transition-metal complexes and enzymes were the two main classes of very efficient asymmetric catalysts. Synthetic chemists have scarcely used small organic molecules as catalysts throughout the last century, even though some of the very first asymmetric catalysts were purely organic molecules. Indeed, already in 1912, Bredig reported a modestly enantioselective alkaloid-catalysed cyanohydrin synthesis. In the 1960s, Pracejus showed that organocatalysts could give significant enantioselectivities. The 1970s brought a milestone in the area of asymmetric organocatalysis, when two industrial groups led by Hajos and Wiechert published the first and highly enantioselective catalytic aldol reactions using the simple amino acid, proline, as the catalyst. The cinchona alkaloids and proline stood as the only familiar organocatalysts for some time. In contrast to the relative inattention paid to organocatalysts by chemists, biological evolution has led to metal catalysis and organocatalysis in equal measure. While the end of the last century has been dominated by the use of metal catalysts,[1] a change in perception has occurred during the last few years, when several reports confirmed that relatively simple organic molecules could be highly effective and remarkably enantioselective catalysts of a variety of fundamentally important transformations. This rediscovery has initiated an explosive growth of research activities in organocatalysis, both in industry and in academia. As the realisation grows that organic molecules not only have ease of manipulation and a green advantage, but also can be very efficient catalysts, asymmetric organocatalysis may begin to catch up with the spectacular advances in enantioselective transition-metal catalysis. Thus, it was demonstrated that, besides the well-established asymmetric metal-complex-catalysed

RSC Catalysis Series No. 3
Recent Developments in Asymmetric Organocatalysis
By Hélène Pellissier
© Hélène Pellissier 2010
Published by the Royal Society of Chemistry, www.rsc.org

syntheses and biocatalysis, the use of pure organic catalysts turned out to be an additional efficient tool for the synthesis of chiral building blocks. Although the first examples were reported several decades ago,[2] the area of enantioselective organocatalysis became a main focus of research only recently. The last decade has seen an exponential growth in the field of asymmetric organocatalysis, and iminium-, enamine- and phosphoramide-based organocatalysis now allows cycloadditions, Michael additions, aldol reactions, nucleophilic substitutions and many other reactions with excellent enantioselectivities. The last few years have witnessed an explosive and impressive growth in the field, with new catalysts, novel methodologies for epoxidation, imine reduction or acylation, and mechanistic studies of aldol condensation, Mannich-type reactions, Michael addition, aza-Henry and Baylis–Hillman reactions and phase-transfer processes.

Organocatalysis, by now, has definitively matured to a recognised third methodology of potentially equal status to organometallic and enzymatic catalysis. Organocatalysts have several important advantages, since they are usually robust, inexpensive, readily available and non-toxic. Because of their inertness towards moisture and oxygen, demanding reaction conditions, *e.g.* inert atmospheres, low temperatures, absolute solvents, *etc.*, are, in many instances, not required. Because of the absence of transition metals, organocatalytic methods seem to be especially attractive for the preparation of compounds that do not tolerate metal contamination such as pharmaceutical products. A collection of chiral organocatalysts has already been involved in asymmetric synthesis, such as the prototypical example of proline, cinchona alkaloids such as quinine and various sugar-, amino acid- or peptide-derived compounds. Indeed, this type of catalysis has several serious advantages when compared to biocatalysis or to the use of transition-metal complexes as catalysts. The catalysts are usually more stable, less expensive and readily available, and can be applied in less-demanding reaction conditions. This type of catalyst can be easily incorporated onto a support, facilitating their recovery and recycling. Moreover, the absence of using a transition metal makes this type of reaction an attractive tool for the synthesis of agrochemical and pharmaceutical compounds, in which the presence of hazardous metallic traces is inadmissible in the final product. In this strategy, these factors contribute to superior atom efficiency, avoiding the protection of the substrate and deprotection of the products, and allowing the direct synthesis of structurally complex molecules, even through asymmetric multicomponent reactions[3] as well as domino,[4] tandem[5] or cascade transformations.[6] Consequently, this methodology will be important in industry, due to its versatility and its favourable environmental impact. This review is an update of the important use of such chiral metal-free organic catalysts in numerous reaction types, such as nucleophilic substitutions, and addition reactions as well as cycloadditions and redox reactions. Asymmetric organocatalysis has regularly been the subject of several excellent reviews by authors, such as List,[7] Dalko[8] and others.[9] This field has been reviewed by Dondoni and Massi, covering the literature up to the end of 2007.[10] In addition, Melchiorre *et al.* reported, in 2008, a review focusing on asymmetric aminocatalysis,[11] and a critical review was also published by

Palomo *et al.*[12] In a more specific context, asymmetric catalysis using chiral primary amine-based organocatalysts has been successively reviewed by Lu *et al.*,[13] and by Shao and Peng.[14] In addition, the organocatalytic formation of quaternary stereocentres was the subject of a review reported in 2009 by Bella and Gasperi.[15] The goal of this book is to cover the recent advances not yet updated in the use of chiral organocatalysts in asymmetric synthesis, focusing on those published since the beginning of 2008 and until the first weeks of 2009. The book is subdivided into ten chapters, according to the different types of reactions based on the use of chiral organocatalysts, such as nucleophilic additions to electron-deficient C=C double bonds, nucleophilic additions to C=O double bonds, nucleophilic additions to C=N double bonds, nucleophilic additions to unsaturated nitrogen, nucleophilic substitutions at aliphatic carbon, cycloaddition reactions, oxidations, reductions, kinetic resolutions and desymmetrisations, and miscellaneous reactions. The vast majority of organocatalytic reactions are amine-based reactions.[16] In this asymmetric aminocatalysis, amino acids, peptides, alkaloids and synthetic nitrogen-containing molecules have been used as chiral catalysts.

References

1. D. J. Ramon and M. Yus, *Chem. Rev.*, 2006, **106**, 2126–2208.
2. (a) Z. G. Hajos and D. R. Parrish, *J. Org. Chem.*, 1974, **39**, 1615–1621; (b) U. Eder, G. Sauer and R. Wiechert, *Ang. Chem., Int. Ed. Engl.*, 1971, **10**, 496–497.
3. D. J. Ramon and M. Yus, *Ang. Chem., Int. Ed. Engl.*, 2005, **44**, 1602–1634.
4. H. Pellissier, *Tetrahedron*, 2006, **62**, 1619–1665.
5. J.-C. Wasilke, S. J. Obrey, R. T. Baker and G. C. Bazan, *Chem. Rev.*, 2005, **105**, 1001–1020.
6. A. Padwa and M. D. Weingarten, *Chem. Rev.*, 1996, **96**, 223–269.
7. J. Seayad and B. List, *Org. Biomol. Chem.*, 2005, **3**, 719–724.
8. (a) P. I. Dalko and L. Moisan, *Ang. Chem., Int. Ed. Engl.*, 2001, **40**, 3726–3748; (b) P. Dalko and L. Moisan, *Ang. Chem., Int. Ed. Engl.*, 2004, **43**, 5138–5175; (c) P. I. Dalko, in *Enantioselective Organocatalysis*, Wiley-VCH, Weinheim, Germany, 2007; (d) P. I. Dalko, *Chimia*, 2007, **61**, 213–218.
9. (a) A. Berkessel and H. Gröger, in *Asymmetric Organocatalysis – From Biomimetic Concepts to Powerful Methods for Asymmetric Synthesis*, Wiley-VCH, Weinheim, 2005; (b) M. S. Taylor and E. N. Jacobsen, *Ang. Chem., Int. Ed. Engl.*, 2006, **45**, 1520–1543; (c) H. Pellissier, *Tetrahedron*, 2007, **63**, 9267–9331; (d) A. G. Doyle and E. N. Jacobsen, *Chem. Rev.*, 2007, **107**, 5713–5743; (e) M. G. Gaunt, C. C. C. Johansson, A. McNally and N. C. Vo, *Drug Discovery Today*, 2007, **2**, 8–27; (f) D. W. C. MacMillan, *Nature*, 2008, **455**, 304–308; (g) X. Yu and W. Wang, *Chem. Asian J.*, 2008, **3**, 516–532.
10. A. Dondoni and A. Massi, *Ang. Chem., Int. Ed. Engl.*, 2008, **47**, 4638–4660.

11. P. Melchiorre, M. Marigo, A. Carlone and G. Bartoli, *Ang. Chem., Int. Ed. Engl.*, 2008, **47**, 6138–6171.
12. C. Palomo, M. Oiarbide and R. Lopez, *Chem. Soc. Rev.*, 2009, **38**, 632–653.
13. L.-W. Xu, J. Luo and Y. Lu, *Chem. Commun.*, 2009, 1807–1821.
14. F. Peng and Z. Shao, *J. Mol. Catal. A.*, 2008, **285**, 1–13.
15. M. Bella and T. Gasperi, *Synthesis*, 2009, **10**, 1583–1614.
16. (a) B. List, *Tetrahedron*, 2002, **58**, 5573–5590; (b) E. R. Jarvo and L. Miller, *Ang. Chem., Int. Ed. Engl.*, 2002, **58**, 2481–2495; (c) B. Westermann, *Ang. Chem., Int. Ed. Engl.*, 2003, **42**, 151–153; (d) B. List, *Chem. Commun.*, 2006, 819–824; (e) G. Lelais and D. W. C. MacMillan, *Aldrichimica Acta*, 2006, **39**, 79–87.

Contents

Chapter 1 Nucleophilic Additions to Electron-deficient C=C Double Bonds 1

   1.1 Intermolecular Michael Additions of C-Nucleophiles 3
      1.1.1 Intermolecular Michael Additions of C-Nucleophiles Catalysed by Proline Derivatives 3
      1.1.2 Intermolecular Michael Additions of C-Nucleophiles Catalysed by Non-proline Derivatives 13
   1.2 Intermolecular Domino Michael Additions of C-Nucleophiles 26
   1.3 Intermolecular Nitro-Michael Additions of C-Nucleophiles 37
   1.4 Intermolecular Domino Nitro-Michael Additions of C-Nucleophiles 54
   1.5 Intermolecular Michael Additions of Other-than-C-Nucleophiles 58
   1.6 Intermolecular Domino Michael Additions of Other-than-C-Nucleophiles 62
   1.7 Intramolecular Michael Additions 68
   1.8 Conclusions 70

Chapter 2 Nucleophilic Additions to C=O Double Bonds 77

   2.1 Aldol Reactions 77
      2.1.1 Aldol Reactions Catalysed by Proline Derivatives 77

RSC Catalysis Series No. 3
Recent Developments in Asymmetric Organocatalysis
By Hélène Pellissier
© Hélène Pellissier 2010
Published by the Royal Society of Chemistry, www.rsc.org

2.1.2   Aldol Reactions Catalysed by Non-proline
        Derivatives   94
2.2   Miscellaneous Reactions   105
2.3   Conclusions   117

Chapter 3   Nucleophilic Additions to C=N Double Bonds   123

3.1   Mannich Reactions   123
3.2   Aza-Henry Reactions   135
3.3   Aza-Morita–Baylis–Hillman Reactions   139
3.4   Strecker Reactions   141
3.5   Conclusions   145

Chapter 4   Nucleophilic Additions to Unsaturated Nitrogen   150

4.1   Nucleophilic Additions to N=N Double Bonds   150
4.2   Nucleophilic Additions to N=O Double Bonds   152
4.3   Conclusions   156

Chapter 5   Nucleophilic Substitutions at Aliphatic Carbon   158

5.1   α-Halogenations of Carbonyl Compounds   158
5.2   α-Alkylations of Carbonyl Compounds and
      Derivatives   160
5.3   α-Aminations of Carbonyl Compounds   167
5.4   Miscellaneous Reactions   168
5.5   Conclusions   170

Chapter 6   Cycloaddition Reactions   172

6.1   Diels–Alder Reactions   172
6.2   Miscellaneous Cycloadditions   179
6.3   Conclusions   189

Chapter 7   Oxidations   192

7.1   Epoxidations of Alkenes   192
7.2   Epoxidations of α,β-Unsaturated Carbonyl Compounds   192
7.3   Other Oxidations   196
7.4   Conclusions   200

Chapter 8   Reductions   202

8.1   Reductions of Imines and Ketones   202
8.2   Other Reductions   207
8.3   Conclusions   209

Chapter 9   Kinetic Resolutions and Desymmetrisations   **213**

Chapter 10   Miscellaneous Reactions   **220**

General Conclusion   **232**

Subject Index   **234**

Abbreviations

Ac	acetyl
Ad	adamantyl
Aib	2-aminoisobutyric acid
Ala	alanine
Ar	aryl
BINAM	1,1'-binaphthalenyl-2,2'-diamine
BINOL	1,1'-bi-2-naphthol
Bmim	1-butyl-3-methylimidazolium
Boc	*tert*-butoxycarbonyl
Bu	butyl
Bz	benzoyl
CAN	cerium ammonium nitrate
Cy	cyclohexyl
Cbz	benzyloxycarbonyl
Cp	cyclopentadienyl
Cy	cyclohexyl
DABCO	1,4-diazabicyclo[2.2.2]octane
BarF	tetrakis[3,5-bis(trifluoromethyl)phenyl]borate
DBAD	dibenzylazodicarboxylate
DBSA	dodecylbenzenesulfonic acid
DBU	1,8-diazabicyclo[5.4.0]undec-7-ene
DCC	dicyclohexylcarbodiimide
DCE	dichloroethane
de	diastereomeric excess
DEAD	diethyl azodicarboxylate
Dec	decanyl
DHQD	dihydroquinidine
DIPEA	diisopropylethylamine
DMAP	4-(dimethylamino)pyridine
DME	dimethoxyethane
DMF	dimethylformamide
DMSO	dimethylsulfoxide
DNP	2,4-dinitrophenol
DPEN	1,2-diphenylethylenediamine
dr	diastereomeric ratio

ee	enantiomeric excess
er	enantiomeric ratio
Et	ethyl
FBSM	1-fluorobis(phenylsulfonyl)methane
Fu	furyl
GABA	γ-amino butyric acid
Hept	heptyl
Hex	hexyl
HFE	hydrofluoroether
HMDS	hexamethyldisilazide
Leu	leucine
Me	methyl
Mes	mesyl
MNP	magnetic nanoparticle
MOM	methoxymethyl
M.S.	molecular sieves
MTBE	methyl *tert*-butyl ether
NADH	nicotinamide adenine dinucleotide
Naph	naphthyl
Nf	nonaflate
NFSI	*N*-fluorobenzenesulfonamide
NIS	*N*-iodosuccinimide
NMP	*N*-methylpyrrolidinone
Non	nonyl
Ns	nosyl
Oct	octyl
PCC	pyridinium chlorochromate
Pent	pentyl
PG	protecting group
Ph	phenyl
Phe	phenylalanine
Piv	pivalate
PMB	*para*-methoxybenzyl
PMP	*para*-methoxyphenyl
Pr	propyl
Pro	proline
PS	polystyrene
Py	pyridine
SOMO	single occupied molecular orbital
TBDPS	*tert*-butyldiphenylsilyl
TBHP	*tert*-butylhydrogenperoxide
TBS	*tert*-butyldimethylsilyl
TCCA	trichloro acid isocyanuric
TEA	triethylamine
TEAB	triethylammonium bicarbonate
Tf	trifluoromethanesulfonyl

CHAPTER 1
Nucleophilic Additions to Electron-deficient C=C Double Bonds

The conjugate addition of nucleophiles to electron-poor alkenes is one of the most frequently used C–C and C–heteroatom bond-forming reactions in organic synthesis.[1] The catalytic asymmetric version of this reaction employing chiral catalysts has been widely developed over the last few years.[2] In particular, the use of chiral organocatalysts has been subjected to a spectacular development in recent years.[3] In the Michael addition, a nucleophile Nu⁻ is added to the β-position of an α,β-unsaturated acceptor. The active nucleophile Nu⁻ is usually generated by deprotonation of the precursor NuH. The mechanistic scheme implies that enantioface differentiation in the addition to the β-carbon atom of the acceptor can be achieved in two ways, either deprotonation of NuH with a chiral base, resulting in a chiral ion pair, which can be expected to add to the acceptor asymmetrically, or phase-transfer catalysis, in which deprotonation of NuH is achieved in one phase with an achiral base and the anion Nu⁻ is transported into the organic phase by a chiral phase-transfer catalyst, also resulting in a chiral ion pair from which asymmetric β-addition may proceed. These methods of providing a chiral environment for the attacking nucleophile can be regarded as the classical ways of approaching asymmetric organocatalysis of Michael additions (Scheme 1.1).

Two highly efficient and very practical alternatives have emerged in recent years (Scheme 1.2). One of these approaches consists of activating the acceptors by reversible conversion into a chiral iminium ion. Thus, the reversible condensation of an α,β-unsaturated carbonyl compound with a chiral secondary amine provides a chiral α,β-unsaturated iminium ion. A face-selective reaction with the nucleophile provides an enamine, which can be either reacted with an electrophile and then hydrolysed, or just hydrolysed to a β-chiral carbonyl compound. The second approach is the enamine pathway. If the nucleophile is

RSC Catalysis Series No. 3
Recent Developments in Asymmetric Organocatalysis
By Hélène Pellissier
© Hélène Pellissier 2010
Published by the Royal Society of Chemistry, www.rsc.org

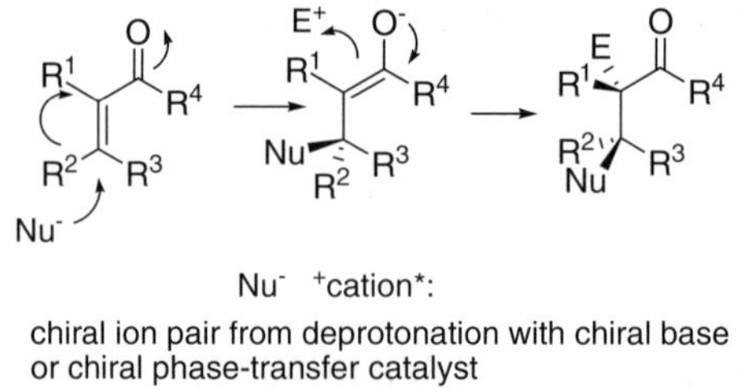

Nu⁻ ⁺cation*:
chiral ion pair from deprotonation with chiral base
or chiral phase-transfer catalyst

Scheme 1.1   Use of chiral bases and phase-transfer catalysis in Michael reactions.

activation of enone by iminium ion formation

activation of carbonyl by enamine formation

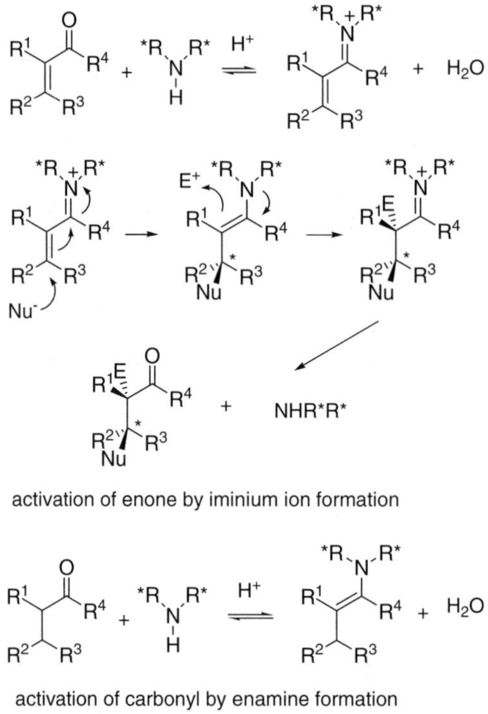

Scheme 1.2   Enamine and iminium catalysis in Michael reactions.

an enolate anion, it can be replaced by a chiral enamine, formed reversibly from the original carbonyl compound and a chiral secondary amine. It is apparent that enamine and iminium catalysis are based on the same origin. Enamine catalysis proceeds *via* iminium ion formation and almost always results in iminium ion formation. In an opposing but complementary fashion, iminium catalysis typically results in the formation of an enamine intermediate. The two catalytic intermediates are opposite, yet interdependent, and they consume and support each other.

1.1 Intermolecular Michael Additions of C-Nucleophiles

1.1.1 Intermolecular Michael Additions of C-Nucleophiles Catalysed by Proline Derivatives

The first part of this chapter deals with Michael additions of C-nucleophiles evolving *via* enamine or iminium ion intermediates.[4] The most successful catalyst for enamine-type reactions is the cheap, natural, simple and readily available amino acid, L-proline, which has been defined in the recent past as a "universal catalyst". Proline can react as a nucleophile with carbonyl groups or Michael acceptors to form iminium ions or enamines. The high enantioselectivities generally observed in proline-mediated reactions can be rationalised by the capacity of this molecule to promote the formation of highly organised transition states with extensive hydrogen-bonding networks.[5] There are several reasons why proline has become an important molecule in asymmetric catalysis, *e.g.* it is an abundant chiral molecule which is inexpensive and available in both enantiomeric forms. A wide series of modifications of the structure of proline have been accomplished with the aim of improving the solubility and/or enhancing the acidity of the directing acid proton. Many successes have been achieved by applying organocatalysts such as proline derivatives to highly reactive Michael donors or acceptors. In contrast, Michael additions of simple aldehydes to enones have received little attention. In this context, Cordova *et al.* have employed an α,α-diarylprolinol ether to promote the first highly enantioselective catalytic conjugate addition of aldehydes to both aliphatic and aromatic alkylidenemalonates.[6] The reaction gave access to β-formyl-substituted malonates with high yields combined with diastereoselectivities of up to 86% de and enantioselectivities of up to 99% ee, as shown in Scheme 1.3.

This organocatalyst and the corresponding trifluoromethyl-substituted derivative have been applied by Zhu and Lu to the asymmetric Michael reaction of aldehydes with other Michael acceptors, such as vinyl sulfones, providing the corresponding Michael products with exceptional enantioselectivities and excellent yields (Scheme 1.4).[7] In order to make this methodology more useful, the authors extended its scope to 2-aryl-substituted vinyl sulfones as acceptors, which yielded the corresponding Michael products in excellent yields, good diastereoselectivities and nearly perfect enantioselectivities, as shown in Scheme 1.4.

In order to explain the stereoselectivity of this reaction, the authors have proposed a plausible transition-state model depicted in Scheme 1.5, in which the bulky biaryl silyl ether moiety exerted steric shielding, resulting in the formation of the observed stereoisomer. This enantioselective addition to vinyl sulfones, in combination with desulfonation, offered a unique, asymmetric entry to α-alkylated aldehydes and their derivatives.

Similar Michael additions of various aldehydes to vinyl sulfones to some of those depicted above have also been developed by Alexakis *et al.* in the presence of an aminal-pyrrolidine organocatalyst derived from proline, albeit leading to

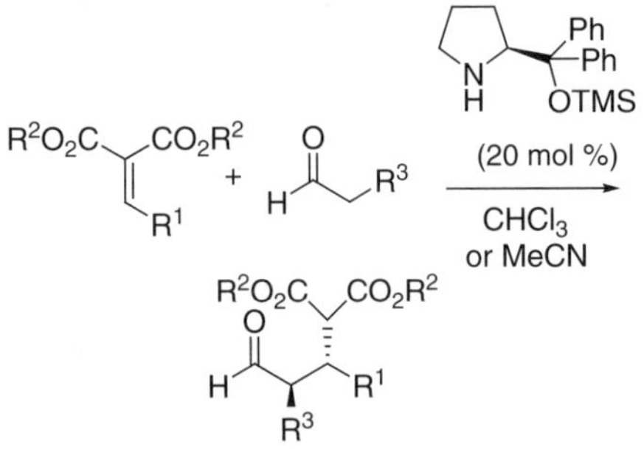

R^1 = p-NO$_2$C$_6$H$_4$, R^2 = R^3 = Me: 88% de = 82% ee = 95%
R^1 = m-NO$_2$C$_6$H$_4$, R^2 = R^3 = Me: 81% de = 66% ee = 97%
R^1 = p-ClC$_6$H$_4$, R^2 = R^3 = Me: 79% de = 66% ee = 95%
R^1 = p-BrC$_6$H$_4$, R^2 = R^3 = Me: 76% de = 74% ee = 94%
R^1 = p-BrC$_6$H$_4$, R^2 = Bn, R^3 = Me: 68% de = 74% ee = 95%
R^1 = Ph, R^2 = R^3 = Me: 96% de = 74% ee > 99%
R^1 = p-NO$_2$C$_6$H$_4$, R^2 = Me, R^3 = Et: 77% de = 86% ee = 99%
R^1 = p-NO$_2$C$_6$H$_4$, R^2 = Me, R^3 = CH$_2$CH=CH$_2$:
86% de = 72% ee = 98%
R^1 = n-Pr, R^2 = R^3 = Me: 84% de = 60% ee = 95%

Scheme 1.3 Silylated biphenylprolinol-catalysed Michael additions of aldehydes to alkylidenemalonates.

both lower enantioselectivities (ee $\leq 91\%$) and yields (Scheme 1.6).[8] In general, the best enantioselectivities were obtained with the more substituted aldehydes with the exception of the more bulky 3,3-dimethylbutyraldehyde, which gave only 75% ee probably due to a strong interaction between the *tert*-butyl group and the catalyst. Moreover, a poor enantiocontrol (16% ee) was observed in the case of using an α,α-disubstituted aldehyde.

In addition, aldehydes have been added by Jorgensen *et al.* to ethyl 2-(diethoxyphosphoryl)acrylate in the presence of 2-[bis(3,5-bistrifluoromethyl-phenyl)trimethylsilylanyloxymethyl]pyrrolidine as the organocatalyst.[9] The obtained Michael products were not purified, but submitted directly to a chemoselective reduction with NaBH$_4$ in methanol, affording the corresponding δ-hydroxyalkanoates, which were cyclised to the corresponding α-diethoxy-phosphoryl-δ-lactones under acidic conditions. These lactones, obtained in good yields as mixtures of epimers, were finally submitted to a Horner–Wadsworth–Emmons olefination process by treatment with formaldehyde, affording the target α-methylene-δ-lactones in good yields and excellent enantioselectivities (Scheme 1.7). As an extension of this methodology, the Michael products could be converted into the corresponding α-methylene-δ-lactams by a reductive amination with benzylamine. The formed δ-aminoalkanoates underwent a spontaneous lactamisation, yielding α-diethoxyphosphoryl-δ-lactams as mixtures of epimers. These products led to the final α-methylene-δ-lactams *via* a Horner–Wadsworth–Emmons olefination process with formaldehyde in high enantioselectivities of up to 94% ee (Scheme 1.7).

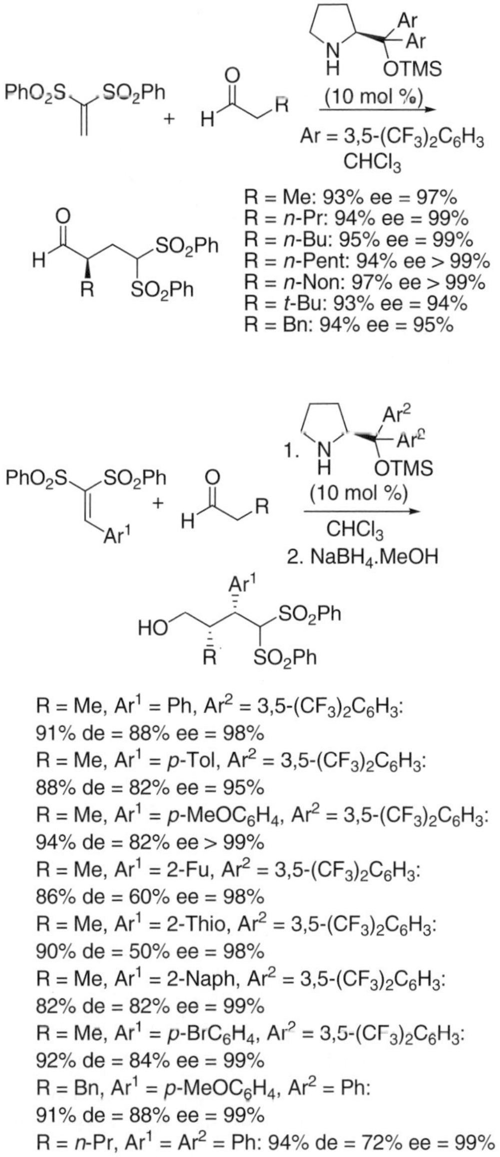
Ar
Ar
N
H
OTMS
PhO2S
SO2Ph
O
+ H
R
(10 mol %)
Ar = 3,5-(CF3)2C6H3
CHCl3
O
H
SO2Ph
R
SO2Ph
R = Me: 93% ee = 97%
R = n-Pr: 94% ee = 99%
R = n-Bu: 95% ee = 99%
R = n-Pent: 94% ee > 99%
R = n-Non: 97% ee > 99%
R = t-Bu: 93% ee = 94%
R = Bn: 94% ee = 95%
Ar2
Ar2
1. N
H
OTMS
PhO2S
SO2Ph
+
Ar1
O
H
R
(10 mol %)
CHCl3
2. NaBH4.MeOH
Ar1
HO
SO2Ph
R
SO2Ph
R = Me, Ar1 = Ph, Ar2 = 3,5-(CF3)2C6H3:
91% de = 88% ee = 98%
R = Me, Ar1 = p-Tol, Ar2 = 3,5-(CF3)2C6H3:
88% de = 82% ee = 95%
R = Me, Ar1 = p-MeOC6H4, Ar2 = 3,5-(CF3)2C6H3:
94% de = 82% ee > 99%
R = Me, Ar1 = 2-Fu, Ar2 = 3,5-(CF3)2C6H3:
86% de = 60% ee = 98%
R = Me, Ar1 = 2-Thio, Ar2 = 3,5-(CF3)2C6H3:
90% de = 50% ee = 98%
R = Me, Ar1 = 2-Naph, Ar2 = 3,5-(CF3)2C6H3:
82% de = 82% ee = 99%
R = Me, Ar1 = p-BrC6H4, Ar2 = 3,5-(CF3)2C6H3:
92% de = 84% ee = 99%
R = Bn, Ar1 = p-MeOC6H4, Ar2 = Ph:
91% de = 88% ee = 99%
R = n-Pr, Ar1 = Ar2 = Ph: 94% de = 72% ee = 99%

Scheme 1.4 Silylated biarylprolinol-catalysed Michael additions of aldehydes to vinyl sulfones.

On the other hand, most organocatalysed Michael additions of stabilised carbon nucleophiles have involved either nucleophiles or electrophiles that are highly activated. As an example, Michael additions of highly activated nucleophiles, such as malonates,[10] to α,β-unsaturated aldehydes have been reported. Therefore, Ma *et al.* have reported the Michael addition of malonates

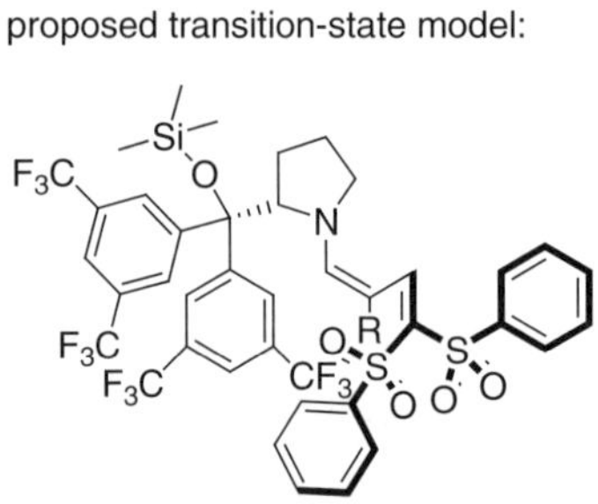
proposed transition-state model:
Si
O
F3C
N
F3C
F3C
CF3
O
R
S
O
S
O
O

Scheme 1.5 Plausible transition-state model for Michael addition of aldehydes to vinyl sulfones.

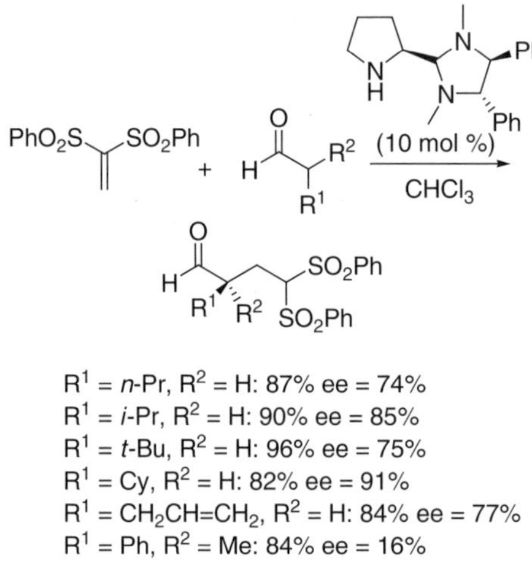
N
Ph
N
H
N
Ph
PhO2S
SO2Ph
O
+
H
R2
R1
(10 mol %)
CHCl3
O
H
SO2Ph
R1
R2
SO2Ph
R1 = n-Pr, R2 = H: 87% ee = 74%
R1 = i-Pr, R2 = H: 90% ee = 85%
R1 = t-Bu, R2 = H: 96% ee = 75%
R1 = Cy, R2 = H: 82% ee = 91%
R1 = CH2CH=CH2, R2 = H: 84% ee = 77%
R1 = Ph, R2 = Me: 84% ee = 16%

Scheme 1.6 Michael additions of aldehydes to vinyl sulfones catalysed by an aminal-pyrrolidine.

to α,β-unsaturated aldehydes catalysed by *O*-TMS protected diphenylprolinol combined with acetic acid in water.[11] A wide range of aldehydes including β-aryl, β-alkyl and β-alkenyl acroleins were found to be compatible with these conditions, providing the corresponding adducts in good yields and with good to excellent enantioselectivities, as shown in Scheme 1.8. It must be noted that the short reaction time (less than 24 h) used was remarkable. These advantages presumably resulted from the combination of Brønsted acids as promoters and water as the reaction medium.[12]

In 2008, Barbas *et al.* developed the first enantioselective thioester Michael addition of simple trifluoroethyl thioesters, thereby establishing a new class of nucleophiles for direct catalytic reactions.[13] Indeed, these nucleophiles were condensed onto a series of α,β-unsaturated aldehydes in the presence of 2-[bis(3,5-bistrifluoromethylphenyl)trimethylsilylanyloxymethyl]pyrrolidine as an

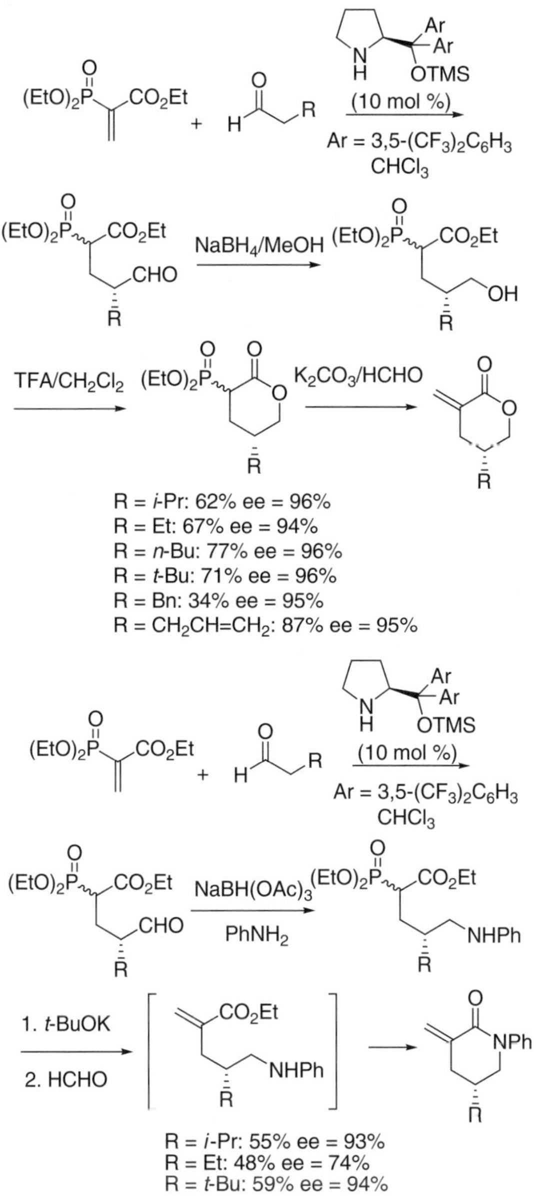

Scheme 1.7 Syntheses of α-methylene-δ-lactones and α-methylene-δ-lactams.

organocatalyst, and benzoic acid as a co-catalyst, providing the corresponding Michael products in good yields and moderate to high enantioselectivities of up to 98% ee, albeit in modest diastereoselectivities in favour of the *anti*-product (Scheme 1.9).

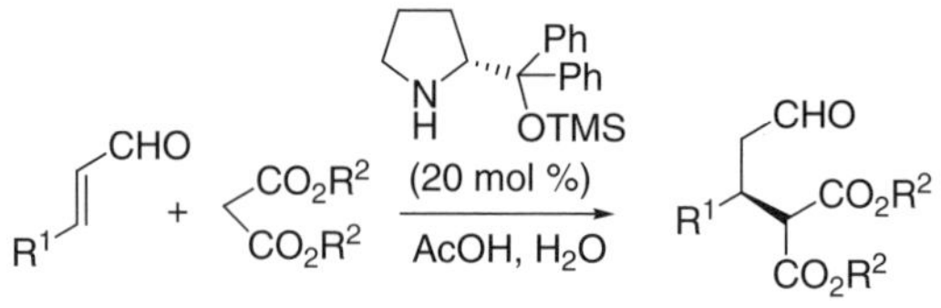

R^1 = Ph, R^2 = Me: 90% ee = 96%
R^1 = Ph, R^2 = Bn: 93% ee = 95%
R^1 = 2-Fu, R^2 = Me: 92% ee = 88%
R^1 = 2-Fu, R^2 = Bn: 85% ee = 90%
R^1 = *p*-MeOC$_6$H$_4$, R^2 = Me: 81% ee = 91%
R^1 = *p*-MeOC$_6$H$_4$, R^2 = Bn: 88% ee = 96%
R^1 = *p*-FC$_6$H$_4$, R^2 = Me: 78% ee = 97%
R^1 = *p*-BrC$_6$H$_4$, R^2 = Me: 75% ee = 94%
R^1 = *p*-BrC$_6$H$_4$, R^2 = Bn: 95% ee = 92%
R^1 = *p*-BrC$_6$H$_4$, R^2 = Me: 74% ee = 90%
R^1 = *p*-NO$_2$C$_6$H$_4$, R^2 = Me: 78% ee = 92%
R^1 = Me, R^2 = Bn: 55% ee = 79%
R^1 = *i*-Pr, R^2 = Bn: 70% ee = 95%
R^1 = (*E*)-CH=CH(Me), R^2 = Bn: 62% ee = 82%
R^1 = (*E*)-CH=CH(*n*-Pent), R^2 = Bn: 67% ee = 82%

Scheme 1.8 Michael additions of malonates to α,β-unsaturated aldehydes catalysed by *O*-TMS protected diphenylprolinol.

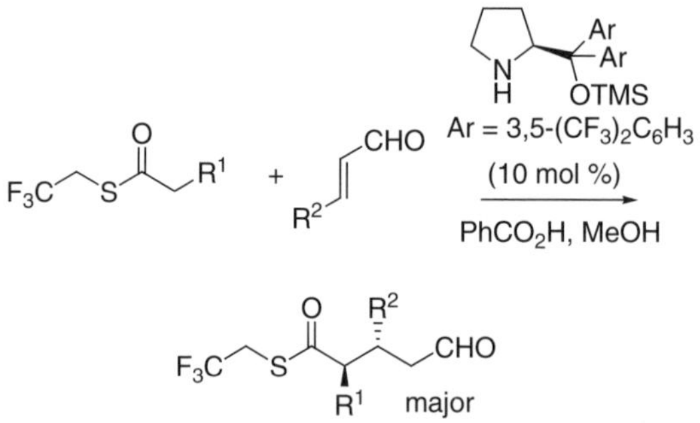

R^1 = *p*-ClC$_6$H$_4$, R^2 = Ph: 70% de = 30% ee = 90%
R^1 = *o*-ClC$_6$H$_4$, R^2 = Ph: 75% de = 30% ee = 89%
R^1 = *p*-CF$_3$C$_6$H$_4$, R^2 = Ph: 74% de = 48% ee = 93%
R^1 = *p*-NO$_2$C$_6$H$_4$, R^2 = Ph: 84% de = 22% ee = 91%
R^1 = R^2 = Ph: 71% de = 34% ee = 82%
R^1 = *p*-MeOC$_6$H$_4$, R^2 = Ph: 48% de = 28% ee = 84%
R^1 = 1-Naph, R^2 = Ph: 66% de = 26% ee = 67%
R^1 = 2-thienyl, R^2 = Ph: 57% de = 62% ee = 84%
R^1 = PhCH=CH, R^2 = Ph: 70% de = 56% ee = 98%
R^1 = *p*-ClC$_6$H$_4$, R^2 = *o*-MeOC$_6$H$_4$: 63% de = 42% ee = 96%
R^1 = *p*-ClC$_6$H$_4$, R^2 = Me: 51% de = 26% ee = 54%

Scheme 1.9 Silylated biarylprolinol-catalysed Michael additions of thioesters to α,β-unsaturated aldehydes.

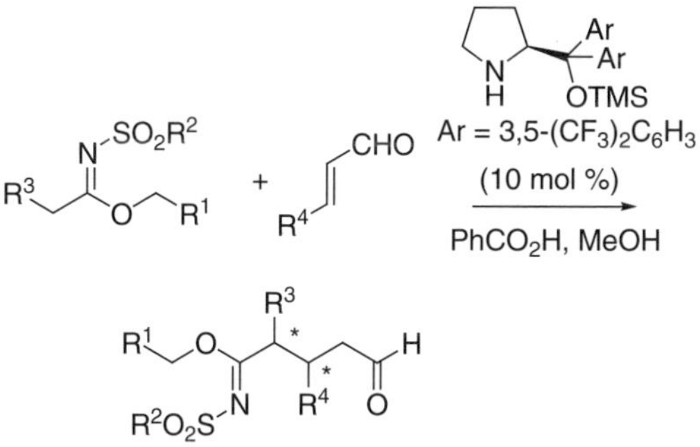

$R^1 = R^3 = R^4 = Ph, R^2 = p$-Tol:
85% de = 10% ee = 98%
$R^1 = H, R^3 = R^4 = Ph, R^2 = p$-Tol:
38% de = 10% ee = 93%
$R^1 = R^4 = Ph, R^3 = p$-$CF_3C_6H_4, R^2 = p$-Tol:
87% de = 20% ee = 94%
$R^1 = R^4 = Ph, R^2 = R^3 = p$-Tol:
50% de = 14% ee = 95%
$R^1 = R^3 = Ph, R^2 = p$-Tol, $R^4 = Me$:
66% de = 16% ee = 52%

Scheme 1.10 Silylated biarylprolinol-catalysed Michael additions of N-tosylimidates to α,β-unsaturated aldehydes.

In 2009, these authors applied this methodology to other nucleophiles, such as N-tosylimidates.[14] Indeed, N-tosylimidates were highly enantioselectively added to α,β-unsaturated aldehydes in the presence of 2-[bis(3,5-bistri-fluoromethylphenyl)trimethylsilylanyloxymethyl]pyrrolidine as the organocatalyst, yielding the corresponding Michael products in high enantioselectivities of up to 98% ee, albeit in low diastereoselectivities, as shown in Scheme 1.10. In particular, α-phenyl-substituted N-tosylimidate showed the best reactivity. In addition, these authors have demonstrated that the kinetic acidity of the α-proton of α-phenyl N-tosylimidate as measured by proton/deuterium NMR exchange experiments was correlated with the potential of N-tosylimidates to act as nucleophiles in organocatalytic reactions.

Other nucleophiles, such as dicyanoolefins, have been submitted to organocatalysed Michael additions with α,β-unsaturated aldehydes by Loh *et al.*, in 2008.[15] Therefore, a pool of water-compatible catalysts, namely dialkyl-(S)-prolinols, was developed for the enantioselective direct vinylogous Michael addition of vinylmalononitriles to α,β-unsaturated aldehydes in aqueous medium. In all the reactions tested, only the *anti*-Michael addition products were obtained. Notably, the best results in both yields and enantioselectivities were obtained by using the catalyst bearing a hexyl group, as shown in Scheme 1.11.

In addition, Jorgensen *et al.* have shown that racemic oxazolones were excellent reagents for the synthesis of chiral quaternary amino acids by

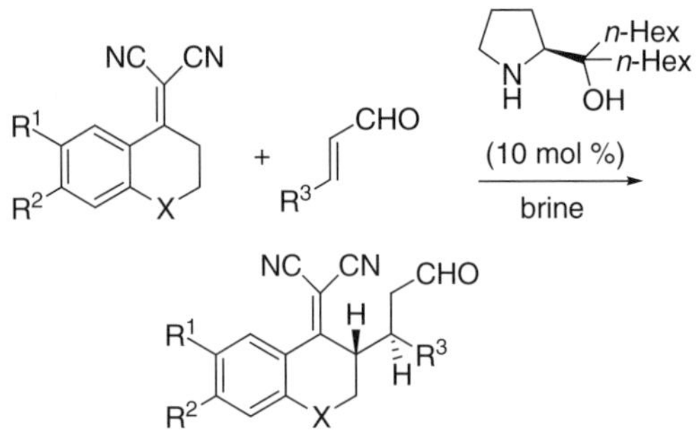

R^1 = R^2 = H, R^3 = Me, X = CH$_2$: 82% ee = 90%
R^1 = R^2 = H, R^3 = Me, X = O: 79% ee = 92%
R^1 = R^2 = H, R^3 = Me, X = S: 90% ee = 91%
R^1 = H, R^2 = OMe, R^3 = Me, X = CH$_2$: 54% ee = 88%
R^1 = OMe, R^2 = H, R^3 = Me, X = CH$_2$: 46% ee = 87%
R^1 = R^2 = H, R^3 = Et, X = CH$_2$: 81% ee = 84%
R^1 = R^2 = H, R^3 = Et, X = O: 85% ee = 88%
R^1 = R^2 = H, R^3 = Et, X = S: 58% ee = 90%
R^1 = H, R^2 = OMe, R^3 = Et, X = CH$_2$: 64% ee = 84%
R^1 = R^2 = H, R^3 = n-Pr, X = CH$_2$: 80% ee = 85%
R^1 = R^2 = H, R^3 = n-Pr, X = O: 84% ee = 87%
R^1 = R^2 = H, R^3 = n-Pr, X = S: 86% ee = 84%
R^1 = R^2 = H, R^3 = Ph, X = CH$_2$: 83% ee = 85%
R^1 = R^2 = H, R^3 = Ph, X = O: 81% ee = 87%
R^1 = R^2 = H, R^3 = p-ClC$_6$H$_4$, X = O: 76% ee = 88%
R^1 = R^2 = H, R^3 = p-BrC$_6$H$_4$, X = O: 74% ee = 85%

Scheme 1.11   Dihexyl-prolinol-catalysed Michael additions of dicyanoolefins to α,β-unsaturated aldehydes.

nucleophilic addition to α,β-unsaturated aldehydes catalysed by diarylprolinol silyl ethers.[16] This novel organocatalytic reaction proceeded with a good diastereoselectivity and an excellent enantioselectivity of up to 96% ee for a broad range of aldehydes and oxazolones, as shown in Scheme 1.12. The synthetic potential of this process was demonstrated by the conversion of the formed chiral products into various chiral products, such as α,α-disubstituted α-amino acids, α-quaternary proline derivatives, amino alcohols, lactams and tetrahydropyranes. Furthermore, the authors have shown by DFT calculations of transition states that the stereoselectivity for one class of compounds was due to hydrogen-bonding interactions between an acceptor in the *ortho*-position of the aromatic α,β-unsaturated aldehyde interacting with the enolate-form of oxazolone, and, in a more general manner, the selectivity could be controlled by a benzhydryl-protecting group in the oxazolone.

On the other hand, a number of asymmetric Michael additions of C-nucleophiles involving acceptors other than α,β-unsaturated aldehydes and catalysed by a proline derivative have recently been reported. As an example,

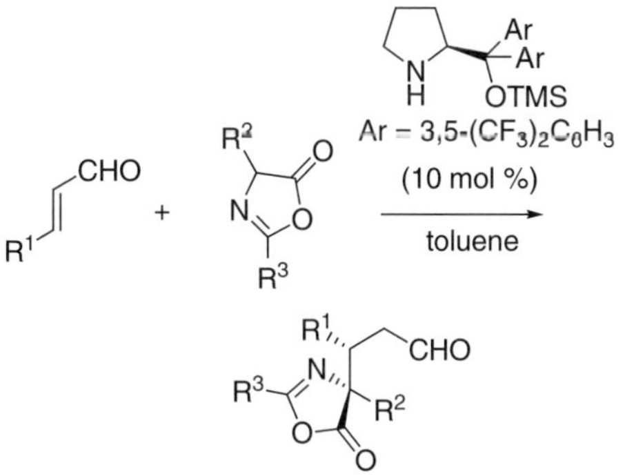

R^1 = Et, R^2 = *i*-Pr, R^3 = Ph: 81% dr = 1:2:1
ee (major isomer) = 96%
R^1 = El, R^2 = *i*-Pr, R^3 – *p*-Tol: 85% dr = 7:3
ee (major isomer) = 93%
R^1 = Et, R^2 = Bn, R^3 = Ph: 75% dr = 3:1
ee (major isomer) = 96%
R^1 = Et, R^2 = Me, R^3 = Ph: 65% dr = 5:1
ee (major isomer) = 93%
R^1 = Et, R^2 = MePr, R^3 = CHPh$_2$: 64% dr > 10:1
ee (major isomer) = 96%
R^1 = Et, R^2 = Ph, R^3 = *t*-Bu: 71% dr = 5:2
ee (major isomer) = 91%
R^1 = Et, R^2 = *i*-Bu, R^3 = *o*-ClC$_6$H$_4$: 54% dr = 2:1
ee (major isomer) = 88%
R^1 = Et, R^2 = MeS(CH$_2$)$_2$, R^3 = Ph: 73% dr = 3:1
ee (major isomer) = 94%
R^1 = Et, R^2 = Me, R^3 = CHPh$_2$: 40% dr > 10:1
ee (major isomer) = 93%
R^1 = Ph, R^2 = Me, R^3 = CHPh$_2$: 47% dr = 6:1
ee (major isomer) = 94%
R^1 = R^2 = Me, R^3 = CHPh$_2$: 59% dr = 5:1
ee (major isomer) = 90%
R^1 = *n*-Hex, R^2 = Me, R^3 = CHPh$_2$: 52% dr > 10:1
ee (major isomer) = 94%
R^1 = *(E)*-MeCH=CH, R^2 = Me, R^3 = CHPh$_2$: 55% dr > 10:1
ee (major isomer) = 92%

Scheme 1.12 Silylated biarylprolinol-catalysed Michael additions of oxazolones to α,β-unsaturated aldehydes.

Ley *et al.* have demonstrated that the pyrrolidinyl tetrazole catalyst, depicted in Scheme 1.13, could be used to efficiently induce the enantioselective Michael addition of malonates to α,β-unsaturated enones.[17] Cyclic, acyclic and aromatic enones could be involved in this process, and the reaction with the most efficient ethyl malonate provided, in the presence of piperidine as an additive, the corresponding Michael products in high yields with good to excellent enantioselectivities, as shown in Scheme 1.13.

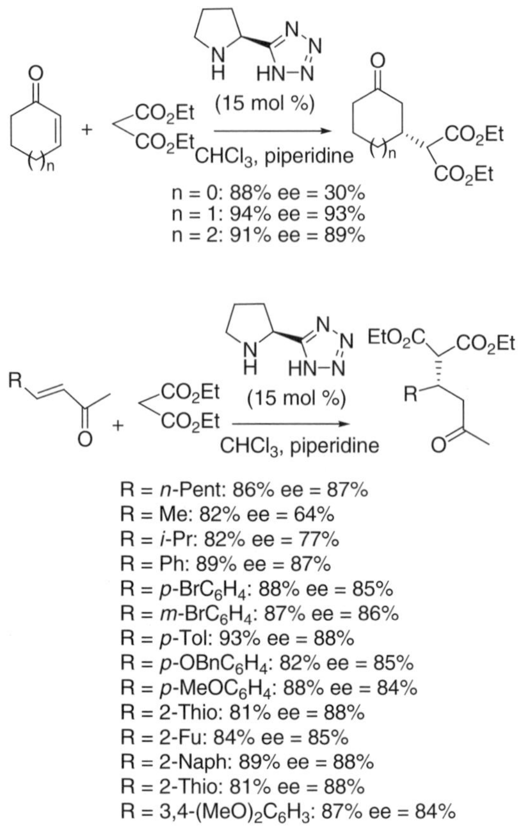

Scheme 1.13   Pyrrolidinyl tetrazole-catalysed Michael additions of ethyl malonate to
α,β-unsaturated enones.

Nitroalkanes are a particularly useful source of stabilised carbanions for the
asymmetric addition to electron-poor alkenes.[18] This type of nucleophiles has
been added to cyclic and acyclic α,β-unsaturated enones in the presence of a
novel class of organocatalysts, such as chiral α-aminophosphonates.[19] This
study revealed that the hydrate salt of a pyrrolidine-based catalyst bearing a
phosphonate group, depicted in Scheme 1.14, was found to be the best catalytic
species, providing, in the presence of *trans*-2,5-dimethylpiperazine as an additive, moderate to good results for a range of substrates, as summarised in
Scheme 1.14.

In the course of synthesising enantioenriched γ-keto *gem*-bisphosphonates
having anti-arthritic and anti-inflammatory activities, Barros and Phillips have
finalised organocatalytic asymmetric Michael additions of cyclic ketones to
vinyl *gem*-bisphosphonates.[20] The reactions were performed in the presence of
(*S*)-(+)-1-(2-pyrrolidininylmethyl)pyrrolidine as an organocatalyst and benzoic acid as an additive, leading to the expected Michael products in high yields,
excellent diastereoselectivities (≥ 98% de) combined with enantioselectivities

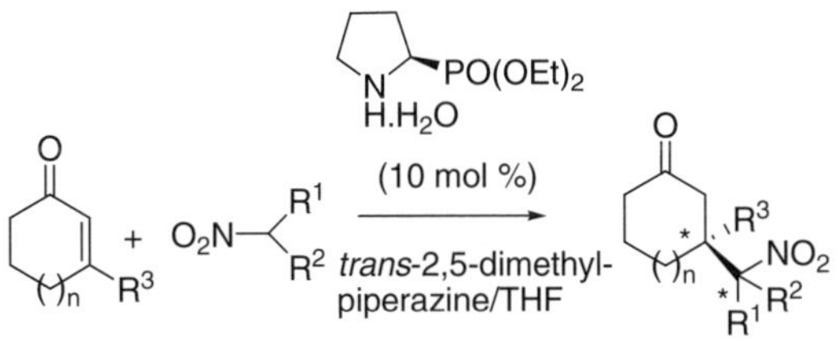

R^1 = R^2 = R^3 = H, n = 1: 18% ee = 80%
R^1 = R^3 = H, R^2 = Me, n = 0: 64% de = 4% ee = 68%
R^1 = R^3 = H, R^2 = Me, n = 1: 85% de = 6% ee = 76%
R^1 = R^3 = H, R^2 = Et, n = 0: 72% de = 2% ee = 74%
R^1 = R^3 = H, R^2 = Et, n = 1: 80% de = 2% ee = 78%
R^1 = R^2 = Me, R^3 = H, n = 0: 68% ee = 72%
R^1 = R^2 = Me, R^3 = H, n = 1: 78% ee = 76%
R^1 = H, R^2 = R^3 = Me, n = 1: 54% de = 26% ee = 76%
R^1 = H, R^2 = Et, R^3 = Me, n = 1: 29% de = 44% ee = 82%

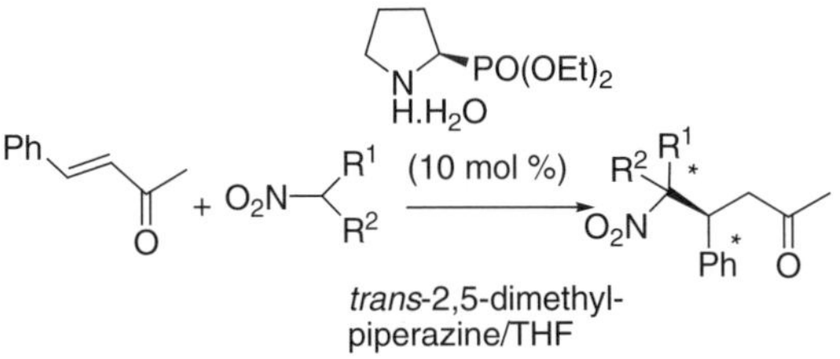

R^1 = H, R^2 = Me: 95% de = 6% ee = 52%
R^1 = H, R^2 = Et: 97% de = 20% ee = 52%
R^1 = R^2 = Me: 98% ee = 36%

Scheme 1.14 α-Aminophosphonate-catalysed Michael additions of nitroalkanes to α,β-unsaturated enones.

of up to 99% ee, as shown in Scheme 1.15. In order to extend the scope of this methodology, these authors have tried to involve linear ketones, β-keto esters or β-ketophosphonates as the Michael donors albeit unsuccessfully, since no reaction occurred with propiophenone, while the formed Michael products derived from either β-keto esters or β-ketophosphonates were obtained in a racemic form.

1.1.2 Intermolecular Michael Additions of C-Nucleophiles Catalysed by Non-proline Derivatives

A number of other-than-proline-derived organocatalysts have been involved to induce chirality in Michael additions of various C-nucleophiles onto a range of Michael acceptors. Among them, bifunctional organocatalysts possessing a thiourea moiety and a tertiary amino group were designed by Takemoto *et al.*

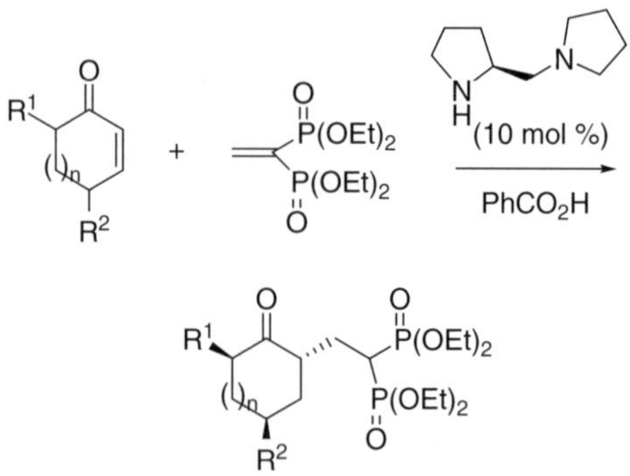

R^1 = R^2 = H, n = 1: 61% ee = 40%
R^1 = H, R^2 = Me, n = 1: 86% de = 64%
ee (major) = 71% ee (minor) = 83%
R^1 = H, R^2 = Ph, n = 1: 78% de = 84%
ee (major) = 76% ee (minor) = 62%
R^1 = Ph, R^2 = H, n = 1: 80% de = 72%
ee (major) = ee (minor) = 0%
R^1 = R^2 = H, n = 0: 58% de = 98% ee = 99%

Scheme 1.15   (*S*)-(+)-1-(2-pyrrolidininylmethyl)pyrrolidine-catalysed Michael additions of cyclic ketones to vinyl *gem*-bisphosphonate.

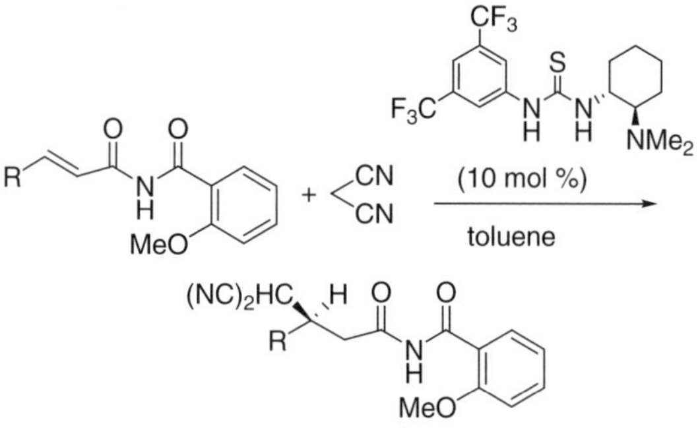

R = Ph: 95% ee = 91%
R = *p*-FC$_6$H$_5$: 99% ee = 92%
R = *p*-MeOC$_6$H$_5$: 92% ee = 90%
R = Me: 96% ee = 90%
R = TBSO(CH$_2$)$_5$: 95% ee = 93%

Scheme 1.16   Michael additions of malononitrile to α,β-unsaturated imides catalysed by a tertiary amine thiourea.

in order to be investigated as catalysts for the asymmetric Michael addition of malononitrile to α,β-unsaturated carboxylic acid derivatives having an imide moiety.[21] As shown in Scheme 1.16, the corresponding Michael adducts bearing various β-substituents were obtained in high yields combined with high enantioselectivities of up to 93% ee.

Among other-than-proline-derived organocatalysts, the modified cinchona alkaloids[22] have been the most employed in the last two years. These tunable bifunctional organocatalysts have emerged in the last four years as robust and tunable bifunctional organocatalysts for a range of synthetically useful transformations. In contrast with chiral secondary amine catalysts, little attention has been paid to the development of chiral primary amine catalysts. In 2009, Ye *et al.* developed a novel type of primary amine thiourea organocatalysts derived from 1,2-diaminocyclohexane and 9-amino (9-deoxy) cinchona alkaloid for the asymmetric Michael addition of malonates to α,β-unsaturated enones.[23] A series of cyclic and acyclic enones could react very well with different malonates in the presence of the catalyst derived from epiquinine, affording the corresponding Michael products with excellent yields and enantioselectivities, as shown in Scheme 1.17.

A closely related catalyst to the above was applied by the same workers to the asymmetric Michael addition of nitroalkanes to both cyclic and acyclic α,β-unsaturated enones, allowing the corresponding Michael products to be obtained in good yields and excellent enantioselectivities of up to 98% ee (Scheme 1.18).[24] This process offered a new way to construct quaternary stereocentres from enones and nitroalkanes.

Another member of the *epi*-cinchona-based thiourea organocatalyst family, depicted in Scheme 1.19, was applied by Vakulya *et al.* to the asymmetric Michael addition of nitroalkanes to chalcones, giving excellent yields and enantioselectivities of up to 98% ee.[25] The extension of this methodology was further explored to encompass α,β-unsaturated *N*-acylpyrroles, as a chalcone mimic. The corresponding Michael products were obtained in high yields and enantioselectivities of up to 94% ee, as shown in Scheme 1.19. This simple novel approach allowed a concise stereoselective synthesis of (*R*)-rolipram to be accomplished.

In 2009, Zhu and Lu reported organocatalytic asymmetric Michael additions of nitroalkanes to another Michael acceptor such as vinyl sulfone mediated by another cinchona alkaloid-derived thiourea catalyst, which afforded the desired Michael products with good enantioselectivities of up to 84% ee (Scheme 1.20).[26] This method in combination with a ready desulfonation represented a new approach to access α-alkylated chiral amines.

On the other hand, several cinchona alkaloid-derived primary amines have been successfully investigated as organocatalysts for asymmetric Michael additions of ketones to Michael acceptors. As an example, Lu *et al.* have described the first Michael addition of cyclic ketones to vinyl sulfone catalysed by a catalyst of this type, providing an easy access to chiral α-alkylated carbonyl compounds with high yields and enantioselectivities of up to 96% ee, albeit with moderate diastereoselectivities (≤72% de), as shown in Scheme 1.21.[27] This novel methodology was applied to the synthesis of sodium cyclamate, an important compound in the artificial sweeteners industry.

This catalyst was also applied to the asymmetric Michael addition of aromatic ketones to alkylidenemalononitriles by Chen *et al.*[28] The expected Michael products were generally isolated in modest to good enantioselectivities

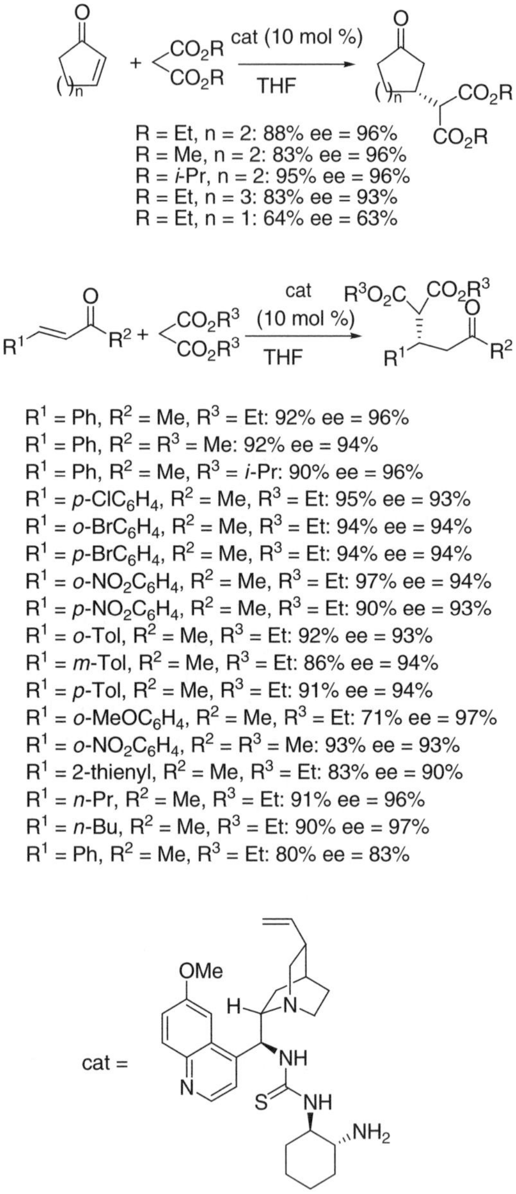
cat (10 mol %)
THF
CO2R
CO2R
R = Et, n = 2: 88% ee = 96%
R = Me, n = 2: 83% ee = 96%
R = i-Pr, n = 2: 95% ee = 96%
R = Et, n = 3: 83% ee = 93%
R = Et, n = 1: 64% ee = 63%
cat
(10 mol %)
THF
R1 = Ph, R2 = Me, R3 = Et: 92% ee = 96%
R1 = Ph, R2 = R3 = Me: 92% ee = 94%
R1 = Ph, R2 = Me, R3 = i-Pr: 90% ee = 96%
R1 = p-ClC6H4, R2 = Me, R3 = Et: 95% ee = 93%
R1 = o-BrC6H4, R2 = Me, R3 = Et: 94% ee = 94%
R1 = p-BrC6H4, R2 = Me, R3 = Et: 94% ee = 94%
R1 = o-NO2C6H4, R2 = Me, R3 = Et: 97% ee = 94%
R1 = p-NO2C6H4, R2 = Me, R3 = Et: 90% ee = 93%
R1 = o-Tol, R2 = Me, R3 = Et: 92% ee = 93%
R1 = m-Tol, R2 = Me, R3 = Et: 86% ee = 94%
R1 = p-Tol, R2 = Me, R3 = Et: 91% ee = 94%
R1 = o-MeOC6H4, R2 = Me, R3 = Et: 71% ee = 97%
R1 = o-NO2C6H4, R2 = R3 = Me: 93% ee = 93%
R1 = 2-thienyl, R2 = Me, R3 = Et: 83% ee = 90%
R1 = n-Pr, R2 = Me, R3 = Et: 91% ee = 96%
R1 = n-Bu, R2 = Me, R3 = Et: 90% ee = 97%
R1 = Ph, R2 = Me, R3 = Et: 80% ee = 83%
cat =
OMe
NH
NH
NH2

Scheme 1.17 Michael additions of malonates to α,β-unsaturated enones catalysed by a primary amine thiourea derived from epiquinine.

(71–84% ee) and acceptable yields (48–85%) when the reaction was performed in the presence of TsOH as an additive (Scheme 1.22).

The use of another cinchona alkaloid-derived primary amine in combination with TFA has allowed highly enantioselective Michael additions of

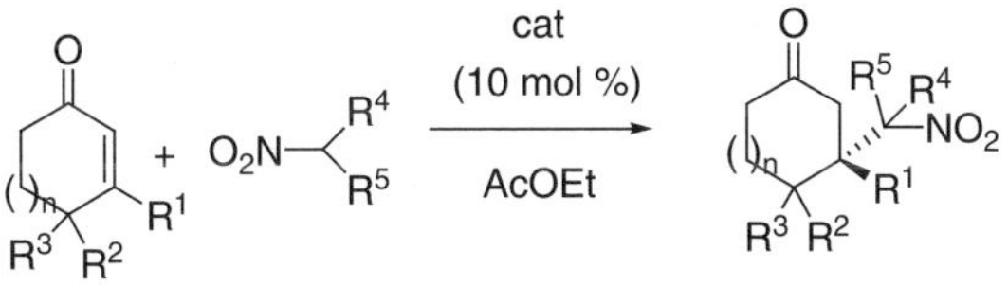

$R^1 = R^2 = R^3 = R^4 = R^5 = H$, n = 1: 70% ee = 96%
$R^1 = R^4 = R^5 = H$, $R^2 = R^3 = Me$, n = 1: 85% ee = 87%
$R^1 = R^2 = R^3 = R^4 = R^5 = H$, n = 0: 70% ee = 80%
$R^1 = R^2 = R^3 = R^4 = R^5 = H$, n = 2: 25% ee = 92%
$R^1 = Me$, $R^2 = R^3 = R^4 = R^5 = H$, n = 1: 82% ee = 94%
$R^1 = n\text{-}Pr$, $R^2 = R^3 = R^4 = R^5 = H$, n = 1: 70% ee = 94%
$R^1 = n\text{-}Bu$, $R^2 = R^3 = R^4 = R^5 = H$, n = 1: 72% ee = 93%
$R^1 = n\text{-}Pent$, $R^2 = R^3 = R^4 = R^5 = H$, n = 1: 79% ee = 93%
$R^1 = R^2 = R^3 = H$, $R^4 = R^5 = Me$, n = 1: 80% ee = 97%

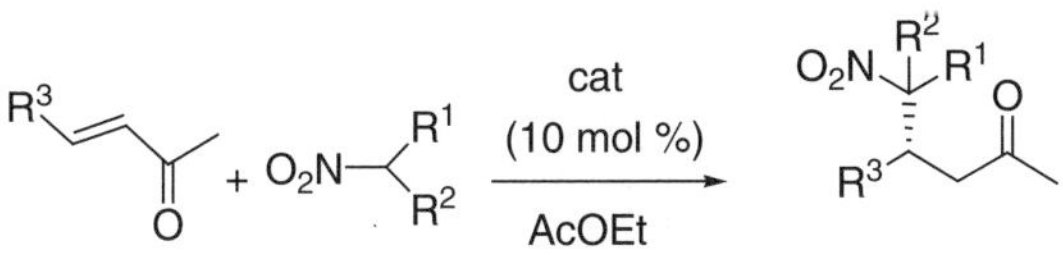

$R^1 = R^2 = H$, $R^3 = Ph$: 90% ee = 85%
$R^1 = R^2 = H$, $R^3 = p\text{-}NO_2C_6H_4$: 60% ee = 73%
$R^1 = R^2 = H$, $R^3 = p\text{-}Tol$: 82% ee = 82%
$R^1 = R^2 = H$, $R^3 = o\text{-}Tol$: 91% ee = 84%
$R^1 = R^2 = H$, $R^3 = p\text{-}MeOC_6H_4$: 65% ee = 80%
$R^1 = R^2 = H$, $R^3 = o\text{-}MeOC_6H_4$: 70% ee = 85%
$R^1 = R^2 = H$, $R^3 = p\text{-}ClC_6H_4$: 73% ee = 86%
$R^1 = R^2 = H$, $R^3 = 2\text{-}thienyl$: 68% ee = 78%
$R^1 = R^2 = H$, $R^3 = n\text{-}Pr$: 93% ee = 84%
$R^1 = R^2 = H$, $R^3 = n\text{-}Bu$: 84% ee = 84%
$R^1 = R^2 = Me$, $R^3 = Ph$: 96% ee = 82%

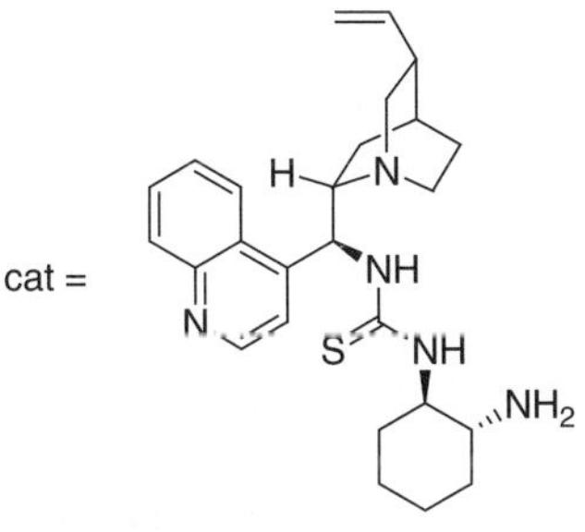

cat =

Scheme 1.18 Michael additions of nitroalkanes to α,β-unsaturated enones catalysed by primary amine thiourea derived from epiquinine.

malononitrile to α,β-unsaturated ketones to be achieved.[29] Indeed, the corresponding Michael products were isolated with excellent enantioselectivities (83–97% ee), as shown in Scheme 1.23.

A variety of cinchona alkaloids bearing a free hydroxyl group have been employed as efficient chiral organocatalysts for asymmetric Michael additions

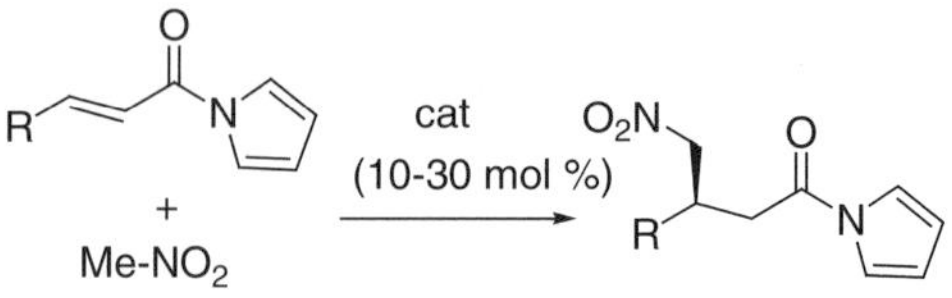

R = Ph: 93% ee = 93%
R = *p*-MeOC$_6$H$_4$: 90% ee = 93%
R = *p*-ClC$_6$H$_4$: 88% ee = 89%
R = 2-Fu: 81% ee = 93%
R = 2-thienyl: 59% ee = 92%
R = (*E*)-PhCH=CH$_2$: 95% ee = 93%
R = Cy: 93% ee = 94%
R = CH$_2$=CH(CH$_2$)$_8$: 87% ee = 90%
R = 3-CpO-4-MeOC$_6$H$_3$: 78% ee = 93%

cat =

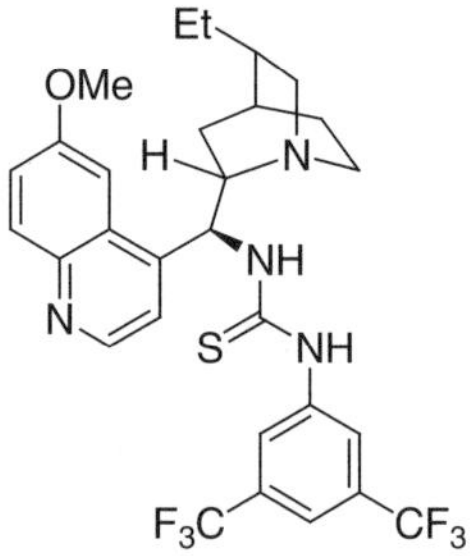

Scheme 1.19 Michael additions of nitroalkanes to α,β-unsaturated *N*-acylpyrroles catalysed by *epi*-cinchona-based thiourea.

of several types of C-nucleophiles to various Michael acceptors. As an example, Jorgensen *et al.* has reported a highly enantioselective approach to geminal bisphosphonates on the basis of Michael additions of cyclic β-keto esters to vinyl bisphosphonate esters catalysed by cheap and commercially available dihydroquinine.[30] High yields and enantioselectivities of up to 99% ee were achieved under simple conditions for a wide range of indanone-based β-keto esters, as well as various unprecedented 5-*tert*-butyloxycarbonyl cyclopentenones (Scheme 1.24). It was shown that the bulkiness of the ester substituent had a pronounced influence on the enantiocontrol, since with the bulkier *tert*-butyl ester, the enantioselectivity was remarkably enhanced. Further elaborations of the formed Michael products to the corresponding biologically important bisphosphonic acids or vinyl phosphonates have been successfully performed with conservation of the optical purity.

In 2009, Garcia Ruano *et al.* reported the synthesis of optically pure cyano *tert*-alkyl sulfones on the basis of enantioselective Michael additions of α-substituted cyanosulfones to vinyl ketones using another cinchona alkaloid bearing a free hydroxyl group as the catalyst.[31] As shown in Scheme 1.25, this methodology was applicable to cyclic as well as acyclic ketones, providing the

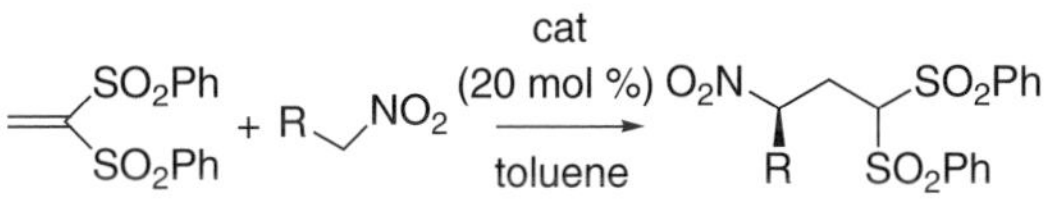

R = Me: 82% ee = 84%
R = Et: 86% ee = 78%
R = *n*-Pr: 82% ee = 80%
R = *n*-Bu: 71% ee = 84%
R = *n*-Hept: 81% ee = 74%
R = *n*-Non: 82% ee = 80%
R = Bn: 87% ee = 74%
R = *p*-MeOC$_6$H$_4$: 82% ee = 72%
R = CH$_2$Cy: 75% ee = 78%

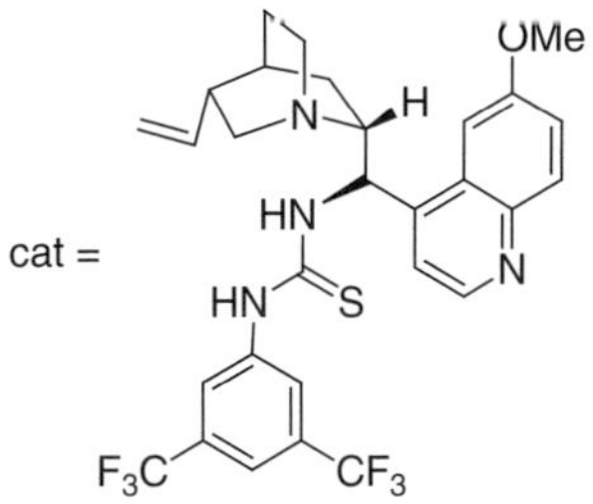

cat =

Scheme 1.20 Michael additions of nitroalkanes to vinyl sulfone catalysed by cinchona alkaloid-derived thiourea.

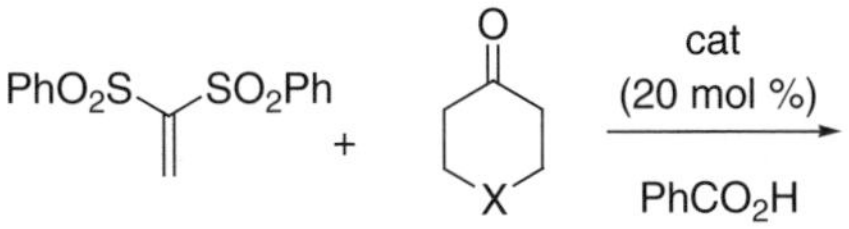

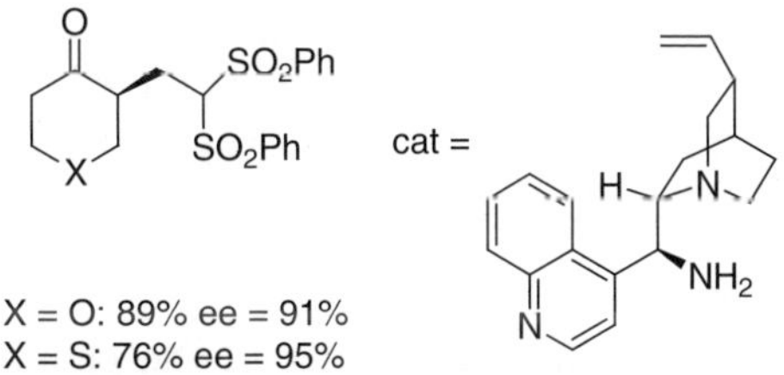

X = O: 89% ee = 91%
X = S: 76% ee = 95%
X = CH(Me): 84% de = 66% ee (major) = 96%
X = CH(Ph): 92% de = 72% ee (major) = 97%
X = CH(*t*-Bu): 93% de = 50% ee (major) = 95%
X = CH(*t*-Pr): 90% de = 60% ee (major) = 94%
X = CH(*n*-Pent): 78% de = 50% ee (major) = 90%
X = CH(CO$_2$Et): 85% de = 66% ee (major) = 88%
X = C(O(CH$_2$)$_2$O): 87% ee = 90%

Scheme 1.21 Michael additions of cyclic ketones to vinyl sulfone catalysed by cinchona alkaloid-derived primary amine.

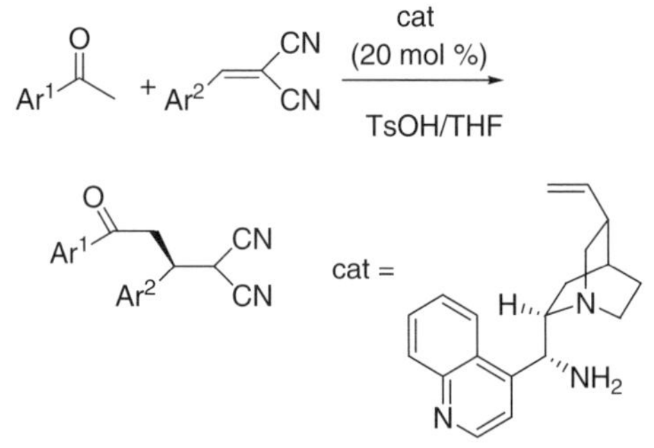

Ar^1 = Ar^2 = Ph: 73% ee = 71%
Ar^1 = *p*-FC_6H_4, Ar^2 = Ph: 72% ee = 75%
Ar^1 = *p*-ClC_6H_4, Ar^2 = Ph: 72% ee = 78%
Ar^1 = *p*-BrC_6H_4, Ar^2 = Ph: 69% ee = 81%
Ar^1 = *m*-BrC_6H_4, Ar^2 = Ph: 79% ee = 78%
Ar^1 = *p*-Tol, Ar^2 = Ph: 77% ee = 74%
Ar^1 = *p*-$MeOC_6H_4$, Ar^2 = Ph: 76% ee = 71%
Ar^1 = *p*-$NO_2C_6H_4$, Ar^2 = Ph: 85% ee = 72%
Ar^1 = 2-Naph, Ar^2 = *p*-$NO_2C_6H_4$: 48% ee = 72%
Ar^1 = *p*-BrC_6H_4, Ar^2 = *p*-FC_6H_4: 65% ee = 80%
Ar^1 = *p*-BrC_6H_4, Ar^2 = *p*-$NO_2C_6H_4$: 75% ee = 84%
Ar^1 = *p*-BrC_6H_4, Ar^2 = *p*-$MeOC_6H_4$: 69% ee = 71%

Scheme 1.22 Michael additions of aromatic ketones to alkylidenemalononitriles catalysed by cinchona alkaloid-derived primary amine.

corresponding Michael products in high yields with moderate to good enantioselectivities of up to 80% ee.

In the same context, Loh *et al.* have developed the first organocatalytic and enantioselective direct vinylogous Michael addition of α,α-dicyanoolefins to maleimides performed in the presence of the cinchona alkaloid depicted in Scheme 1.26.[32] This novel procedure generated the corresponding Michael *anti*-products in good yields with excellent diastereo- and enantioselectivities of up to 99% ee, as shown in Scheme 1.26. In this study, the authors have demonstrated that the free hydroxyl group of the catalyst played a key role in the substrate activation.

Acrylic esters, thioesters and *N*-acryloyl pyrrole have been identified by Dixon and Rigby as effective electrophiles in the enantioselective Michael addition reaction with β-keto esters catalysed by a cinchona alkaloid bearing a bulky phenanthrene group (Scheme 1.27).[33] High yields combined with excellent enantioselectivities of up to 96% ee were obtained in almost all cases of substrates.

Since quinones have an important role in nature in biological processes, a large number of reactions have been performed with these compounds. In this context, Jorgensen *et al.* have developed the first organocatalytic asymmetric Michael addition of dicyanoalkylidenes to quinones catalysed by cinchona

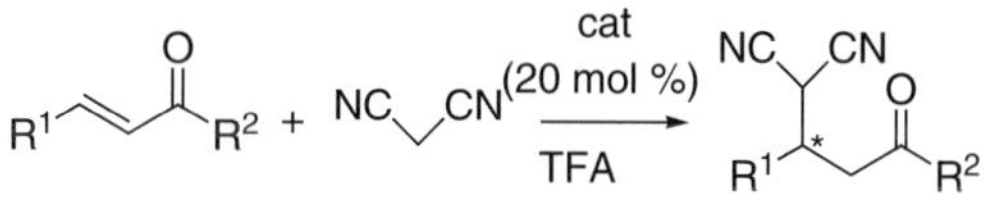

R^1 = Ph, R^2 = Me: 93% ee = 95%
R^1 = *p*-Tol, R^2 = Me: 87% ee = 93%
R^1 = *p*-MeOC$_6$H$_4$, R^2 = Me: 94% ee = 94%
R^1 = *p*-ClC$_6$H$_4$, R^2 = Me: 85% ee = 96%
R^1 = *p*-BrC$_6$H$_4$, R^2 = Me: 84% ee = 93%
R^1 = *p*-NO$_2$C$_6$H$_4$, R^2 = Me: 70% ee = 87%
R^1 = 2-Naph, R^2 = Me: 87% ee = 88%
R^1 = 2-Fu, R^2 = Me: 99% ee = 95%
R^1 = *n*-Pr, R^2 = Me: 35% ee = 93%
R^1 = Ph, R^2 = Et: 85% ee = 97%
R^1 = R^2 = Ph: 60% ee = 88%
R^1 = *p*-MeOC$_6$H$_4$, R^2 = Ph: 80% ee = 85%
R^1 = *p*-FC$_6$H$_4$, R^2 = Ph: 78% ee = 80%
R^1 = Ph, R^2 = *p*-MeOC$_6$H$_4$: 79% ee = 87%

cat =

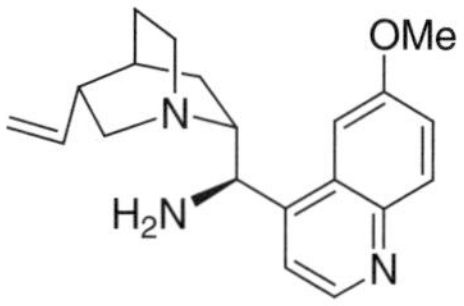

Scheme 1.23 Michael additions of malononitrile to α,β-unsaturated ketones catalysed by cinchona alkaloid-derived primary amine.

alkaloids, which led to the formation of 1,4-diketone derivatives with high
diastereomeric ratios of up to 96% de and enantioselectivities of up to 99%
ee.[34] A screening of the appropriate conditions and catalysts has demonstrated
that the dimeric cinchona alkaloid depicted in Scheme 1.28 was more efficient
than the corresponding monomeric catalysts. The formed products were useful
for a number of transformations, such as the synthesis of optically active α-aryl
ketones.

Phase transfer catalysis is an attractive alternative for organic reactions in
which charged intermediates are involved. Reactions are usually carried out in
two- or three-phase systems, most commonly in vigorously stirred aqueous/
apolar solvent mixtures. An inorganic base, such as K$_2$CO$_3$ or Cs$_2$CO$_3$, is used
to form the reactive enolate. The role of the catalyst is primarily that of an ion
shuttle. Chiral organocatalysts have been used to act as templates to direct the
approach of the reagent. Asymmetric phase transfer catalysed reactions were
firstly carried out with cinchona alkaloids. A recent example was reported by
Shibata *et al.*, who used the ammonium salts of sterically demanding cinchona
alkaloids to promote the first asymmetric Michael addition of 1-fluorobis-
(phenylsulfonyl)methane (FBSM), a potential monofluoromethide equivalent,

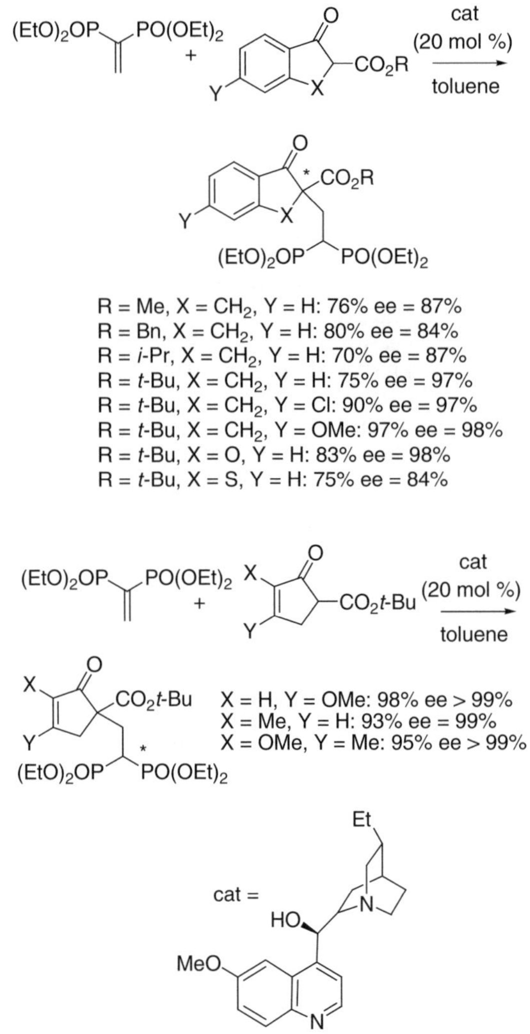

Scheme 1.24   Michael additions of β-keto esters to vinyl bisphosphonate esters catalysed by dihydroquinine.

to α,β-unsaturated ketones, providing versatile and enantiomerically enriched
products (Scheme 1.29).[35] A wide substrate scope, a high level of enantioselectivity and the flexibility to generate either enantiomer of the product have
been achieved. Moreover, these products constituted useful intermediates for
the synthesis of chiral monofluoromethylated molecules and biologically
interesting targets.

Another phase-transfer catalyst based on the structural motif of dihy-
drocinchonine and bearing a sterically demanding 9-anthracenylmethyl substituent at the quinuclidine nitrogen atom was employed by Jorgensen *et al.* to

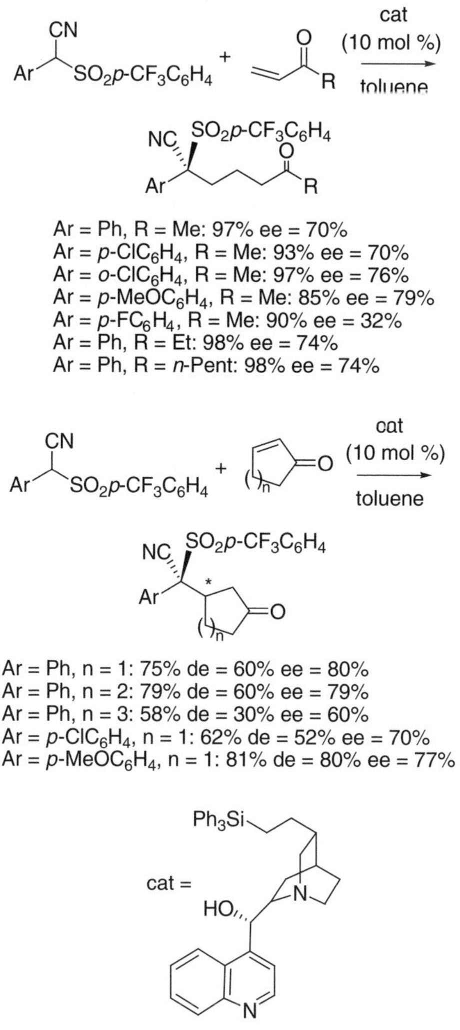
CN
Ar
SO2p-CF3C6H4
+
O
R
cat
(10 mol %)
toluene
NC
SO2p-CF3C6H4
O
Ar
R
Ar = Ph, R = Me: 97% ee = 70%
Ar = p-ClC6H4, R = Me: 93% ee = 70%
Ar = o-ClC6H4, R = Me: 97% ee = 76%
Ar = p-MeOC6H4, R = Me: 85% ee = 79%
Ar = p-FC6H4, R = Me: 90% ee = 32%
Ar = Ph, R = Et: 98% ee = 74%
Ar = Ph, R = n-Pent: 98% ee = 74%
CN
Ar
SO2p-CF3C6H4
+
O
n
cat
(10 mol %)
toluene
NC
SO2p-CF3C6H4
Ar
*
O
n
Ar = Ph, n = 1: 75% de = 60% ee = 80%
Ar = Ph, n = 2: 79% de = 60% ee = 79%
Ar = Ph, n = 3: 58% de = 30% ee = 60%
Ar = p-ClC6H4, n = 1: 62% de = 52% ee = 70%
Ar = p-MeOC6H4, n = 1: 81% de = 80% ee = 77%
Ph3Si
cat =
HO
N
N

Scheme 1.25 Michael additions of α-substituted cyanosulfones to vinyl ketones catalysed by cinchona alkaloid.

promote the first example of catalytic asymmetric Michael addition to electron-deficient allenes.[36] Therefore, the condensation of various cyclic β-keto esters onto allenes, having a ketone or an ester motif as the electron-withdrawing group as well as different substituents in the 3-position, led to the formation of the corresponding chiral β,γ-unsaturated carbonyl compounds in high yields and excellent diastereo- and enantioselectivities, as shown in Scheme 1.30. The scope of the reaction could be extended to the use of glycine imine derivatives as nucleophiles, which underwent the asymmetric conjugate addition to allenes in high yields and with enantioselectivities in the range of 60–88% ee, thus providing a rapid entry to chiral α-vinyl-substituted α-amino acid derivatives.

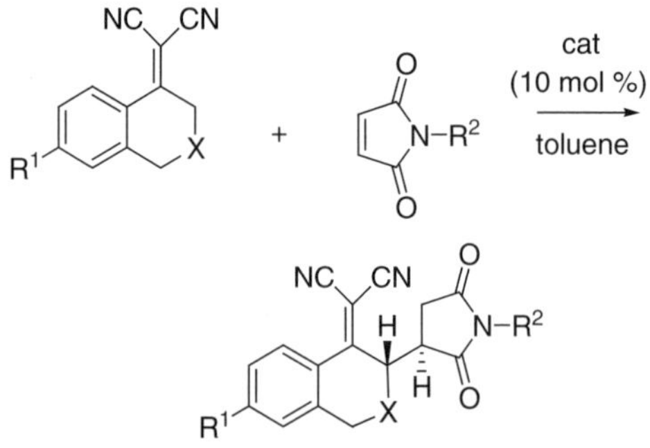

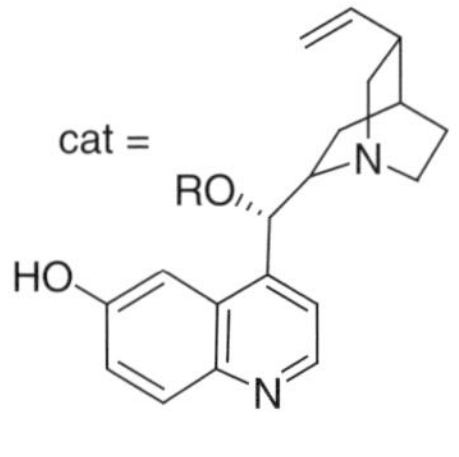

Scheme 1.26 Michael additions of α,α-dicyanoolefins to maleimides catalysed by cinchona alkaloid.

with R = Bn:
R^1 = H, X = O, R^2 = Ph: 80% ee = 85%
R^1 = H, X = CH$_2$, R^2 = Ph: 80% ee = 85%
R^1 = H, X = S, R^2 = Ph: 90% ee = 85%
R^1 = OMe, X = CH$_2$, R^2 = Ph: 35% ee = 81%
R^1 = H, X = O, R^2 = Bn: 81% ee > 99%
R^1 = H, X = S, R^2 = Bn: 89% ee = 92%
with R = 2-C$_{10}$H$_7$CH$_2$:
R^1 = H, X = O, R^2 = Ph: 86% ee = 92%
R^1 = H, X = CH$_2$, R^2 = Ph: 82% ee = 96%
R^1 = H, X = S, R^2 = Ph: 88% ee = 82%
R^1 = OMe, X = CH$_2$, R^2 = Ph: 36% ee = 82%
R^1 = H, X = S, R^2 = Bn: 88% ee = 95%

In 2009, Shimizu and Shirakawa reported the synthesis of inherently chiral calix[4]arenes containing a quaternary ammonium moiety, which were applied as chiral phase-transfer organocatalysts for Michael addition reactions.[37] The Michael products derived from the addition of benzylmalonate to benzylacetone and from the addition of a glycine derivative to methyl vinyl ketone, respectively, were isolated in almost quantitative yields, albeit with only low enantioselectivities (≤6% ee).

In addition, several organocatalysts other than cinchona alkaloid derivatives have been developed very recently. As an example, a chiral bicyclic guanidine was found by Tan *et al.* to be an excellent catalyst for enantioselective Michael additions of malonates or ethyl benzoylacetates to cyclopentenone or

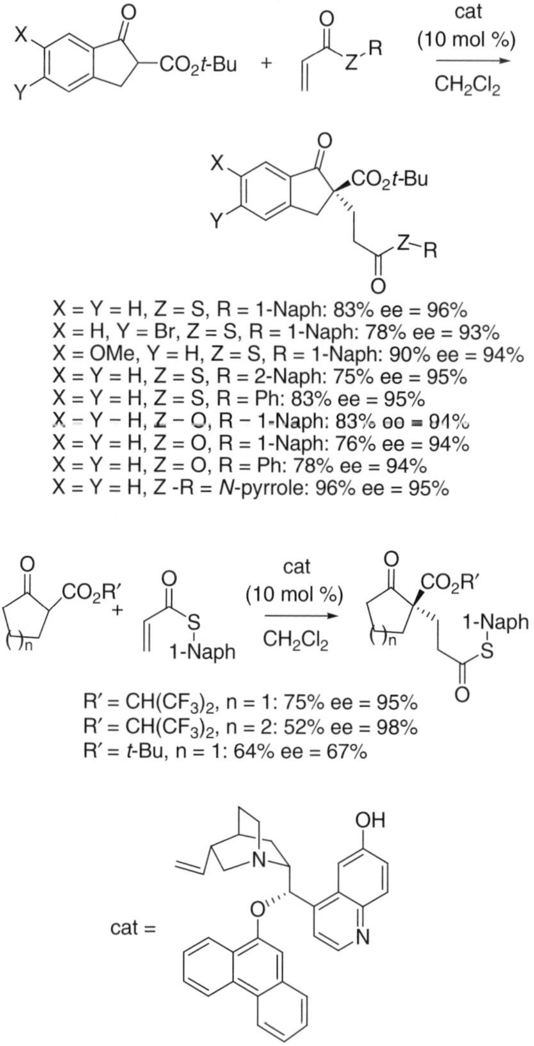
cat
(10 mol %)
CH2Cl2
CO2t-Bu
X = Y = H, Z = S, R = 1-Naph: 83% ee = 96%
X = H, Y = Br, Z = S, R = 1-Naph: 78% ee = 93%
X = OMe, Y = H, Z = S, R = 1-Naph: 90% ee = 94%
X = Y = H, Z = S, R = 2-Naph: 75% ee = 95%
X = Y = H, Z = S, R = Ph: 83% ee = 95%
X = Y = H, Z = O, R = 1-Naph: 83% ee = 91%
X = Y = H, Z = O, R = 1-Naph: 76% ee = 94%
X = Y = H, Z = O, R = Ph: 78% ee = 94%
X = Y = H, Z -R = N-pyrrole: 96% ee = 95%
CO2R′
1-Naph
R′ = CH(CF3)2, n = 1: 75% ee = 95%
R′ = CH(CF3)2, n = 2: 52% ee = 98%
R′ = t-Bu, n = 1: 64% ee = 67%
cat =

Scheme 1.27 Michael additions of β-keto esters to acrylic esters, thioesters and *N*-acryloyl pyrrole catalysed by cinchona alkaloid.

N-alkylmaleimides.[38] It was noted that additives such as TEA could accelerate the rate of the reaction. When TEA was used as the solvent, the reaction rate could be increased up to a thousand times without compromising the enantioselectivity. As shown in Scheme 1.31, the Michael products were obtained in excellent yields of up to 99% and enantioselectivities of up to 96% ee, albeit with no diastereoselectivity in cases of nucleophiles, such as ethyl benzoylacetates or *N*-alkylmaleimides.

Given the frequent occurrence of the indole core structure in biologically active substances and natural products, Rueping *et al.* have developed a highly

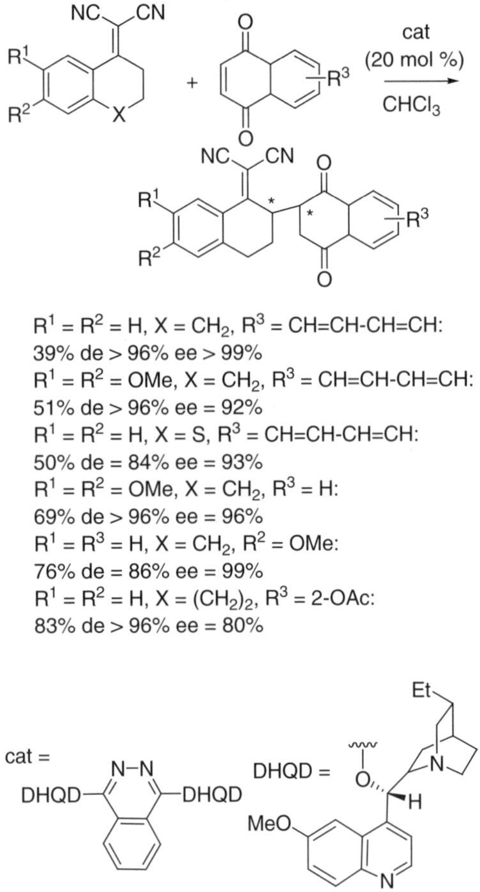
NC CN
R1
R2
X
O
O
R3
cat
(20 mol %)
CHCl3
NC CN
R1
R2
*
*
O
O
R3
R1 = R2 = H, X = CH2, R3 = CH=CH-CH=CH:
39% de > 96% ee > 99%
R1 = R2 = OMe, X = CH2, R3 = CH=CH-CH=CH:
51% de > 96% ee = 92%
R1 = R2 = H, X = S, R3 = CH=CH-CH=CH:
50% de = 84% ee = 93%
R1 = R2 = OMe, X = CH2, R3 = H:
69% de > 96% ee = 96%
R1 = R3 = H, X = CH2, R2 = OMe:
76% de = 86% ee = 99%
R1 = R2 = H, X = (CH2)2, R3 = 2-OAc:
83% de > 96% ee = 80%
cat =
N–N
DHQD
DHQD
DHQD =
Et
O
N
H
MeO
N

Scheme 1.28 Michael additions of dicyanoalkylidenes to quinones catalysed by dimeric cinchona alkaloid.

enantioselective organocatalytic Michael addition of indoles to α,β-unsaturated α-keto esters, providing the corresponding α-keto esters in good yields and high enantioselectivities of up to 92% ee (Scheme 1.32).[39] These novel reactions were catalysed by a silylated *N*-triflylphosphoramide, providing a direct access to optically pure α-keto and α-amino acids.

1.2 Intermolecular Domino Michael Additions of C-Nucleophiles

The great potential of asymmetric domino processes to generate chemical efficiency through the formation of multiple new bonds and stereocentres in a one-pot system is amply documented. This strategy avoids time-consuming and

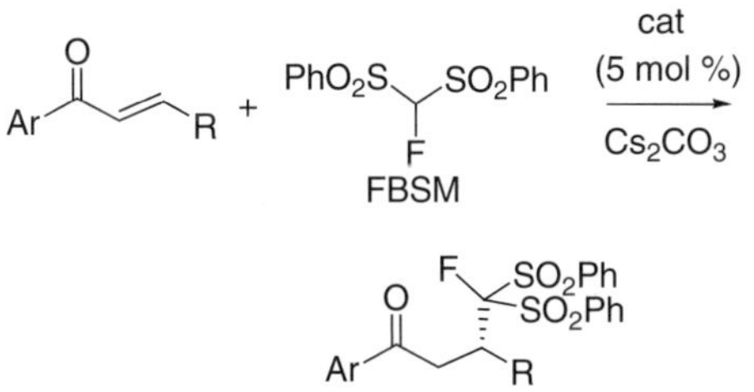

Ar = R = Ph: 80% ee = 97%
Ar = Ph, R = p-ClC$_6$H$_4$: 76% ee = 97%
Ar = Ph, R = m-ClC$_6$H$_4$: 85% ee = 98%
Ar = Ph, R = p-BrC$_6$H$_4$: 86% ee = 97%
Ar = Ph, R = p-Tol: 77% ee = 94%
Ar = p-ClC$_6$H$_4$, R = Ph: 52% ee = 91%
Ar = p-BrC$_6$H$_4$, R = Ph: 82% ee = 95%
Ar = p-BocOC$_6$H$_4$, R = p-BrC$_6$H$_4$: 32% ee = 95%
Ar = Ph, R = Me: 91% ee = 85%
Ar = Ph, R = Et: 69% ee = 90%

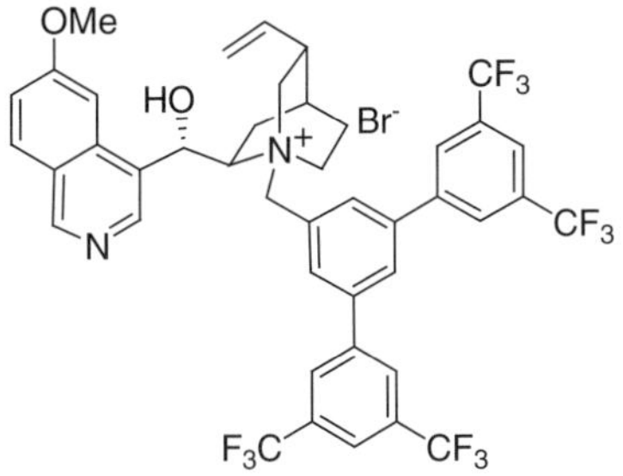

Scheme 1.29   Michael additions of FBSM to α,β-unsaturated ketones catalysed by cinchona-based phase-transfer catalyst.

costly processes, such as the purification of intermediates and the protection or deprotection of functional groups. These favourable features have stimulated the development of a range of asymmetric organocatalytic domino reactions. While reports on the area of organocatalysis appeared in the early 2000s, it was only at the end of 2005 that this organocatalytic strategy began to be intensively investigated. Ever since, the use of domino reactions in asymmetric synthesis is increasing constantly.[40] The first catalytic asymmetric domino Michael reaction was reported by Shibasaki *et al.*, in 1996.[41] This domino reaction was promoted by the catalytic use of a heterobimetallic multifunctional asymmetric complex. Since this work, a number of asymmetric organocatalytic domino Michael addition reactions have been successfully developed.[42] In the last year, biaryl-prolinol ethers have been the most used organocatalysts for asymmetric domino Michael additions of C-nucleophiles. In particular, silylated biphenylprolinol has been successfully involved in a range of asymmetric domino Michael reactions, such as domino Michael-aldol reactions developed by Hong *et al.*[43] These reactions, performed in the presence of this catalyst and AcOH,

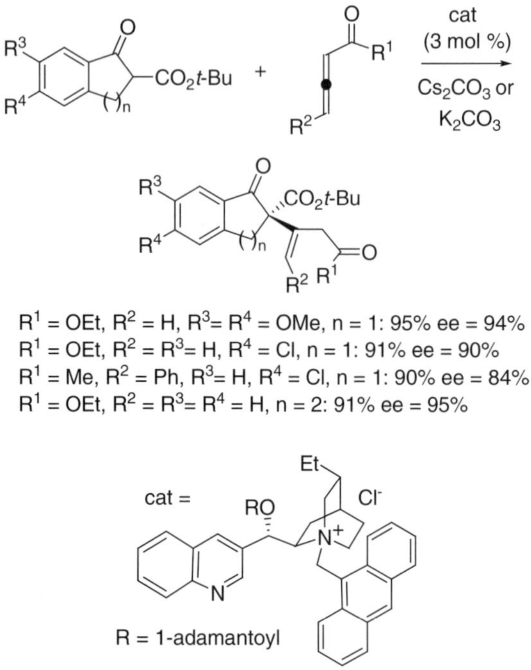

R^1 = OEt, R^2 = H, R^3= R^4 = OMe, n = 1: 95% ee = 94%
R^1 = OEt, R^2 = R^3= H, R^4 = Cl, n = 1: 91% ee = 90%
R^1 = Me, R^2 = Ph, R^3= H, R^4 = Cl, n = 1: 90% ee = 84%
R^1 = OEt, R^2 = R^3= R^4 = H, n = 2: 91% ee = 95%

cat =

R = 1-adamantoyl

Scheme 1.30 Michael additions of cyclic β-keto esters to electron-deficient allenes catalysed by cinchona-based phase-transfer catalyst.

occurred between 5-oxohexanal or glutaraldehyde and 3-arylpropenal, providing an expedited access towards highly functionalised cyclohexenedicarbaldehydes in excellent diastereo- and enantioselectivities of up to 99% ee (Scheme 1.33). The structure of the products was confirmed by X-ray analysis.

This catalyst was also used by Ma *et al.* to promote a domino Michael addition-cyclisation process of aldehydes with α-keto-α,β-unsaturated esters, proceeding smoothly in water in the presence of AcOH as an additive.[44] The corresponding chiral cyclic hemiacetals were formed in both high yields and diastereoselectivities and almost complete enantioselectivities, and were subsequently oxidised by treatment with PCC to furnish highly functionalised 3,4,5,6-tetrasubstituted dihydropyrones (Scheme 1.34).

In 2009, another application of this organocatalyst was reported by Franzen and Fisher, dealing with a short enantioselective one-pot, two-step synthesis of the indolo[2,3a]quinolizidine and the benzo[a]quinolizidine skeleton, which are found in a large number of naturally occurring compounds having diverse biological activities.[45] The key step of the synthesis was an asymmetric Michael addition followed by a subsequent acid-catalysed cyclisation of the acyliminium ion, providing the quinolizidine derivatives in good yields, moderate to good diastereoselectivities and high enantioselectivities of up to 95% ee (Scheme 1.35).

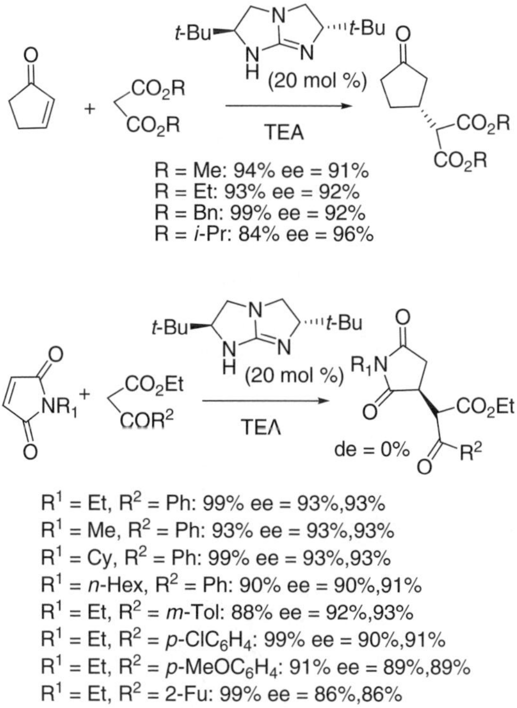

Scheme 1.31 Bicyclic guanidine-catalysed Michael additions of malonates or ethyl benzoylacetates to cyclopentenone or *N*-alkylmaleimides.

Another important application of this catalyst was the development by Jorgensen *et al.* of a novel asymmetric cascade reaction occurring between an α,β-unsaturated aldehyde and a tricarbonyl compound, which led to the formation of important optically active bicyclo[3.3.1]non-2-ene compounds (Scheme 1.36).[46] With this novel reaction, organocatalysis has been taken to a new level, allowing the selective formation of four new carbon–carbon bonds, providing six new stereocentres, one of which was quaternary, leading to the control of one out of sixty-four possible stereoisomers by mixing two simple molecules. Moreover, the Michael products were obtained in high yields combined with excellent diastereo- and enantiocontrol and performing the process on the gram scale. In addition, these products were precursors for important biomolecules with antitumour activity. The proposed mechanism for this process involved a first Michael addition of one molecule of the tricarbonyl compound to the α,β-unsaturated aldehyde, followed by the cyclisation of the thus-formed intermediate, and then the addition of a second molecule of tricarbonyl compound, providing the final bicyclic product.

In 2008, Cordova *et al.* reported the silylated biphenylprolinol-catalysed domino Michael addition-α-alkylation reaction between 2-bromonitromethane and α,β-unsaturated aldehydes.[47] This highly enantioselective process provided

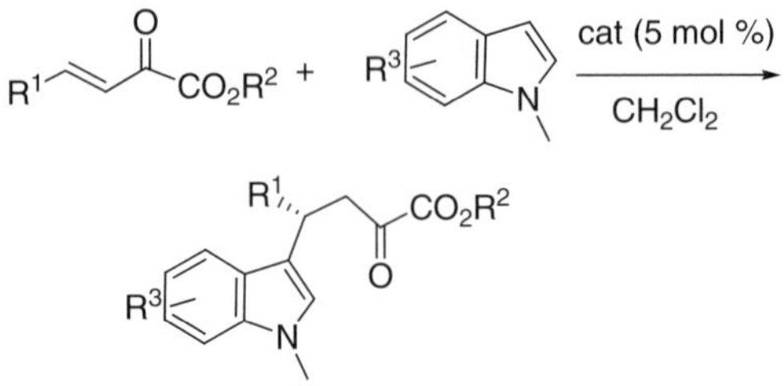

R^1 = Ph, R^2 = Me, R^3 = H: 62% ee = 88%
R^1 = Ph, R^2 = Et, R^3 = H: 81% ee = 87%
R^1 = Ph, R^2 = Me, R^3 = 5-Br: 43% ee = 87%
R^1 = Ph, R^2 = Me, R^3 = 7-Me: 78% ee = 88%
R^1 = p-ClC$_6$H$_4$, R^2 = Me, R^3 = H: 65% ee = 88%
R^1 = p-BrC$_6$H$_4$, R^2 = Me, R^3 = H: 60% ee = 90%
R^1 = p-Tol, R^2 = Me, R^3 = H: 69% ee = 92%
R^1 = p-Tol, R^2 = Me, R^3 = Br: 55% ee = 80%
R^1 = p-MeOC$_6$H$_4$, R^2 = Me, R^3 = H: 88% ee = 86%
R^1 = 2-Naph, R^2 = Me, R^3 = H: 70% ee = 90%

cat =

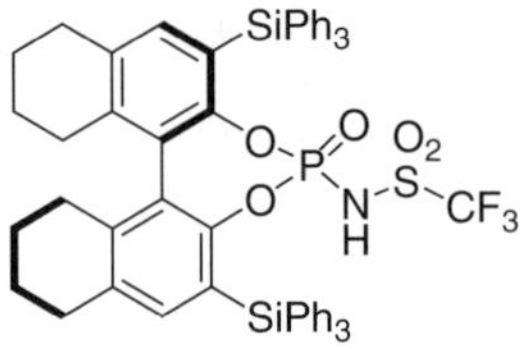

Scheme 1.32 *N*-Triflylphosphoramide-catalysed Michael additions of indoles to α,β-unsaturated α-keto esters.

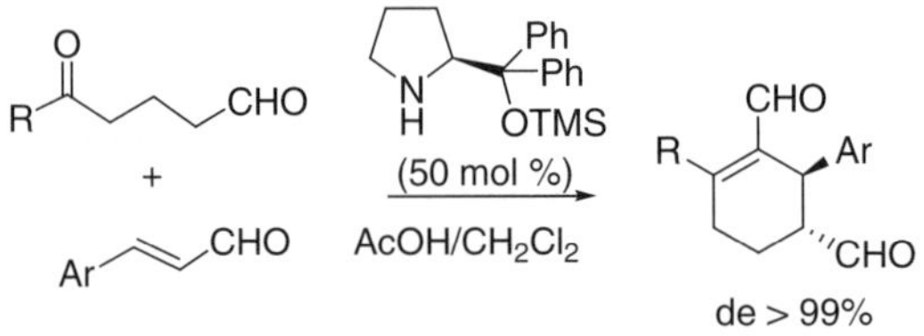

R = H, Ar = Ph: 66% ee = 89%
R = H, Ar = p-MeOC$_6$H$_4$: 65% ee = 93%
R = H, Ar = p-Tol: 63% ee = 93%
R = H, Ar = o-TBSOC$_6$H$_4$: 68% ee = 97%
R = H, Ar = 2-Fu: 55% ee = 74%
R = H, Ar = p-NO$_2$C$_6$H$_4$: 69% ee = 96%
R = H, Ar = p-BrC$_6$H$_4$: 61% ee = 94%
R = Me, Ar = Ph: 60% ee = 95%
R = Me, Ar = 2-Fu: 53% ee = 92%
R = Me, Ar = p-NO$_2$C$_6$H$_4$: 62% ee > 99%
R = Me, Ar = p-BrC$_6$H$_4$: 58% ee > 99%
R = Me, Ar = p-MeOC$_6$H$_4$: 60% ee > 99%

Scheme 1.33 Silylated biphenylprolinol-catalysed domino Michael-aldol reactions.

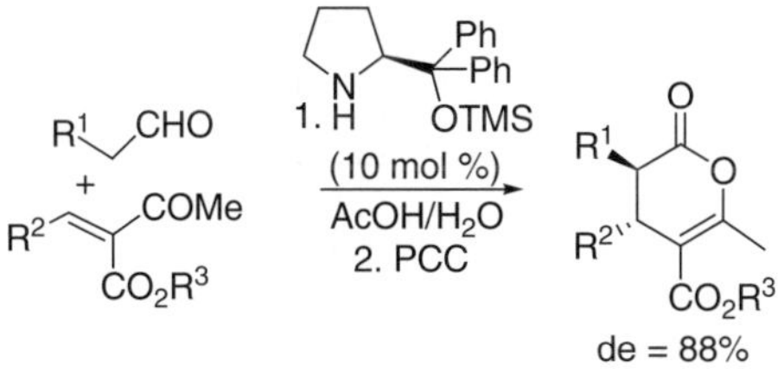

R^1 = (CH$_2$)$_7$CH=CH$_2$, R^2 = *n*-Bu, R^3 = Et: 58% ee > 99%
R^1 = (CH$_2$)$_2$OBn, R^2 = *n*-Bu, R^3 = Et: 54% ee = 98%
R^1 = *i*-Pr, R^2 = *n*-Bu, R^3 = Et: 54% ee = 98%
R^1 = (CH$_2$)$_7$CH=CH$_2$, R^2 = *n*-Bu, R^3 = *t*-Bu: 59% ee = 98%
R^1 = (CH$_2$)$_2$OBn, R^2 = *n*-Bu, R^3 = *t*-Bu: 54% ee = 98%
R^1 = *n*-Pr, R^2 = (CH$_2$)$_3$OBn, R^3 = *t*-Bu: 63% ee = 99%
R^1 = (CH$_2$)$_7$OBn, R^2 = (CH$_2$)$_3$OBn, R^3 = *t*-Bu: 53% ee = 99%
R^1 = Me, R^2 = (CH$_2$)$_3$OBn, R^3 = *t*-Bu: 54% ee = 94%
R^1 = *n*-Pr, R^2 = CO$_2$Et, R^3 = *t*-Bu: 66% ee > 99%
R^1 = *i*-Pr, R^2 = CO$_2$Et, R^3 = *t*-Bu: 68% ee > 99%

Scheme 1.34 Silylated biphenylprolinol-catalysed domino Michael addition-cyclisation reactions.

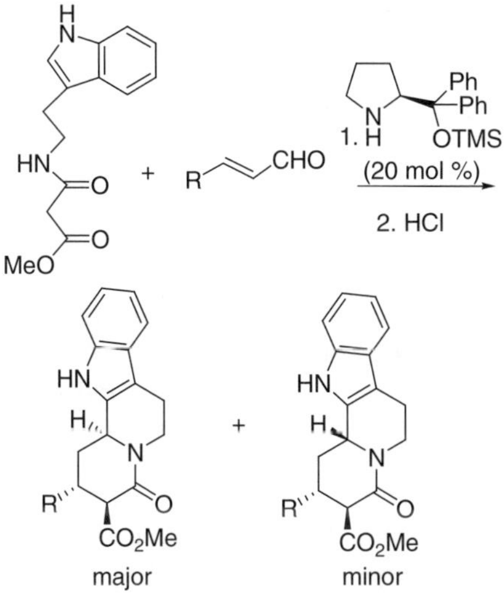

R = Ph: 69 % de = 70% ee (major) = 94%
R = *o*-NO$_2$C$_6$H$_4$: 53 % de = 80% ee (major) = 95%
R = 3-MeO-4-OAcC$_6$H$_3$: 62 % de = 48% ee (major) = 95%
R = *p*-MeOC$_6$H$_4$: 71 % de = 66% ee (major) = 87%
R = 3,4-Cl$_2$C$_6$H$_3$: 56 % de = 66% ee (major) = 89%

Scheme 1.35 Silylated biphenylprolinol-catalysed domino Michael addition-cyclisation reactions.

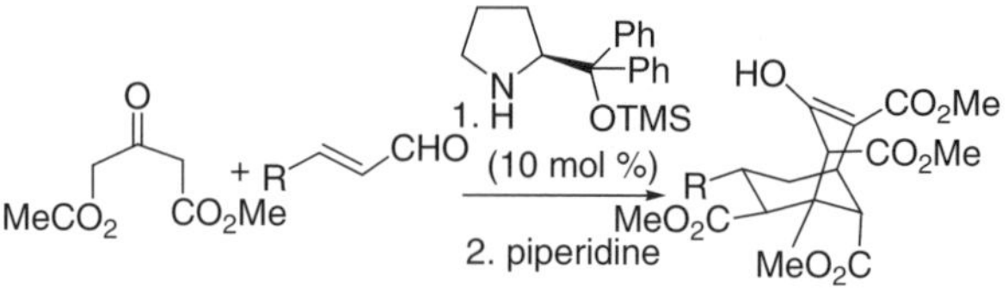

R = Et: 48% de > 98% ee = 94%
R = *i*-Pr: 65% de > 98% ee = 96%
R = *n*-Hept: 69% de = 76% ee = 95%
R = CO_2Et: 38% de > 98% ee = 89%
R = Ph: 70% de > 98% ee = 93%
R = 2-Fu: 86% de = 88% ee = 90%
R = *p*-BrC$_6$H$_4$: 86% de > 98% ee = 96%

Scheme 1.36 Silylated biphenylprolinol-catalysed domino Michael addition-cyclisation reactions.

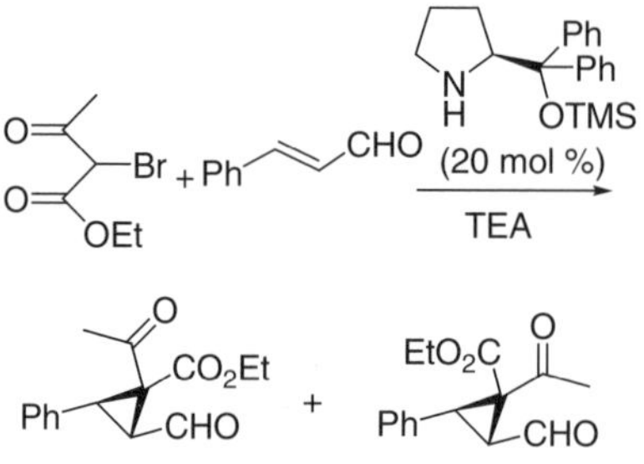

94% de > 92% ee = 94%

Scheme 1.37 Silylated biphenylprolinol-catalysed domino Michael addition-α-alkylation reaction.

the corresponding chiral 2-formyl-cyclopropanes in moderate to good yields (35–63%), high enantioselectivities (83–99% ee) and moderate diastereoselectivities (0–56% de). The cyclopropyl 1-nitro-2-formyl-cyclopropane derivatives synthesised from this domino reaction were further converted to the corresponding β-nitromethyl-acid esters, which are excellent precursors of GABA analogues such as Baclofen. In addition, this catalyst proved to be the most effective catalyst in the domino Michael-α-alkylation reaction depicted in Scheme 1.37.[48] The final cyclopropane derivative was isolated in high diastereo- and enantioselectivity, as shown in Scheme 1.37.

This catalyst was involved in the development of a new asymmetric organocatalytic domino reaction of α,β-unsaturated aldehydes and β-keto esters following a Michael addition-Morita–Baylis–Hillman mechanism.[49] As shown in Scheme 1.38, the formed cyclohexenone products were obtained in moderate to good yields and with high stereoselectivities for a wide range of substrates.

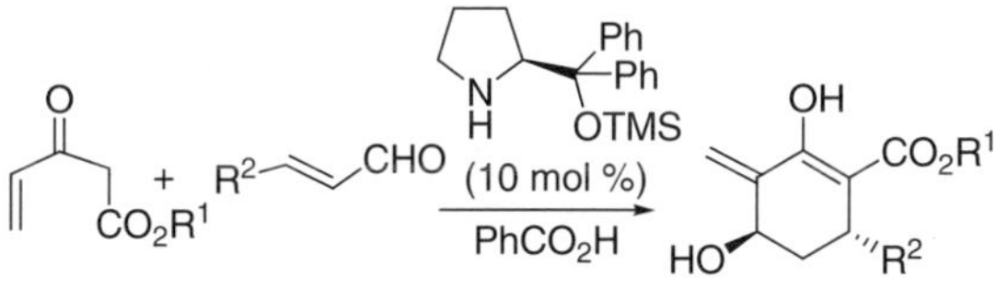

R^1 = Ph, R^2 = Et: 55% de = 76% ee = 94%
R^1 = Ph, R^2 = *t*-Bu: 68% de = 66% ee = 94%
R^1 = Ph, R^2 = CH$_2$CH=CH$_2$: 45% de = 72% ee = 94%
R^1 = *p*-MeOC$_6$H$_4$, R^2 = Et: 69% de = 80% ee = 93%
R^1 = *p*-NO$_2$C$_6$H$_4$, R^2 = Et: 58% de = 60% ee = 96%
R^1 = *p*-NO$_2$C$_6$H$_4$, R^2 = *t*-Bu: 51% de = 60% ee = 95%
R^1 = 2-thienyl, R^2 = Et: 57% de = 72% ee = 95%
R^1 = CO$_2$Et, R^2 = Et: 51% de = 45% ee = 98%

Scheme 1.38 Silylated biphenylprolinol-catalysed domino Michael addition-Morita–Baylis–Hillman reactions.

In 2009, Hayashi *et al.* reported a novel asymmetric domino Michael-Knoevenagel reaction catalysed by silylated biphenylprolinol occurring between an α,β-unsaturated aldehyde and dimethyl 3-oxopentanedioate, affording the corresponding chiral substituted cyclohexenone derivatives, which were further converted into the corresponding alcohols as single diastereomers by treatment with NaBH$_4$ with excellent enantioselectivities of up to 99% ee (Scheme 1.39).[50]

A complete enantioselectivity was achieved by Melchiorre *et al.* for the one-pot synthesis of tri- and tetra-substituted cyclohexene carbaldehydes bearing three or four stereogenic carbon atoms, one of which was quaternary by all-carbon substitution.[51] These enantiopure products were formed *via* a stereoselective three-component domino strategy catalysed by silylated biphenylprolinol occurring between an aldehyde, an α,β-unsaturated aldehyde and a cyanoacrylate. This triple cascade process involved a first Michael addition of the aldehyde to the cyanoacrylate leading to the Michael product, which was then engaged as a nucleophile in a second Michael addition to the α,β-unsaturated aldehyde. The last step was an enamine-promoted intramolecular aldol reaction, in which the less-hindered aldehyde acted as a nucleophile, affording after a subsequent dehydration the final desired cyclohexene carbaldehyde. In general, the best diastereocontrol was obtained with β-substituted cyanoacrylates (up to 90% de). The best results are collected in Scheme 1.40.

Another diarylprolinol ether has been used to catalyse several asymmetric domino Michael reactions. Therefore, Jorgensen *et al.* have demonstrated its efficiency for the asymmetric synthesis of 1,4-dihydropyridines, which are closely related to the NADH system, a biological system of utmost importance, and moreover these molecules are important drugs used in the treatment of a number of diseases, such as cardiovascular diseases and Alzheimer's disease.[52] Thus, a series of chiral 1,4-dihydropyridines were prepared on the basis of a one-pot multicomponent reaction between a α,β-unsaturated aldehyde, a β-diketone

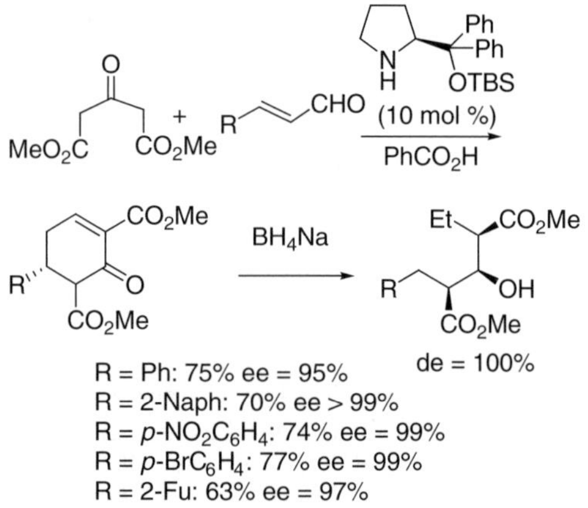

R = Ph: 75% ee = 95%   de = 100%
R = 2-Naph: 70% ee > 99%
R = p-NO$_2$C$_6$H$_4$: 74% ee = 99%
R = p-BrC$_6$H$_4$: 77% ee = 99%
R = 2-Fu: 63% ee = 97%

Scheme 1.39   Silylated biphenylprolinol-catalysed domino Michael addition-Knoevenagel reactions.

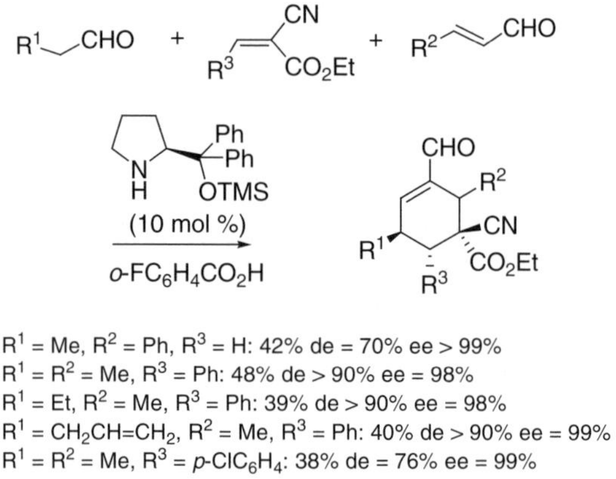

R^1 = Me, R^2 = Ph, R^3 = H: 42% de = 70% ee > 99%
R^1 = R^2 = Me, R^3 = Ph: 48% de > 90% ee = 98%
R^1 = Et, R^2 = Me, R^3 = Ph: 39% de > 90% ee = 98%
R^1 = CH$_2$CH=CH$_2$, R^2 = Me, R^3 = Ph: 40% de > 90% ee = 99%
R^1 = R^2 = Me, R^3 = p-ClC$_6$H$_4$: 38% de = 76% ee = 99%

Scheme 1.40   Silylated biphenylprolinol-catalysed triple cascade Michael addition-Michael addition–aldol reactions.

and a primary amine catalysed by 2-[bis(3,5-bistrifluoromethylphenyl)trimethyl-silylanyloxymethyl]pyrrolidine (Scheme 1.41). The first step of the process was the Michael addition of the β-diketone to the α,β-unsaturated aldehyde, affording the corresponding Michael adduct. This intermediate reacted with the primary amine to form the corresponding enamine, of which the amine moiety then attacked the keto functionality of the intermediate in a 6-*exo*-trig fashion to form the final 1,4-dihydropyridine after dehydration with moderate to good yields and high enantioselectivities of up to 92% ee. The same authors have developed the synthesis of chiral 3,4-dihydropyran derivatives on the basis of an asymmetric domino Michael addition-cyclisation reaction catalysed by the same organocatalyst occurring between an α,β-unsaturated aldehyde and

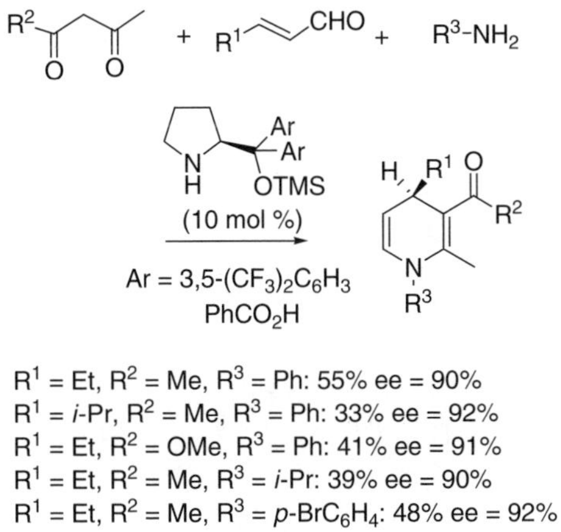

Scheme 1.41 Silylated biarylprolinol-catalysed synthesis of 1,4-dihydropyridines.

1,3-cyclopentanedione.[53] The 3,4-dihydropyrans were obtained with moderate to good yields (48–95%) and diastereoselectivities (34–90% de) combined with high enantioselectivities (82–96% ee) with a broad range of aromatic and aliphatic aldehydes. The scope of the reaction could be extended to other 1,3-cycloalkanediones, such as 1,3-cyclohexanedione and 1,3-cycloheptanedione, providing similar results. On the other hand, the same authors have employed this catalyst to prepare enantiopure 5-(trialkylsilyl)cyclohex-2-enones on the basis of an asymmetric domino Michael reaction between β-keto esters and silicon-substituted α,β-unsaturated aldehydes.[54] The process involved the Michael addition of the β-keto ester to the silicon-substituted α,β-unsaturated aldehyde leading to the corresponding intermediate which, after an acid-catalysed decarboxylation and an aldol condensation, yielded the desired 5-(trialkylsilyl)cyclohex-2-enone, which was isolated in 42–69% yield and enantioselectivity of 99% ee in all cases of substrates.

Chiral functionalised chromenes are important compounds which, due to their biological activities, find wide applications in medicinal chemistry. In this context, Rueping *et al.* have developed an enantioselective access to chromenones based on a domino Michael addition-cyclisation reaction, involving cyclohexane-1,3-dione and an α,β-unsaturated aldehyde, catalysed by silylated biarylprolinol ethers.[55] The corresponding chiral 2-hydroxychromenones were isolated in good yields and with excellent enantioselectivities of up to 98% ee, as shown in Scheme 1.42. Moreover, this efficient method was applied to other diketones such as dimedone with similar results. The resulting chromenones could be readily converted into various pharmacologically interesting compounds, such as the corresponding lactones by oxidation, and the corresponding cyclic ethers, oxadecalinones, by treatment with NaBH$_4$ with no loss of the enantioselectivity in each case.

As an extension of the precedent methodology, these authors have prepared a diverse set of chiral 1,4-pyranonaphthoquinones through the corresponding

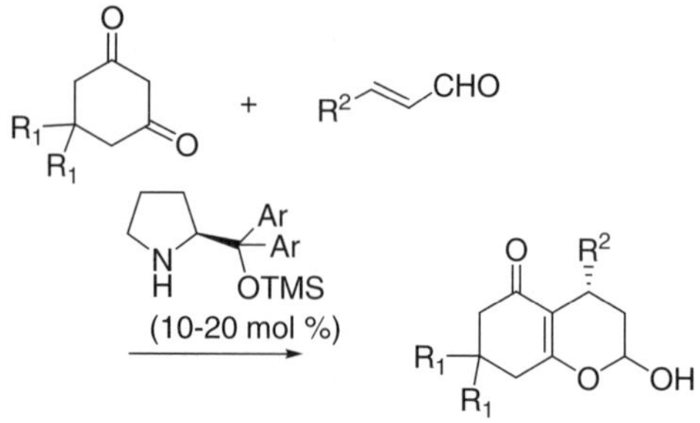

with Ar = 3,5-(CF$_3$)$_2$C$_6$H$_3$:
R^1 = H, R^2 = *n*-Pr: 78% ee = 96%
R^1 = H, R^2 = *n*-Hept: 66% ee = 96%
R^1 = H, R^2 = *n*-Dec: 67% ee = 96%
R^1 = Me, R^2 = Et: 62% ee = 94%
R^1 = Me, R^2 = *n*-Pr: 73% ee = 95%
R^1 = Me, R^2 = *n*-Bu: 68% ee = 96%
R^1 = Me, R^2 = *n*-Hept: 67% ee = 96%
R^1 = Me, R^2 = *n*-Dec: 67% ee = 96%
R^1 = Me, R^2 = Ph: 89% ee = 94%
with Ar = H:
R^1 = H, R^2 = *p*-NO$_2$C$_6$H$_4$: 66% ee = 98%
R^1 = H, R^2 = 2-Fu: 81% ee = 95%

Scheme 1.42   Silylated biarylprolinol-catalysed synthesis of chromenones.

domino reaction of 2-hydroxy-1,4-naphthoquinone with various α,β-unsaturated aldehydes (Scheme 1.43).[56] The biologically interesting formed quinones were isolated in good yields and excellent enantioselectivities (90–99% ee) and could be further converted without loss of the enantioselectivity into various 1,2-pyranonaphthoquinones, lapachones and 1,4-naphthoquinonelactones, which are considered as privileged structures in medicinal chemistry.

In addition, several asymmetric domino Michael reactions have been catalysed by organocatalysts which were not derived from proline. As an example, Gong *et al.* have reported an asymmetric three-component cyclisation of cinnamaldehydes and primary amines with 1,3-dicarbonyl compounds catalysed by a chiral phosphonic acid.[57] In a first step, the cinnamaldehyde reacted with the primary amine to give the corresponding unsaturated imine, which was added by the 1,3-dicarbonyl compound to give an intermediate, which finally cyclised to give after a final dehydration the expected chiral dihydropyridine. Scheme 1.44 collects the best results obtained for this process affording a series of dihydropyridines, which have been recognised as an important class of organic calcium-channel modulators for the treatment of cardiovascular diseases.

In 2008, Fréchet *et al.* reported a one-pot multicomponent asymmetric cascade reaction catalysed by soluble star polymers with highly branched non-interpenetrating catalytic cores.[58] Therefore, in the presence of four

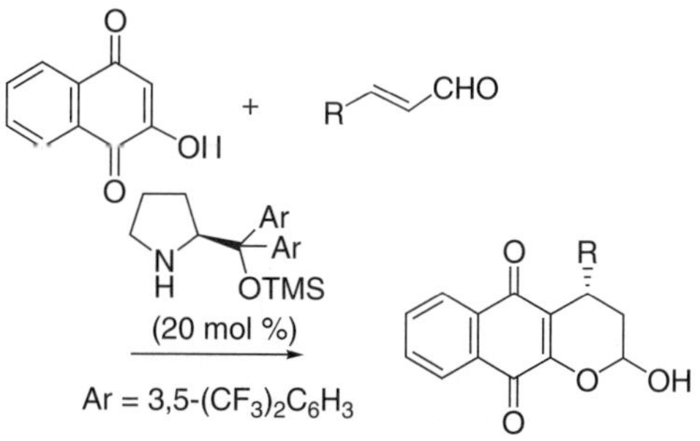

R = *i*-Pr: 77% ee = 99%
R = Et: 55% ee = 98%
R = *n*-Pr: 75% ee = 99%
R = *n*-Bu: 52% ee = 98%
R = *n*-Hept: 46% ee = 98%
R = *n*-Dec: 43% ee = 99%
R = Ph: 56% ee = 91%
R = *o*-ClC_6H_4: 57% ee = 99%
R = *o*-BrC_6H_4: 84% ee = 99%
R = *o*-Tol: 57% ee = 96%
R = *o*-$CF_3C_6H_4$: 57% ee = 96%
R = *p*-BrC_6H_4: 87% ee = 92%

Scheme 1.43 Silylated biarylprolinol-catalysed synthesis of 1,4-pyranonaphthoquinones.

incompatible catalyst components, the sophisticated domino Michael reaction occurred between *N*-methyl indole, 2-hexenal and methyl vinyl ketone, providing the corresponding cascade product in 89% yield, 86% de, and >99% ee.

1.3 Intermolecular Nitro-Michael Additions of C-Nucleophiles

Organocatalytic Michael methodologies involving nitroalkenes as acceptors are among the most widely studied,[59] in part because of the versatility of the nitro group for further synthetic manipulation and, in part, because of the optimum performance of nitroalkenes in these types of reactions. In the last two years, a large number of results have been reported dealing with the asymmetric organocatalysed conjugate addition of C-nucleophiles to nitroolefins. Among them, the Michael addition of aldehydes to nitroalkenes catalysed by silylated biphenylprolinol has been widely investigated by a number of workers, providing generally excellent enantioselectivities. Therefore, almost complete enantioselectivities combined with excellent yields were observed by the groups of Hayashi,[60] List,[61] Vicario,[62] Ma,[63] Gellman[64] and Alexakis,[65] who involved a wide range of aldehydes and nitroalkenes in the presence of this organocatalyst

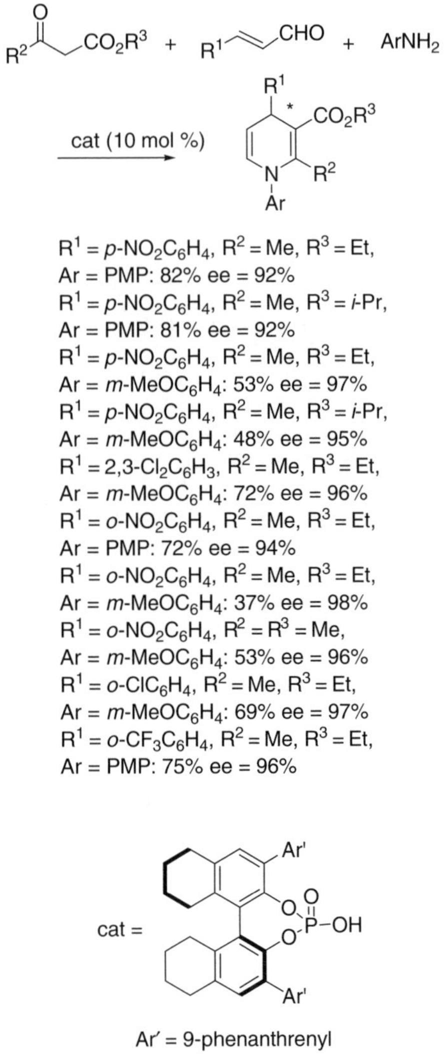

R2 CO2R3 + R1 CHO + ArNH2
O
cat (10 mol %)
R1
*
CO2R3
N R2
Ar
R1 = p-NO2C6H4, R2 = Me, R3 = Et,
Ar = PMP: 82% ee = 92%
R1 = p-NO2C6H4, R2 = Me, R3 = i-Pr,
Ar = PMP: 81% ee = 92%
R1 = p-NO2C6H4, R2 = Me, R3 = Et,
Ar = m-MeOC6H4: 53% ee = 97%
R1 = p-NO2C6H4, R2 = Me, R3 = i-Pr,
Ar = m-MeOC6H4: 48% ee = 95%
R1 = 2,3-Cl2C6H3, R2 = Me, R3 = Et,
Ar = m-MeOC6H4: 72% ee = 96%
R1 = o-NO2C6H4, R2 = Me, R3 = Et,
Ar = PMP: 72% ee = 94%
R1 = o-NO2C6H4, R2 = Me, R3 = Et,
Ar = m-MeOC6H4: 37% ee = 98%
R1 = o-NO2C6H4, R2 = R3 = Me,
Ar = m-MeOC6H4: 53% ee = 96%
R1 = o-ClC6H4, R2 = Me, R3 = Et,
Ar = m-MeOC6H4: 69% ee = 97%
R1 = o-CF3C6H4, R2 = Me, R3 = Et,
Ar = PMP: 75% ee = 96%
cat =
Ar'
O
O
P–OH
O
Ar'
Ar′ = 9-phenanthrenyl

Scheme 1.44 Phosphonic acid-catalysed synthesis of dihydropyridines.

combined or not with an additive, such as benzoic acid or 3-nitrobenzoic acid. The best results are collected in Scheme 1.45.

This catalyst was also used by Rodriguez *et al.* to promote a Michael addition between an *in situ* generated silylnitronate and an unsaturated aldehyde (Scheme 1.46).[66] The resulting γ-nitroaldehyde, obtained in excellent yield and enantioselectivity, was further converted into diversely substituted enantiopure fused isoxazolines.

On the other hand, several other analogues of proline[67] were also capable of inducing chirality for the Michael addition of various aldehydes to nitroalkenes.

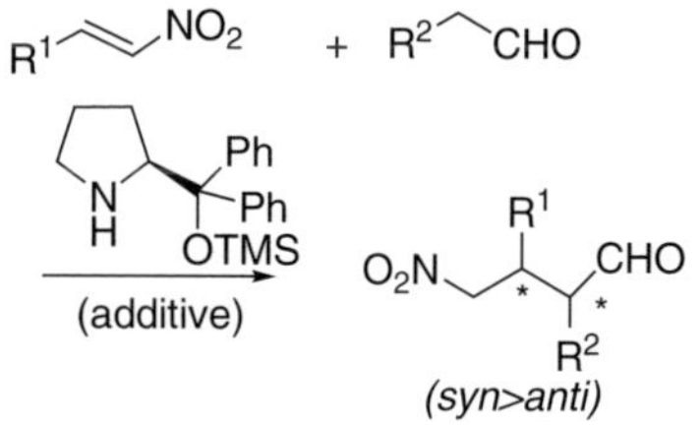

with 5 mol % of catalyst:
R^1 = (E)-PhCH=CH, R^2 = i-Pr: 91% de = 88% ee > 99%
R^1 = (E)-p-ClC$_6$H$_4$CH=CH, R^2 = i-Pr: 93% de = 84% ee = 98%
R^1 = PhC C, R^2 = i-Pr: 83% de = 88% ee = 94%
with 10 mol % of catalyst:
R^1 = Ph, R^2 = H: 75% ee = 96%
R^1 = p-ClC$_6$H$_4$, R^2 = H: 75% ee = 92%
R^1 = 2-Fu, R^2 = H: 72% ee = 97%
R^1 = t-Bu, R^2 = H: 61% ee > 99%
R^1 = CH(OMe)$_2$, R^2 = i-Pr: 75% de > 90% ee > 99%
R^1 = CH(OMe)$_2$, R^2 = n-Hex: 85% de > 90% ee > 99%
with 20 mol % of catalyst:
R^1 = Cy, R^2 = H: 61% ee = 88%
with 20 mol % of catalyst and ClCH$_2$CO$_2$H as additive:
R^1 = CO$_2$t-Bu, R^2 = Oi-Pr: 100% ee = 96%
with 0.5–2 mol % of catalyst and PhCO$_2$H as additive in water:
R^1 = CO$_2$t-Bu, R^2 = n-Pr: 90% de = 94% ee > 99%
R^1 = CO$_2$Bn, R^2 = n-Pr: 86% de = 96% ee > 99%
R^1 = CO$_2$Me, R^2 = n-Bu: 83% de = 96% ee > 99%
R^1 = Ph, R^2 = n-Pr: 96% de = 96% ee > 99%
R^1 = p-BrC$_6$H$_4$, R^2 = n-Pr: 98% de = 90% ee > 99%
R^1 = 2-Fu, R^2 = n-Pr: 96% de = 96% ee > 99%
with 2 mol % of catalyst and m-NO$_2$C$_6$H$_4$CO$_2$H as additive:
R^1 = H, R^2 = Me: 95% ee = 98%
R^1 = H, R^2 = Et: 96% ee = 98%
R^1 = H, R^2 = i-Bu: 94% ee > 99%
R^1 = H, R^2 = n-Bu: 95% ee = 99%
R^1 = H, R^2 = (CH$_2$)$_4$N(Boc)$_2$: 92% ee = 98%
R^1 = H, R^2 = CH$_2$CO$_2$Me: 92% ee = 96%

Scheme 1.45 Silylated biphenylprolinol-catalysed Michael additions of aldehydes to nitroalkenes.

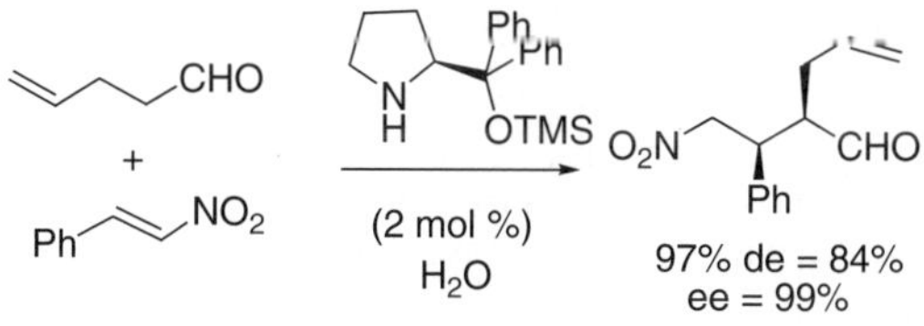

Scheme 1.46 Silylated biphenylprolinol-catalysed Michael addition of unsaturated aldehyde to nitrostyrene.

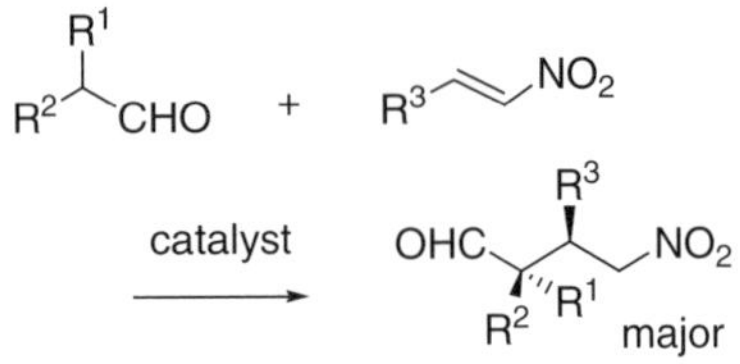

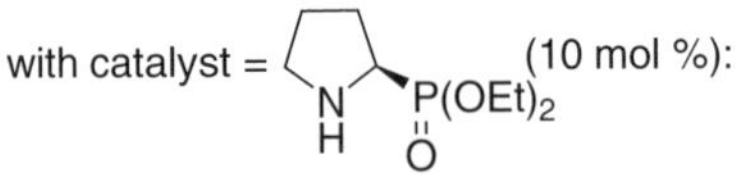

R^1 = R^2 = Me, R^3 = Ph: 78% ee = 91%
R^1 = H, R^2 = Et, R^3 = Ph: 80% de = 72% ee = 87%
R^1 = H, R^2 = Et, R^3 = p-BrC$_6$H$_4$: 78% de = 50% ee = 96%
R^1 = H, R^2 = n-Hex, R^3 = p-FC$_6$H$_4$: 78% de > 98% ee = 87%

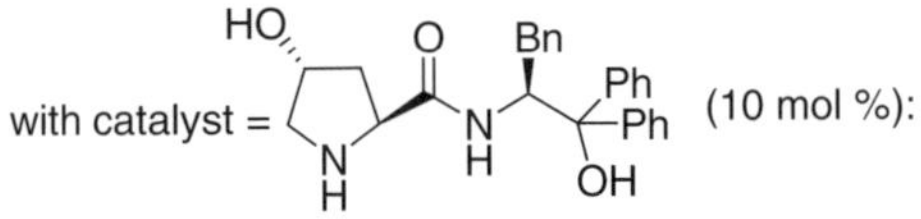

R^1 = H, R^2 = Me, R^3 = Ph: 71% de = 98% ee = 93%
R^1 = H, R^2 = Me, R^3 = p-ClC$_6$H$_4$: 65% de = 92% ee = 94%
R^1 = H, R^2 = n-Pr, R^3 = Bn: 72% de = 90% ee = 93%
R^1 = H, R^2 = n-Pr, R^3 = 2-thienyl: 96% ee = 95%

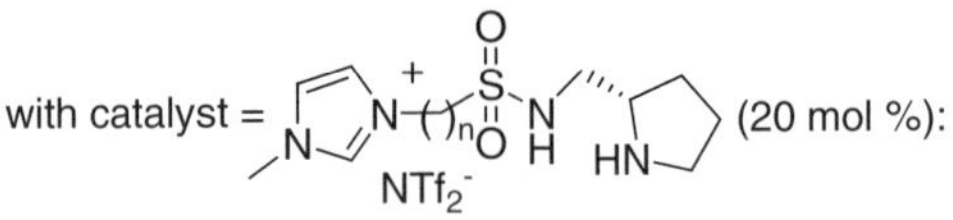

R^1 = R^2 = Me, R^3 = Ph, n = 1: 90% ee = 84%
R^1,R^2 = (CH$_2$)$_4$, R^3 = Ph, n = 1: 82% ee = 85%
R^1 = R^2 = Me, R^3 = p-Tol, n = 1: 43% ee = 80%
R^1 = H, R^2 = Me, R^3 = Ph, n = 1: 99% de = 45% ee = 88%
R^1 = H, R^2 = n-Bu, R^3 = Ph, n = 3: 64% de = 68% ee = 94%
R^1 = H, R^2 = n-Pr, R^3 = p-MeOC$_6$H$_4$, n = 3: 29% de = 67%
ee = 84%

Scheme 1.47   Michael additions of aldehydes to nitroalkenes catalysed by various pyrrolidines.

As an example, Tang *et al.* have reported good yields and stereoselectivities for these types of reactions when a chiral pyrrolidine-phosphonate such as that depicted in Scheme 1.47 was applied as the organocatalyst.[68] Similar reactions were also performed by Takeshita *et al.* in the presence of a 4-hydroxy-prolinamide alcohol having one covalent site and different non-covalent coordination sites (Scheme 1.47).[69] Enantioselectivities of up to 95% ee combined with diastereoselectivities of up to 98% ee and good yields could be achieved and explained by DFT calculations. It was proposed that the hydroxyl group at the 4-position on the pyrrolidine ring made a hydrogen bond with a substrate molecule and the amide moiety at the 2-position on the pyrrolidine ring acted to control the equilibrium between enamine conformers and blocked one enamine face to afford a high enantioselectivity. In addition, a new class of recyclable pyrrolidine-based functionalised chiral ionic liquids in combination with TfOH as an additive has been developed by Headley *et al.* for the asymmetric Michael addition of aldehydes to nitroalkenes, providing the expected Michael products in good enantioselectivities of up to 85% ee, high yields and diastereoselectivities of up to 94% de in favour of the *syn* adduct (Scheme 1.47).[70]

A series of pyrrolidine catalysts has been investigated for the asymmetric Michael addition of cyclohexanone to nitroalkenes, which is the classical model commonly used. When an aminal-pyrrolidine, developed by Alexakis *et al.*,[8] was employed as an organocatalyst for the conjugate addition of cyclohexanone to nitrostyrene, it led quantitatively to the Michael product with 84% de and 87% ee (Scheme 1.48). Better stereoselectivities for this reaction were

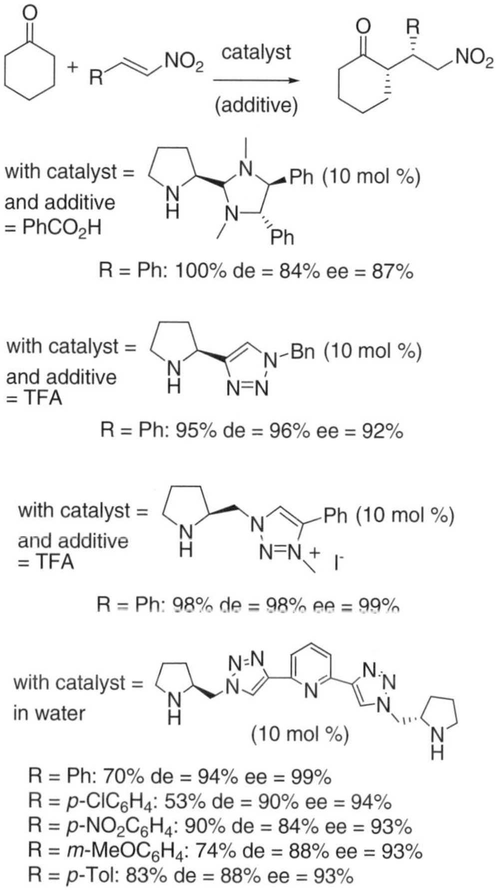

with catalyst = (10 mol %)
and additive = PhCO$_2$H

R = Ph: 100% de = 84% ee = 87%

with catalyst = (10 mol %)
and additive = TFA

R = Ph: 95% de = 96% ee = 92%

with catalyst = (10 mol %)
and additive = TFA

R = Ph: 98% de = 98% ee = 99%

with catalyst = (10 mol %)
in water

R = Ph: 70% de = 94% ee = 99%
R = *p*-ClC$_6$H$_4$: 53% de = 90% ee = 94%
R = *p*-NO$_2$C$_6$H$_4$: 90% de = 84% ee = 93%
R = *m*-MeOC$_6$H$_4$: 74% de = 88% ee = 93%
R = *p*-Tol: 83% de = 88% ee = 93%

Scheme 1.48 Michael additions of cyclohexanone to nitrostyrenes catalysed by various pyrrolidines.

obtained by Chandrasekhar *et al.* by using a novel chiral pyrrolidine-triazole as the organocatalyst in combination with TFA as an additive, as shown in Scheme 1.47.[71] Almost complete diastereo- and enantioselectivities combined with an almost quantitative yield were attained by Liebscher *et al.* for this reaction by employing new recyclable task specific ionic liquids assembled from a chiral pyrrolidin-2-ylmethyl moiety and a triazolium salt as phase tag (Scheme 1.48).[72] In addition, the efficiency of a novel pyrrolidine-triazole-based C_2-symmetric organocatalyst to induce chirality for the Michael addition of cyclohexanone was extended to a series of substituted β-nitrostyrenes, providing the corresponding Michael adducts in high yields, diastereoselectivities of up to 94% ee and enantioselectivities of up to 99% ee (Scheme 1.48).[73]

A novel and effective organocatalytic system consisting in chiral pyrrolidinyl-thioimidazole and a thioureido acid additive was demonstrated by Xu *et al.* to efficiently promote the asymmetric Michael addition of ketones to nitroolefins, affording the products with high diastereoselectivities of up to 98% de and excellent enantioselectivities of up to 99% ee (Scheme 1.49).[74] Previously, these authors have shown that the use of salicylic acid as an additive to this catalyst in these reactions gave similar results but at a higher catalyst loading of 20 mol % instead of 5 mol % with the thioureido acid additive.[75]

In 2009, Peng *et al.* reported the synthesis of chiral 4-trifluoromethanesulfonamidyl prolinol *tert*-butyldiphenylsilyl ether and its highly efficient application as an organocatalyst for the Michael addition of cyclohexanone to various nitroalkenes, providing almost quantitative yields combined to

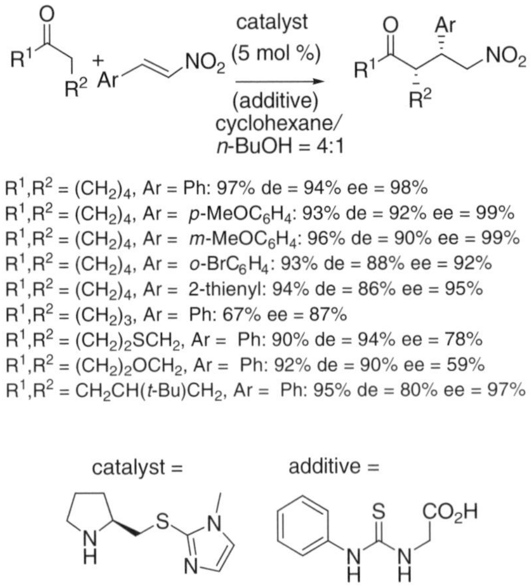

$R^1, R^2 = (CH_2)_4$, Ar = Ph: 97% de = 94% ee = 98%
$R^1, R^2 = (CH_2)_4$, Ar = p-MeOC$_6$H$_4$: 93% de = 92% ee = 99%
$R^1, R^2 = (CH_2)_4$, Ar = m-MeOC$_6$H$_4$: 96% de = 90% ee = 99%
$R^1, R^2 = (CH_2)_4$, Ar = o-BrC$_6$H$_4$: 93% de = 88% ee = 92%
$R^1, R^2 = (CH_2)_4$, Ar = 2-thienyl: 94% de = 86% ee = 95%
$R^1, R^2 = (CH_2)_3$, Ar = Ph: 67% ee = 87%
$R^1, R^2 = (CH_2)_2SCH_2$, Ar = Ph: 90% de = 94% ee = 78%
$R^1, R^2 = (CH_2)_2OCH_2$, Ar = Ph: 92% de = 90% ee = 59%
$R^1, R^2 = CH_2CH(t\text{-Bu})CH_2$, Ar = Ph: 95% de = 80% ee = 97%

catalyst =   additive =

Scheme 1.49   Pyrrolidinyl-thioimidazole-catalysed Michael additions of ketones to nitroalkenes.

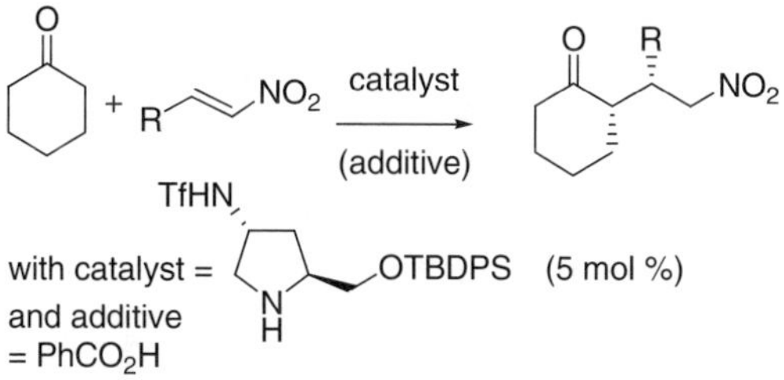

R = Ph: 100% de = 90% ee = 98%
R = *p*-FC_6H_4: 100% de = 92% ee = 98%
R = *p*-ClC_6H_4: 100% de = 94% ee = 99%
R = *o*-ClC_6H_4: 100% de = 98% ee = 99%
R = *o*-BrC_6H_4: 96% de = 88% ee = 98%
R = *m*-$NO_2C_6H_4$: 100% de = 86% ee = 99%
R = *m*-BrC_6H_4: 99% de = 90% ee = 97%
R = *p*-Tol: 100% de = 92% ee = 95%
R = *o*-Tol: 90% de = 98% ee = 98%
R = 2-Naph: 95% de = 92% ee = 99%

Scheme 1.50 4-Trifluoromethanesulfonamidyl prolinol *tert*-butyldiphenylsilyl ether-catalysed Michael additions of cyclohexanone to nitroalkenes.

excellent diastereo- and enantioselectivities (Scheme 1.50).[76] Control experiments suggested that the *trans*-configuration relationship between the bulky group (CH_2OTBDPS) and the sulfonamido group at the 4-position of the pyrrolidine ring was important to achieve high yield and stereoselectivity. The condensation of cyclohexanone to styrene was also studied by Kokotos in the presence of the *p*-toluenesulfonamide of the non-natural amino acid homoproline as the organocatalyst but, however, the Michael product was obtained in only 33% yield, 92% ee and a similar diastereoselectivity of 90% de compared when prepared by using the precedent catalyst, but at a higher catalyst loading (10 mol % instead of 5 mol %).[77]

An alternative approach to the attainment of high stereoselectivity by means of substrate control could be achieved by using a proline derivative in which the amino acid moiety was a part of a chiral cavity in which the reaction took place. In this context, Iuliano and Puleo have synthesised prolinamide derivatives of bile acids by linking a proline moiety to the different functionalised positions of cholic and deoxycholic acids, which were successfully used as organocatalysts in the Michael addition of cyclohexanone to various nitroolefins.[78] It was shown that the derivative bearing a D-prolinamide moiety linked at the 7-position of the cholic acid emerged as the most efficient, giving the Michael adducts in satisfactory yields and enantioselectivities of up to 95% ee (Scheme 1.51).The corresponding system having a free OH group at the 3-position of the cholestanic backbone afforded the opposite enantiomer of the product, suggesting that the transition state was developed at the inner part of the cholestanic cavity, which was responsible for the substrate control determining the stereochemical outcome of the reaction.

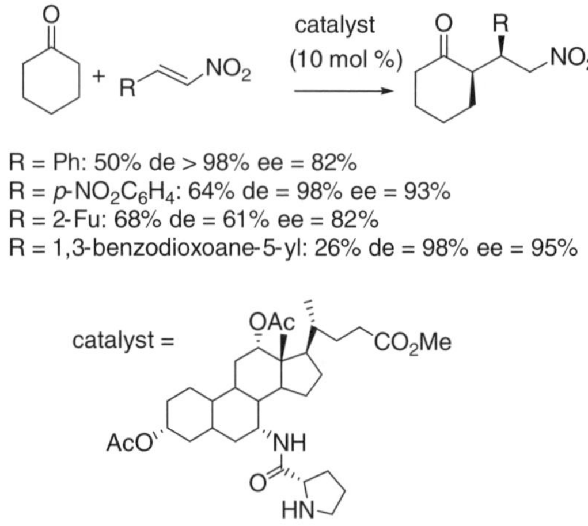

R = Ph: 50% de > 98% ee = 82%
R = p-NO$_2$C$_6$H$_4$: 64% de = 98% ee = 93%
R = 2-Fu: 68% de = 61% ee = 82%
R = 1,3-benzodioxoane-5-yl: 26% de = 98% ee = 95%

Scheme 1.51 Prolinamide cholic acid-catalysed Michael additions of cyclohexanone to nitroalkenes.

In 2008, chiral cyclic α-aminophosphonates were found to be novel organocatalysts for the Michael addition of cyclohexanone to various nitroolefins. Using a catalyst loading of 10 mol % of the isopropyl pyrrolidine-phosphonate, Tang *et al.* prepared different Michael products in high yields and diastereoselectivities of up to 98% de combined with moderate to good enantioselectivities of up to 72% ee (Scheme 1.52).[68] Higher enantioselectivities of up to 96% ee, albeit combined with lower yields and diastereoselectivities, were obtained by using the corresponding ethylphosphonate at a catalyst loading of 20 mol % and in the presence of TFA as an additive (Scheme 1.52).[79] In 2009, a remarkably efficient organocatalyst allowed excellent yields (87–99%) and stereoselectivities (96–98% de and 92–99% ee) to be obtained for similar reactions.[80] Indeed, this new catalyst containing a pyrrolidine unit and a phosphine oxide moiety was proven to be very important in controlling the stereochemistry of the adducts (Scheme 1.52).

In recent years, several studies on asymmetric Michael additions using immobilised proline derivatives on various supports have been reported. To develop highly enantioselective and efficient chiral organocatalysts with applicability and easy recyclability in this field is essential. In 2008, Ma *et al.* explored a silica-supported triazole-pyrrolidine connecting through an amine tether as a recyclable catalyst for the asymmetric Michael addition of cyclohexanone to a range of nitroalkenes.[81] As shown in Scheme 1.53, this supported catalyst exhibited a good activity with high diastereo- and enantioselectivities of up to 90% de and 93% ee, respectively. Furthermore, this procedure permitted recycling of the catalyst up to four times without loss of the activity. Triazole-pyrrolidine has been anchored by Gao *et al.* to dendrimers and applied as an organocatalyst in the presence of TFA as an additive for the Michael

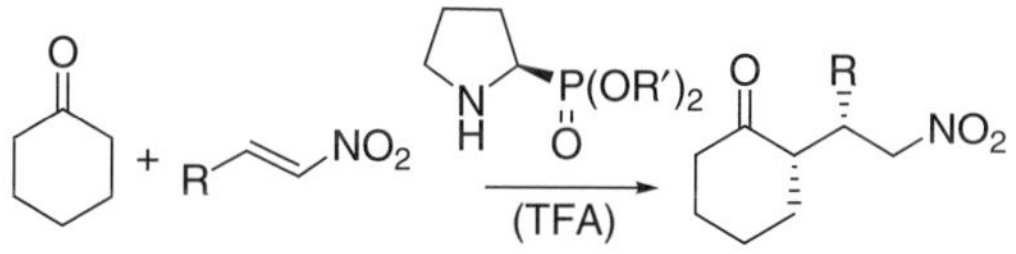

at 10 mol %:
R′ = *i*-Pr, R = Ph: 94% de = 96% ee = 79%
R′ = *i*-Pr, R = *p*-ClC$_6$H$_4$: 97% de = 94% ee = 72%
R′ = *i*-Pr, R = 2,4-Cl$_2$C$_6$H$_3$: 95% de > 98% ee = 72%
at 20 mol % and with TFA:
R′ = Et, R = Ph: 92% de = 86% ee = 90%
R′ = Et, R = *p*-NO$_2$C$_6$H$_4$: 92% de = 86% ee = 90%
R′ = Et, R = 2,4-Cl$_2$C$_6$H$_3$: 95% de = 90% ee = 90%

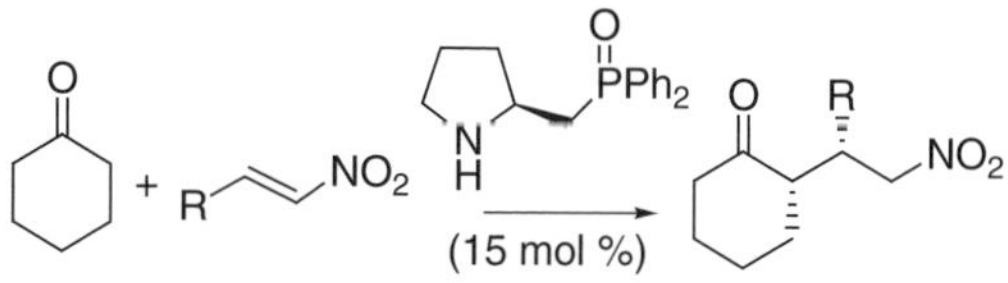

R = Ph: 99% de > 98% ee > 99%
R = *p*-MeOC$_6$H$_4$: 96% de > 98% ee = 97%
R = *m*-Tol: 98% de > 98% ee = 98%
R = 2-Fu: 97% de > 98% ee > 99%
R = 3-Fu: 98% de = 98% ee = 98%
R = *p*-BrC$_6$H$_4$: 97% de > 98% ee = 99%
R = *p*-ClC$_6$H$_4$: 96% de = 96% ee = 98%
R = *o*-ClC$_6$H$_4$: 93% de = 98% ee = 99%

Scheme 1.52 Michael additions of cyclohexanone to nitroalkenes catalysed by α-aminophosphonates or pyrrolidine phosphine oxide.

additions of cyclohexanone to nitroalkenes, providing good yields, high diastereoselectivities of up to 96% de and enantioselectivities of up to 95% ee (Scheme 1.53).[82] Furthermore, this novel catalyst could be reused at least five times without significant loss of either the activity or the stereoselectivity. The use of acyclic ketones as Michael acceptors was also evaluated but, however, the corresponding Michael products were isolated in only moderate yields (≤60%) and with a dramatic drop in both enantioselectivity (≤72% ee) and diastereoselectivity (≤78% de). In addition, Wang and Miao have immobilised triazole-pyrrolidine on polystyrene, and obtained, using this novel catalyst in the presence of TFA as an additive, almost complete enantio- and diastereoselectivity combined with excellent yields for the addition of cyclohexanone to a wide range of nitroalkenes (Scheme 1.53).[83] Furthermore, this highly efficient catalyst could be recovered and recycled by a simple filtration and used for more than ten consecutive trials without loss of the activity.

In addition, several supported chiral ionic liquids have been investigated for these reactions. Therefore, Wang *et al.* have developed a new type of polymer-immobilised pyrrolidine-based chiral ionic liquids, which were capable of inducing the Michael addition of ketones to nitrostyrenes in high yields,

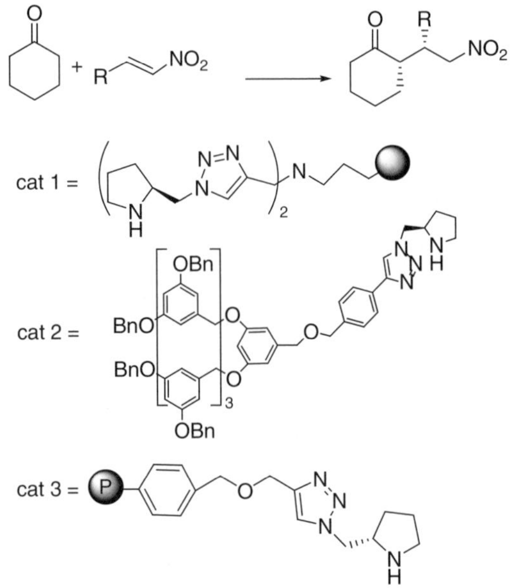

with cat 1 (10 mol %):
R = Ph: 97% de = 92% ee = 91%
R = *p*-$MeOC_6H_4$: 98% de > 90% ee = 84%
R = *o*-ClC_6H_4: 96% de > 90% ee = 89%
R = *p*-ClC_6H_4: 97% de > 90% ee = 93%
R = *p*-BrC_6H_4: 96% de = 92% ee = 89%
with cat 2 (10 mol %, with TFA):
R = Ph: 99% de = 96% ee = 93%
R = *p*-ClC_6H_4: 93% de = 96% ee = 93%
R = *p*-BrC_6H_4: 94% de = 96% ee = 95%
R = *p*-Tol: 96% de = 96% ee = 92%
R = *p*-$MeOC_6H_4$: 94% de = 94% ee = 93%
R = 2-Fu: 93% de = 92% ee = 92%
with cat 3 (10 mol %, with TFA):
R = Ph: 97% de > 99% ee > 98%
R = *p*-FC_6H_4: 94% de = 92% ee > 98%
R = *p*-BrC_6H_4: 90% de > 99% ee = 96%
R = *p*-ClC_6H_4: 92% de = 99% ee > 98%
R = *p*-$CF_3C_6H_4$: 91% de = 96% ee = 98%
R = *p*-$AcOC_6H_4$: 85% de > 99% ee > 98%
R = *o*-ClC_6H_4: 92% de = 90% ee = 96%

Scheme 1.53 Michael additions of cyclohexanone to nitroalkenes catalysed by supported triazole-pyrrolidine.

excellent enantioselectivities of up to 99% ee and diastereoselectivities of up to 98% de (Scheme 1.54).[84] Moreover, this catalyst could be reused at least eight times without significant loss of the activity. The Michael reactions were evaluated with ketones other than cyclohexanone, and it was found that tetrahydrothiopyran-4-one and tetrahydro-4*H*-pyran-4-one were also suitable substrates as Michael donors, providing enantioselectivities of 93% ee and 95%

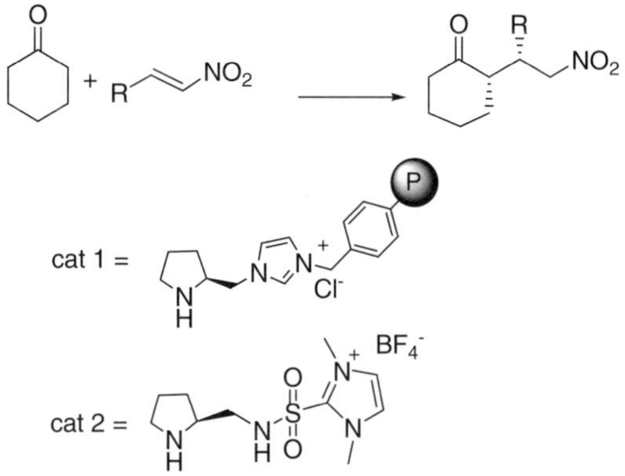

Scheme 1.54 is illustrated with the reaction scheme and catalyst structures, followed by:

with cat 1 (10 mol %).
R = Ph: 97% de = 98% ee = 99%
R = p-Tol: 96% de > 98% ee > 99%
R = p-MeOC$_6$H$_4$: 96% de = 98% ee = 99%
R = p-CF$_3$C$_6$H$_4$: 98% de > 98% ee > 99%
R = p-BrC$_6$H$_4$: 96% de = 96% ee = 96%
R = o-BrC$_6$H$_4$: 96% de = 98% ee = 98%
R = p-ClC$_6$H$_4$: 97% de = 98% ee = 99%
R = 2-thienyl: 90% de = 98% ee > 99%
R = 2-Naph: 86% de = 96% ee = 98%
with cat 2 (10 mol %):
R = Ph: 91% de = 90% ee = 90%
R = p-Tol: 90% de = 98% ee = 99%
R = p-NO$_2$C$_6$H$_4$: 83% de = 98% ee = 98%
R = p-CF$_3$C$_6$H$_4$: 90% de = 96% ee = 95%
R = p-BrC$_6$H$_4$: 87% de = 86% ee = 94%
R = 2,4-Cl$_2$C$_6$H$_3$: 96% de = 98% ee = 99%
R = 2-Fu: 97% de = 84% ee = 80%

Scheme 1.54   Michael additions of cyclohexanone to nitroalkenes catalysed by supported pyrrolidine-based ionic liquids.

ee, respectively, along with high yields and diastereoselectivities. However, acetone and cyclopentanone served as efficient Michael donors to generate the desired products with excellent yields (96–98%), but with moderate to good enantioselectivities ($\leq$57% ee). In 2009, Headley *et al.* applied a new class of ionic liquid supported (*S*)-pyrrolidine sulfonamide organocatalyst to the Michael additions of ketones to nitroolefins.[85] Excellent diastereo- and enantioselectivities of up to 98% de and 99% ee, respectively, were obtained along with excellent yields, as shown in Scheme 1.54. This catalyst could be reused for at least five times without significant loss of the activity. The Michael additions of ketones other than cyclohexanone, such as tetrahydrothiopyran-4-one and tetrahydro-4*H*-pyran-4-one were also investigated, providing enantioselectivities of 92% ee in each case. On the other hand, ketones such as acetone and cyclopentanone gave poor enantioselectivities ($\leq$14% ee).

Other-than-proline derivatives have been successfully employed to catalyse a great number of asymmetric nitro-Michael reactions. Therefore, excellent results have been reported for these reactions performed in the presence of various chiral amine-thioureas as the organocatalysts. The Michael addition of acetylacetone to nitroolefins catalysed by chiral bifunctional amine-thiourea catalysts has been investigated by several authors, providing in each case the corresponding adducts in high yields and stereoselectivities. As shown in Scheme 1.55, a bifunctional thiourea bearing both a tertiary amino group and a saccharide scaffold,[86] bifunctional thioureas bearing a chiral cyclohex-anediamine moiety,[87] as well as a bifunctional amine-thiourea bearing a binaphthyl unit[88] have allowed a range of Michael adducts to be obtained in comparable high yields and stereoselectivities.

Chiral amine-thiourea catalysts have been successfully applied to the Michael addition of other nucleophiles such as malonates to nitroalkenes. As an example, Pedrosa *et al.* have obtained excellent yields and enantioselec-tivities for the addition of diethyl malonate to a series of nitroalkenes by using a thiourea derived from D-valine (Scheme 1.56).[89] In addition, the key step of the total syntheses of the antidepressant (*R*)-rolipram and the selective serotonin reuptake inhibitor (3*S*,4*R*)-paroxetine, reported by Dixon *et al.*, was a highly enantioselective Michael addition of dimethyl malonate to a nitroolefin cata-lysed by a cinchona alkaloid amine-thiourea.[90] Other excellent results were reported independently by Wang *et al.* (Scheme 1.56) and Takemoto and Miyabe[21] for the Michael addition of α-substituted β-keto esters to nitroolefins performed in the presence of a chiral bifunctional amine thiourea bearing a cyclohexane unit.[91] In addition, Michael additions of various ketones have been performed by Mandal and Zhao in the presence of a mixture of two self-assembled chiral organocatalysts, such as a quinidine thiourea and L-phenyl-glycine, which led to the corresponding ketones with moderate to good yields (40–92%) and diastereoselectivities (50–92% de) combined with high enantioselectivities (84–99% ee).[92] The assembly of these two organocatalysts provided the corresponding ammonium salt by reaction between the carboxylic acid of the amino acid and the tertiary amine group of the cinchona alkaloid.

Various other C-nucleophiles have been submitted to the asymmetric nitro-Michael addition in the presence of chiral thiourea catalysts. Thus, nitroalkanes could be added to nitroalkenes with high asymmetric inductions (91–95% ee), good yields (55–94%) and moderate to good diastereoselectivities (48–80% de) in the presence of a BINAM bifunctional thiourea as the organocatalyst.[93] Remarkably, the asymmetric induction increased with decreasing the catalyst loading with the optimal compromise between rate and induction at a loading of 2 mol %. The cinchona alkaloid thiourea organocatalyst depicted in Scheme 1.57 was applied to promote the Michael addition of oxazolones to nitroalk-enes by Jorgensen *et al.*, which provided the Michael adducts in good yields (61–98%), good to high diastereoselectivities (60–90% de) and enantiose-lectvities of up to 92% ee.[94] These products were further converted into the corresponding α,α-quaternary α-amino acid derivatives. On the other hand, asymmetric Michael reactions of fluorinated nucleophiles with nitroolefins catalysed by this cinchona alkaloid thiourea organocatalyst generated the desired Michael products containing vicinal fluorinated quaternary and tertiary

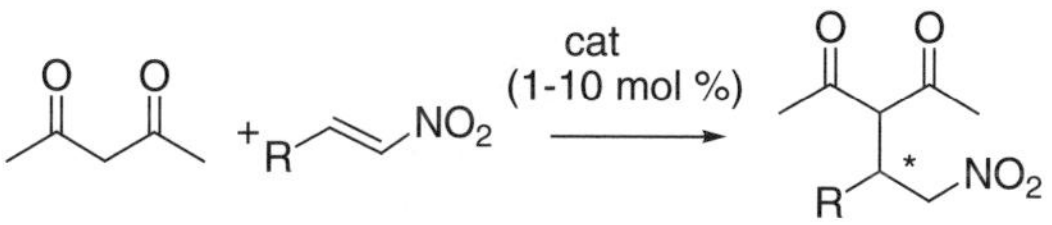

with cat 1 (10 mol %):
R = Ph: 93% ee = 96%
R = *p*-Tol: 80% ee = 89%
R = *m*-MeOC$_6$H$_4$: 99% ee = 92%
R = *o*-MeOC$_6$H$_4$: 89% ee = 92%
R = 1-Naph: 97% ee = 95%
R = 2-Fu: 99% ee = 94%

with cat 2 (1 mol %):
R = Ph: 97% ee = 97%
R = *p*-Tol: 96% ee = 95%
R = *o*-Tol: 95% ee = 98%
R = *p*-MeOC$_6$H$_4$: 93% ee = 98%
R = *m*-MeOC$_6$H$_4$: 96% ee = 97%
R = *o*-BrC$_6$H$_4$: 97% ee = 96%
R = *p*-FC$_6$H$_4$: 96% ee = 99%

with cat 3 (10 mol %):
R = Ph: 82% ee = 93%
R = *p*-Tol: 80% ee = 93%
R = *p*-MeOC$_6$H$_4$: 85% ee = 92%
R = *p*-ClC$_6$H$_4$: 86% ee = 94%
R = *p*-BrC$_6$H$_4$: 84% ee = 94%
R = *o*-BnOC$_6$H$_4$: 78% ee = 96%

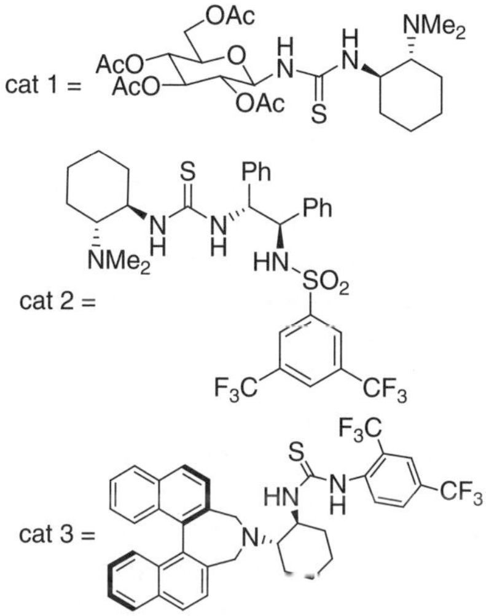

cat 1 =

cat 2 =

cat 3 =

Scheme 1.55 Michael additions of acetylacetone to nitroalkenes catalysed by bifunctional amine-thioureas.

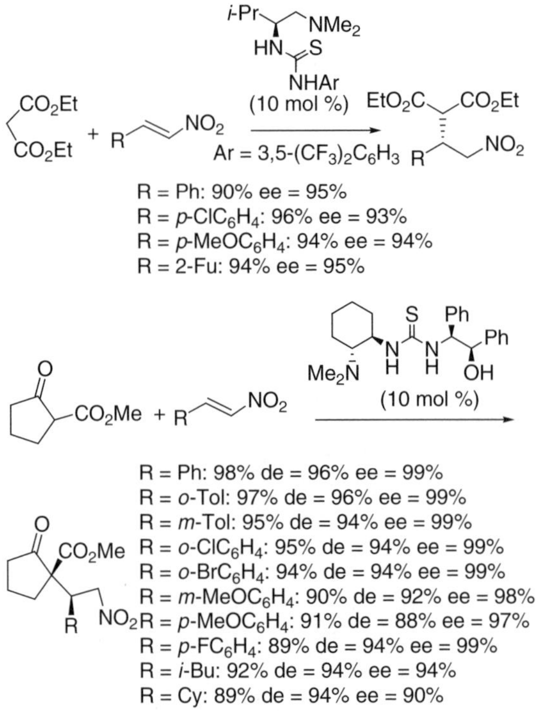

Scheme 1.56 Michael additions of malonate and α-substituted β-keto esters to nitroalkenes catalysed by bifunctional amine-thioureas.

chiral centres with exceptional enantioselectivities of up to 98% ee, as shown in Scheme 1.56.[95] The first organocatalytic enantioselective Michael addition of 2-hydroxy-1,4-naphthoquinones to nitroalkenes was investigated by Du *et al.* in the presence of a chiral amine-thiourea (Scheme 1.57).[96] The formed nitroalkylated naphthoquinone derivatives were obtained in good yields and excellent enantioselectivities of up to 99% ee.

In addition, a collection of other chiral organocatalysts has been developed to promote the Michael additions of various C-nucleophiles to nitroalkenes. In particular, the conjugate addition of ketones to nitroalkenes has been studied in the presence of several organocatalysts, such as a chiral bifunctional sulfonamide primary amine derived from chiral *trans*-cyclohexane, which allowed good yields (33–94%) combined with good levels of enantioselectivity (57–96% ee) to be obtained for the Michael products at 15 mol % of catalyst loading.[97] On the other hand, a highly diastereoselective and enantioselective Michael addition of cyclohexanone, acetone and other ketones to nitroolefins was developed by Hu *et al.* by the use of an amine organocatalyst based on bispidine (Scheme 1.58).[98] Additionally, a theoretical study of transition structures revealed that this bispidine-based primary-secondary amine catalyst could serve through an enamine intermediate and H-bond interaction, which

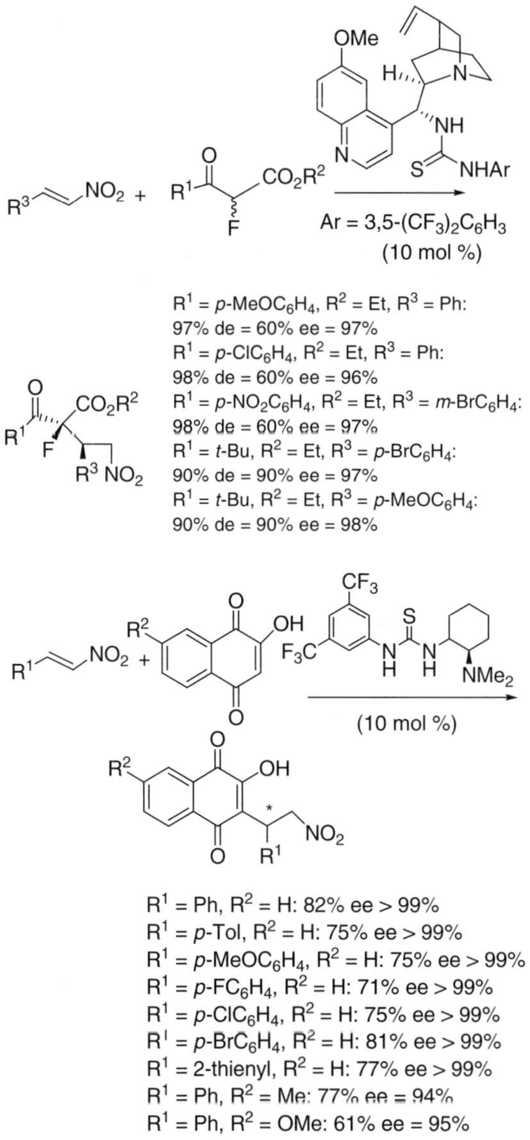

Scheme 1.57 Michael additions of fluorinated methines and naphthoquinones to nitroalkenes catalysed by bifunctional amine-thioureas.

was important for the reactivity and selectivity of this reaction. A heterogeneous version of the asymmetric Michael addition of ketones to nitroalkenes was proposed by Cheng *et al.*, involving a supported organocatalyst derived from the *in situ* complexation of polystyrene/sulfonic acids and simple chiral diamines through acid-base interactions.[99] This novel recyclable catalyst

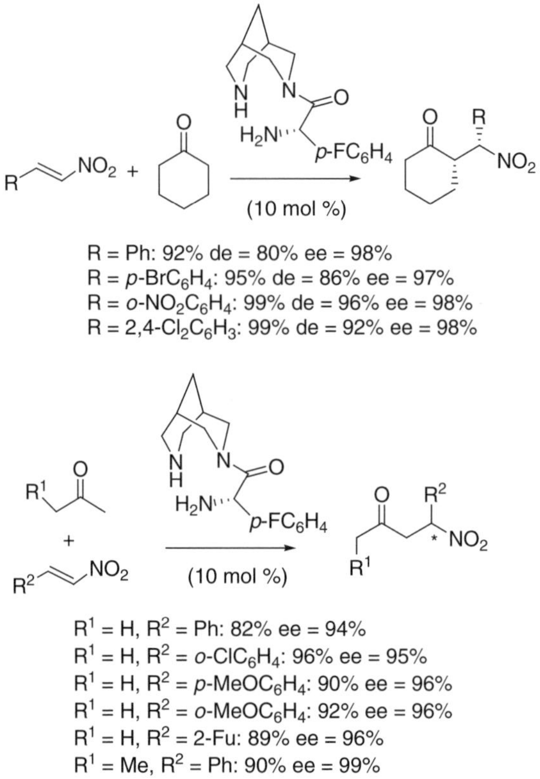

R = Ph: 92% de = 80% ee = 98%
R = p-BrC$_6$H$_4$: 95% de = 86% ee = 97%
R = o-NO$_2$C$_6$H$_4$: 99% de = 96% ee = 98%
R = 2,4-Cl$_2$C$_6$H$_3$: 99% de = 92% ee = 98%

R^1 = H, R^2 = Ph: 82% ee = 94%
R^1 = H, R^2 = o-ClC$_6$H$_4$: 96% ee = 95%
R^1 = H, R^2 = p-MeOC$_6$H$_4$: 90% ee = 96%
R^1 = H, R^2 = o-MeOC$_6$H$_4$: 92% ee = 96%
R^1 = H, R^2 = 2-Fu: 89% ee = 96%
R^1 = Me, R^2 = Ph: 90% ee = 99%

Scheme 1.58   Michael additions of ketones to nitroalkenes catalysed by bispidine-derived amine.

has allowed Michael products derived from a range of ketones and nitro-alkenes to be obtained with moderate to excellent enantioselectivities (20–90% ee), diastereoselectivities of up to 92% de and moderate to high yields (21–98%).

C-Nucleophiles such as aldehydes have been condensed to nitroethylene by Wennemers *et al.* in the presence of peptides, affording synthetically useful chiral monosubstituted γ-nitroaldehydes in high yields and high enantioselectivities, requiring catalyst loadings of only 1 mol %.[100] In particular, peptide H-D-Pro-Pro-Glu-NH$_2$ was found to be the most efficient catalyst for the title reaction, providing high general yields (67–90%) and excellent general enantioselectivities (95–99% ee). The enantiopure γ-nitroaldehydes were readily converted into γ-amino acids. Noyori's Ts-DPEN ligand bearing an amino sulfonamide moiety and with a primary amino group on a chiral scaffold was found to be a simple and efficient bifunctional organocatalyst for the asymmetric Michael addition of 1,3-dicarbonyl indane compounds to nitroolefins, which gave highly functionalised Michael adducts with quaternary centres in

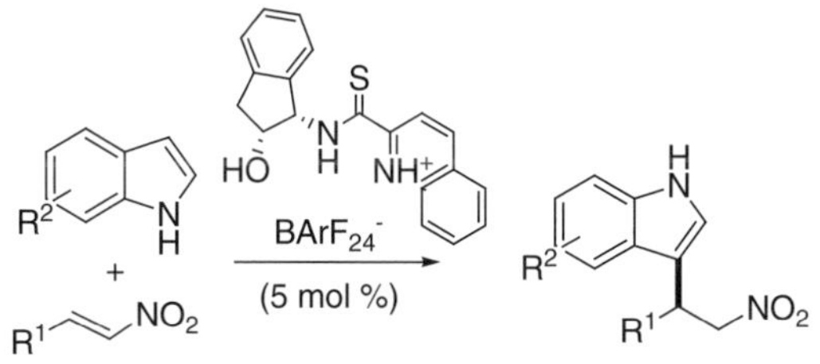

R^1 = Ph, R^2 = H: 92% ee = 94%
R^1 = Ph, R^2 = 5-OMe: 84% ee = 98%
R^1 = Ph, R^2 = 5-Br: 96% ee = 95%
R^1 = p-ClC$_6$H$_4$, R^2 = H: 96% ee = 93%
R^1 = o-ClC$_6$H$_4$, R^2 = H: 90% ee = 94%
R^1 = 2-Naph, R^2 = H: 91% ee = 97%
R^1 = i-Pr, R^2 = H: 92% ee = 94%
R^1 = CO$_2$Et, R^2 = H: 80% ee = 95%

Scheme 1.59 Michael additions of indoles to nitroalkenes catalysed by protonated quinolinium thioamide.

moderate to good enantio- (60–84% ee) and diastereoselectivities (46–70% de), along with general high yields (72–87%).[101] Highly enantioselective Michael additions of indoles to nitroalkenes were achieved by using a new type of chiral protonated quinolinium thioamide catalyst.[102] By using a catalyst loading of only 5 mol%, the Michael adducts were obtained in excellent yields and enantioselectivities, as shown in Scheme 1.59.

Finally, several cinchona alkaloid-derived catalysts were successfully applied to asymmetric Michael additions of various C-nucleophiles to nitroalkenes. In this context, Ricci *et al.* have accomplished the first organocatalysed phase-transfer enantioselective Michael addition of cyanide ion derived from acetone cyanohydrin to β,β′-disubstituted nitroolefins, which led to the formation of an all-carbon quaternary stereogenic centre with moderate to good enantioselec-tivities (33–72% ee) and yields (52–75%).[103] In this process, the most efficient organocatalyst used at 10 mol% of catalyst loading was a cinchoninium salt. Another cinchona alkaloid-derivative was employed by Wang *et al.* to promote the synthesis of chiral fluorinated quaternary carbon containing β-keto esters on the basis of an asymmetric Michael addition of α-fluoroketo esters to nitroolefins.[104] By using a catalyst loading as low as 1 mol%, the Michael adducts were isolated in general high yields (75–98%) and enantioselectivities (86–99% ee), albeit with moderate diastereoselectivities (50–60% de). In 2009, Lu *et al.* introduced a novel type of cinchona alkaloid-derived catalysts incorporating N-sulfonamides into quinidine structural scaffold.[105] Applied to the Michael addition of bicyclic α-substituted β-keto esters to nitroolefins, these catalysts allowed a wide series of Michael adducts to be obtained in high yields (96–99%), excellent enantioselectivities (91–96% ee) and moderate to good diastereoselectivities (42–80% de), as shown in Scheme 1.60.

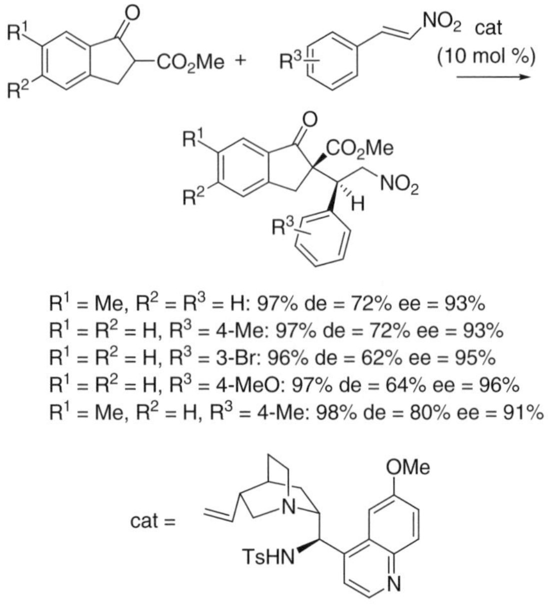
O
R1
R2
CO2Me +
R3
NO2 cat
(10 mol %)
CO2Me
NO2
H
R1 = Me, R2 = R3 = H: 97% de = 72% ee = 93%
R1 = R2 = H, R3 = 4-Me: 97% de = 72% ee = 93%
R1 = R2 = H, R3 = 3-Br: 96% de = 62% ee = 95%
R1 = R2 = H, R3 = 4-MeO: 97% de = 64% ee = 96%
R1 = Me, R2 = H, R3 = 4-Me: 98% de = 80% ee = 91%
cat =
OMe
N
TsHN
N

Scheme 1.60 Michael additions of cyclic β-keto esters to nitroalkenes catalysed by cinchona alkaloid.

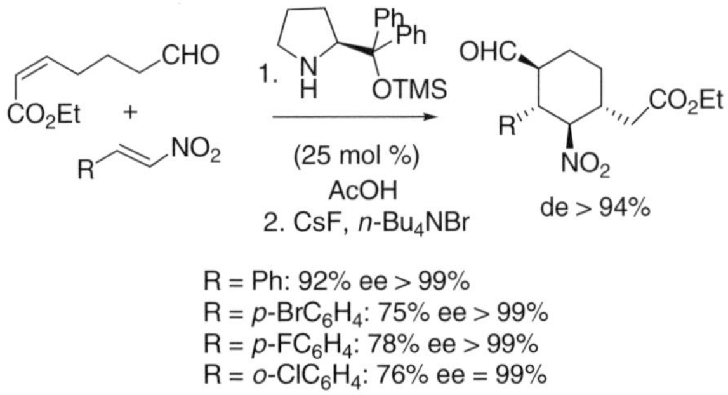
CHO
CO2Et +
R
NO2
1.
N
H
Ph
Ph
OTMS
(25 mol %)
AcOH
2. CsF, n-Bu4NBr
OHC
R
CO2Et
NO2
de > 94%
R = Ph: 92% ee > 99%
R = p-BrC6H4: 75% ee > 99%
R = p-FC6H4: 78% ee > 99%
R = o-ClC6H4: 76% ee = 99%

Scheme 1.61 Domino Michael reactions of 7-oxohept-2-enoate with nitroalkenes.

1.4 Intermolecular Domino Nitro-Michael Additions of C-Nucleophiles

A number of highly efficient asymmetric domino nitro-Michael additions of C-nucleophiles catalysed by organocatalysts were described in 2008. One of the most employed organocatalysts for these types of reactions is diphenylprolinol silyl ether. This catalyst was used by Hong *et al.* to develop a highly diastereo- and enantioselective domino reaction occurring between 7-oxohept-2-enoate and nitroalkenes (Scheme 1.61).[106] The reaction afforded highly functionalised

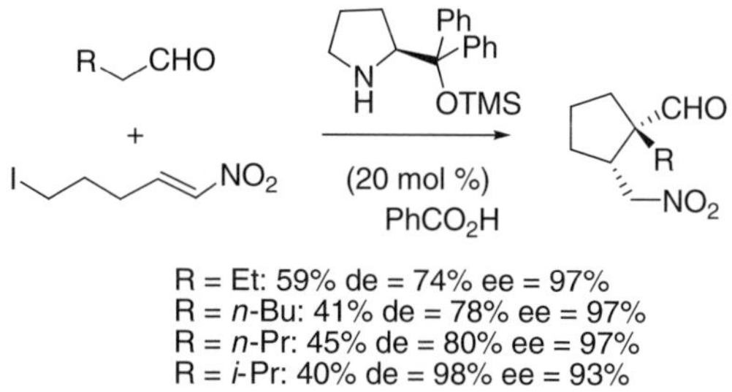

Scheme 1.62   Domino Michael reactions of aldehydes with (*E*)-5-iodo-1-nitropent-1-ene.

cyclohexane carboesters bearing four stereogenic centres with high diastereo- and enantioselectivities of up to 99% ee. Some adducts were transformed into the key intermediates for the synthesis of (−)-α- and (−)-β-lycorane.

An asymmetric domino Michael addition-alkylation reaction between aldehydes and (*E*)-5-iodo-1-nitropent-1-ene was developed by Enders *et al.* by using this catalyst and following an enamine-enamine mechanism.[107] This method provided an access to cyclic γ-nitroaldehydes of the cyclopentane type containing an all-carbon-substituted quaternary stereogenic centre. Enantioselectivities of up to 97% ee could be reached, whereas both yields and diastereoselectivities remained moderate, as shown in Scheme 1.62. Furthermore, a novel cyclic γ-amino acid of potential pharmaceutical relevance could be synthesised from a domino product.

The use of this catalyst allowed the same authors to elaborate an asymmetric domino Michael-Michael-aldol reaction, involving two aldehydes and a nitroalkene on the basis of an enamine-iminium-enamine activation.[108] The corresponding cyclohexene-carbaldehydes were isolated with virtually complete diastereo- and enantioselectivities, as shown in Scheme 1.63.

On the other hand, Chen *et al.* have reported an asymmetric three-component reaction of aldehydes, diethyl α-aminomalonate and nitroalkenes.[109] The Michael adducts were obtained in excellent yields and enantioselectivities of up to 98% ee by using a chiral bifunctional urea-tertiary amine as the organocatalyst (Scheme 1.64).

A cinchona alkaloid-derived organocatalyst was demonstrated by Zhong *et al.* to be a highly efficient catalyst to promote domino Michael–Henry reactions.[110] Therefore, upon catalysis with readily available 9-amino-9-deoxyepiquinine, the domino reaction between a diketo ester and a nitroolefin furnished the corresponding synthetically useful, highly functionalised chiral cyclohexanes with four stereogenic centres containing two quaternary stereocentres in good to excellent yields, excellent enantioselectivities and high diastereoselectivities (Scheme 1.65). This highly efficient methodology could be extended to the synthesis of highly functionalised cyclopentanes through domino reactions of nitroalkenes with ethyl 2-acetyl-4-oxo-4-phenylbutanoate with excellent yields and enantioselectivities along with complete diastereoselectivities (Scheme 1.65).[111]

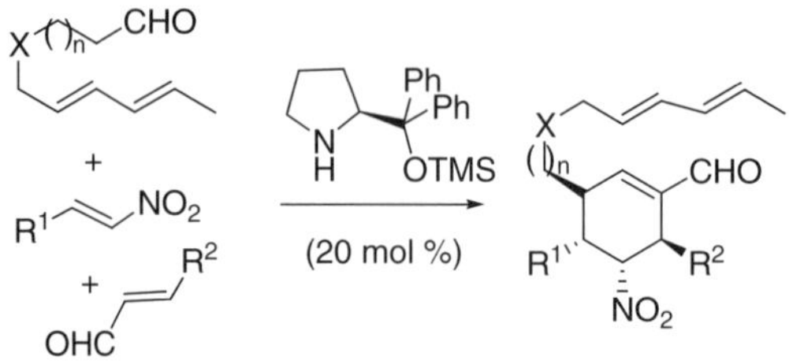

R¹ = R² = Ph, X = CH₂, n = 0: 51% de > 98% ee > 99%
R¹ = o-ClC₆H₄, R² = Ph, X = CH₂, n = 0: 59% de > 96%
ee > 99%
R¹ = R² = Ph, X = CH₂, n = 1: 47% de > 96% ee > 99%
R¹ = o-ClC₆H₄, R² = Ph, X = CH₂, n = 1: 55% de > 96%
ee > 99%
R¹ = Ph, R² = H, X = CH₂, n = 1: 49% de > 96% ee = 98%
R¹ = o-ClC₆H₄, R² = H, X = CH₂, n = 1: 51% de > 96%
ee = 97%
R¹ = R² = Ph, X = O, n = 1: 20% de > 96% ee > 99%
R¹ = R² = Ph, X = O, n = 2: 31% de > 96% ee > 99%

Scheme 1.63 Domino Michael–Michael-aldol reactions.

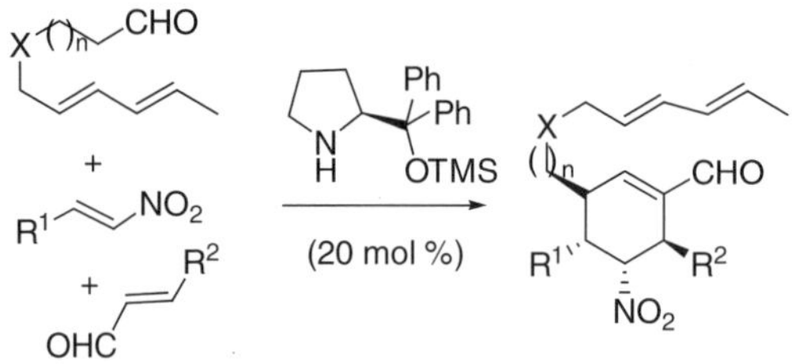

R¹ = R² = Ph, X = CH₂, n = 0: 51% de > 98% ee > 99%
R¹ = o-ClC₆H₄, R² = Ph, X = CH₂, n = 0: 59% de > 96%
ee > 99%
R¹ = R² = Ph, X = CH₂, n = 1: 47% de > 96% ee > 99%
R¹ = o-ClC₆H₄, R² = Ph, X = CH₂, n = 1: 55% de > 96%
ee > 99%
R¹ = Ph, R² = H, X = CH₂, n = 1: 49% de > 96% ee = 98%
R¹ = o-ClC₆H₄, R² = H, X = CH₂, n = 1: 51% de > 96%
ee = 97%
R¹ = R² = Ph, X = O, n = 1: 20% de > 96% ee > 99%
R¹ = R² = Ph, X = O, n = 2: 31% de > 96% ee > 99%

Scheme 1.64 Three-component Michael additions.

In the same area, an asymmetric domino double Michael reaction was initiated by this catalyst, providing an expedited access to chiral multi-functionalised five-membered rings.[112] This time, the reaction occurred between a nitroalkene and ethyl 2-acetyl-5-oxohexanoate, yielding the desired double Michael adducts in high yields and excellent diastereo- and enantio-selectivities (Scheme 1.66). An immense potential of synthetic versatility of the

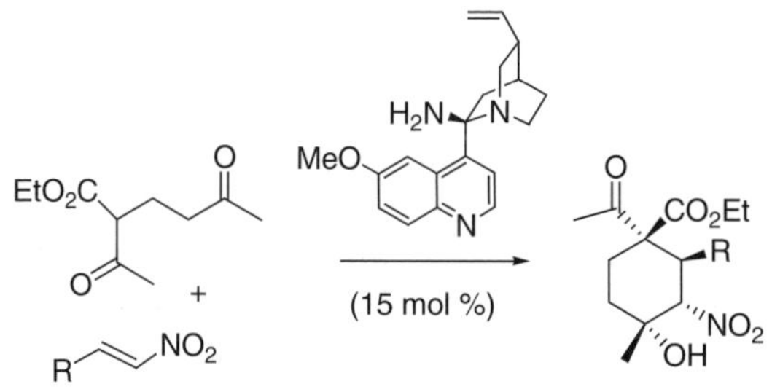

R = Ph: 93% de = 96% ee > 99%
R = p-MeOC₆H₄: 91% de = 90% ee = 98%
R = p-Tol: 89% de = 92% ee = 98%
R = p-BrC₆H₄: 88% de = 86% ee > 99%
R = o-BrC₆H₄: 90% de = 98% ee = 97%
R = p-CF₃C₆H₄: 91% de = 90% ee = 98%

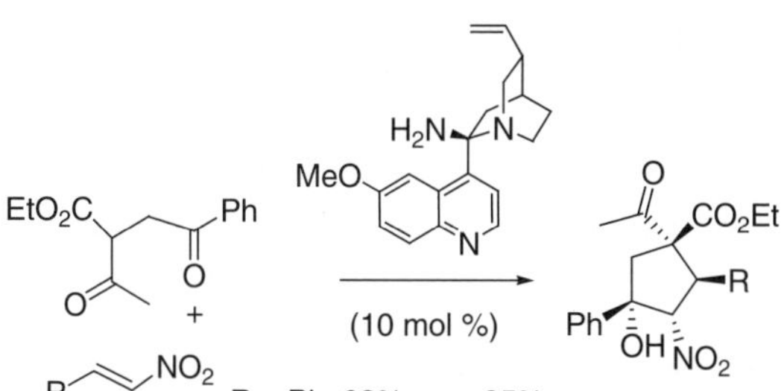

R = Ph: 93% ee = 95%   de = 100%
R = p-Tol: 91% ee = 95%
R = p-ClC₆H₄: 95% ee = 95%
R = 1-Naph: 91% ee = 96%

Scheme 1.65   Domino Michael–Henry reactions.

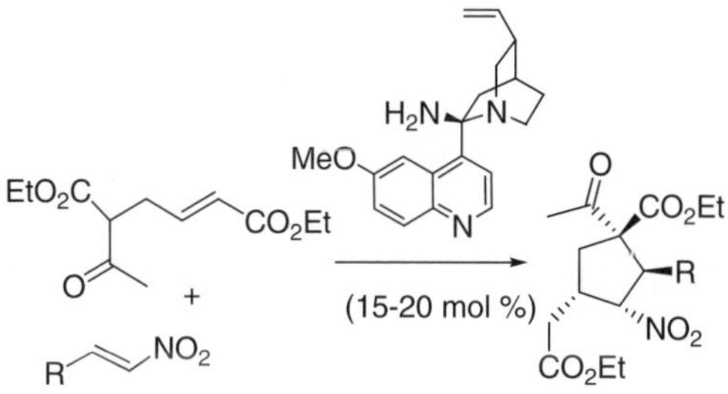

R = Ph: 91% de > 98% ee = 97%
R = p-Tol: 89% de = 94% ee = 95%
R = m-Tol: 85% de = 92% ee = 95%
R = 1-Naph: 87% de = 94% ee = 97%
R = 3-Fu: 86% de = 96% ee = 97%
R = 2-Fu: 87% de = 90% ee = 96%
R = 2-thienyl: 91% de > 98% ee = 96%
R = p-NO₂C₆H₄: 91% de = 92% ee = 95%

Scheme 1.66   Domino double Michael reactions.

products rendered this novel simple procedure highly appealing for asymmetric synthesis.

Finally, Sankararaman *et al.* have reported the synthesis of chiral nitro-chromenes on the basis of an asymmetric domino Michael–Henry reaction followed by dehydration, involving nitrostyrene and *ortho*-hydro-xybenzaldehyde.[73] This reaction was catalysed by a novel C_2-symmetric bis-(pyrrolidine-triazole)-based organocatalyst and provided only moderate yields (22–40%) and poor enantioselectivities (7–24% ee).

1.5 Intermolecular Michael Additions of Other-than-C-Nucleophiles

Enantioselective conjugate addition reactions of nitrogen-centred heterocyclic nucleophiles to electron-deficient olefins constitute a powerful preparative method in the area of asymmetric heterocyclic chemistry but, to date, reports on this reaction are sparse. In 2008, Deng and Lu reported the first highly enantioselective aza-Michael reaction of simple α,β-unsaturated ketones per-formed in the presence of an organocatalyst.[113] Therefore, the reaction of Boc-protected *N*-benzyloxyamine with a broad range of α,β-unsaturated ketones in the presence of a chiral 9-amino cinchona alkaloid combined with TFA led to the expected aza-Michael products in excellent enantioselectivities and good to high yields, as shown in Scheme 1.67. In the same area, the first enantioselective Michael addition of aniline to chalcones was promoted by cinchona alkaloids under solvent-free conditions by Scettri *et al.*, in 2008.[114] The most efficient organocatalyst was cinchonine, which allowed a series of aza-Michael products to be synthesised in moderate to excellent yields (48–100%) and low to moderate enantioselectivities (11–58% ee).

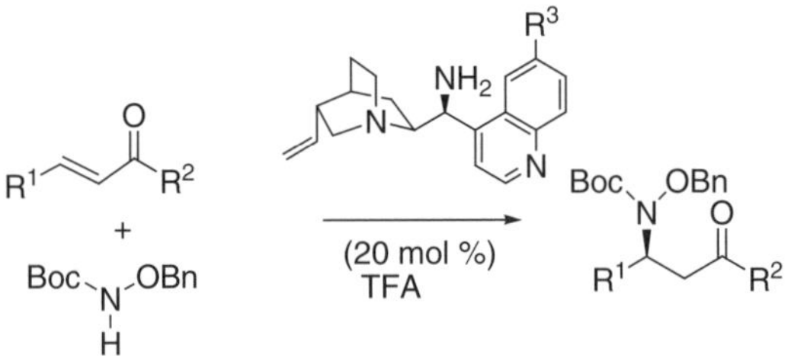

R^1 = BnCH$_2$, R^2 = Et, R^3 = OMe: 80% ee = 93%
R^1 = *n*-Pent, R^2 = Me, R^3 = OMe: 76% ee = 94%
R^1 = Me, R^2 = Et, R^3 = OMe: 83% ee = 90%
R^1 = *n*-Pent, R^2 = Et, R^3 = OMe: 70% ee = 93%
R^1 = *n*-Hex, R^2 = Et, R^3 = OMe: 83% ee = 93%
R^1 = *n*-Hep, R^2 = Et, R^3 = OMe: 70% ee = 96%
R^1 = *p*-FC$_6$H$_4$, R^2 = Me, R^3 = H: 61% ee = 92%
R^1 = *p*-ClC$_6$H$_4$, R^2 = Me, R^3 = H: 71% ee = 93%

Scheme 1.67   Aza-Michael reactions catalysed by cinchona alkaloids.

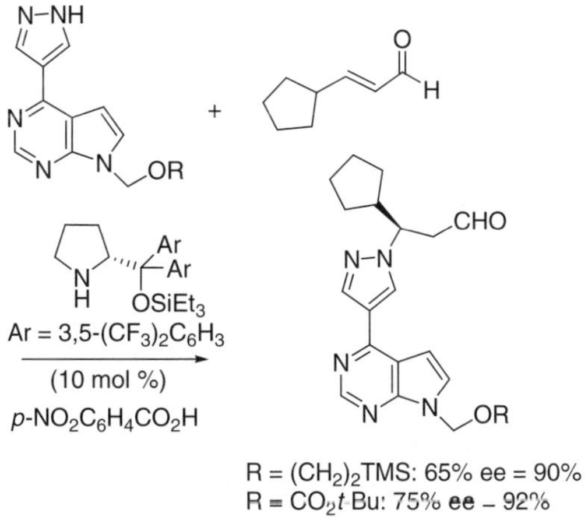

Scheme 1.68 Aza-Michael reactions catalysed by diarylprolinol silyl ether.

In 2009, Lin *et al.* reported an enantioselective synthesis of an important Janus kinase inhibitor, INCB018424, the key step of which was an asymmetric aza-Michael addition of pyrazoles to an α,β-unsaturated aldehyde catalysed by a chiral diarylprolinol silyl ether.[115] The use of benzoic acid or 4-nitrobenzoic acid as an additive was shown to increase the reaction rate. The highest enantioselectivities of up to 93% ee were observed for the reactions using the more sterically hindered organocatalysts (Scheme 1.68).

Using a chiral primary amine salt derived from 9-amino-(9-deoxy)-*epi*-hydroquinine, Melchiorre *et al.* achieved an organocatalytic asymmetric protocol for the highly enantioselective sulfa-Michael addition, which was applicable to a large variety of α,β-unsaturated ketones.[116] The highest enantioselectivities of up to 96% ee combined with high yields were obtained with benzyl mercaptan as a nucleophile, as shown in Scheme 1.69.

In 2009, Enders and Hoffman explored the reactivity of α,β-unsaturated sulfonates and aromatic thiols in sulfa-Michael additions.[117] When the reactions were catalysed by a chiral bifunctional thiourea derived from quinine, the sulfa-Michael adducts were formed in moderate enantioselectivities of up to 64% ee, albeit with generally good yields, as shown in Scheme 1.70.

A chiral bicyclic guanidine derivative was employed as an organocatalyst for the sulfa-Michael addition of thiols to *tert*-butyl 2-phthalimido acrylates by Tan *et al.*, providing the expected chiral Michael adducts in excellent yields and enantioselectivities of up to 92% ee, as shown in Scheme 1.71.[118]

In addition, Shimizu *et al.* have applied novel multifunctional inherently chiral calix[4]arene to the sulfa-Michael addition of thiols to cyclohexenone.[119] The expected Michael adducts were obtained in generally excellent yields (18–99%) but with low enantioselectivities (≤25% ee). The reaction system was also applied to other cyclic and acyclic enones, providing the corresponding products with a comparable degree of enantioselectivity to that obtained with cyclohexenone.

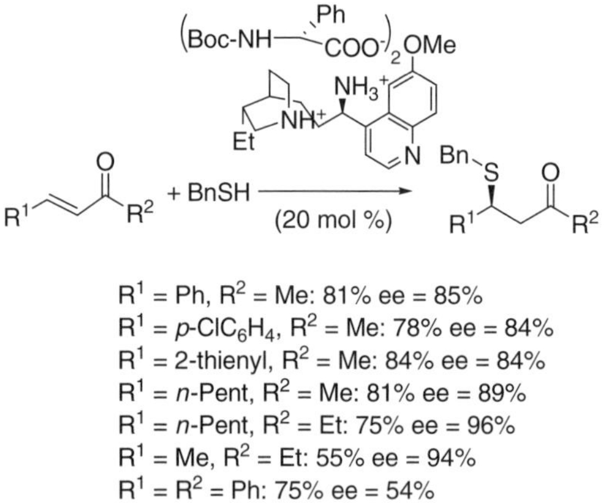

Scheme 1.69   Sulfa-Michael reactions catalysed by cinchona alkaloid ammonium salt.

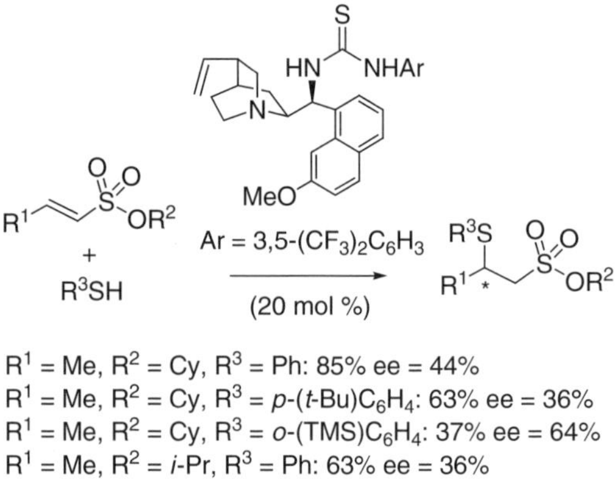

Scheme 1.70   Sulfa-Michael reactions of α,β-unsaturated sulfonates catalysed by cinchona alkaloid.

In 2008, Falck *et al.* developed highly efficient organocatalytic oxa-Michael additions to γ-hydroxy-α,β-unsaturated ketones.[120] The key transformation was the asymmetric conjugate addition triggered by complexation between a boronic acid hemiester, generated *in situ* from the corresponding γ-hydroxy-α,β-unsaturated ketone, and a chiral cinchona alkaloid-derived amine-thiourea catalyst. The methodology could be generalised to a wide range of γ-hydroxy-α,β-enones, allowing the corresponding products to be obtained in both high yields and high enantioselectivities, as shown in Scheme 1.72.

In addition, Cordova *et al.* have developed highly enantioselective organo-catalytic hydrophosphination reactions of α,β-unsaturated aldehydes.[121] These novel reactions were catalysed by protected chiral diarylprolinol derivatives

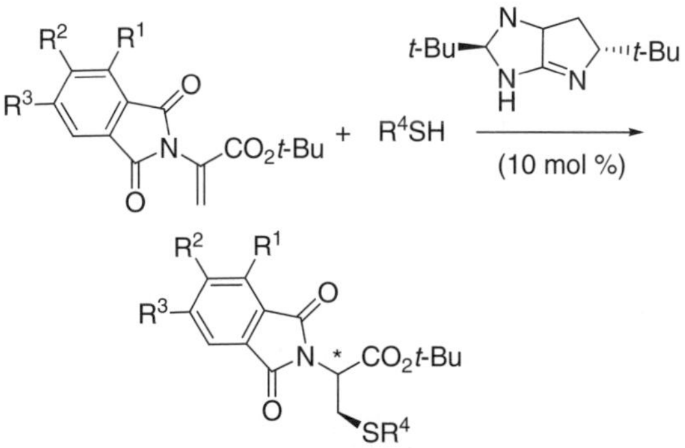

R^1 = F, R^2 = R^3 = H, R^4 = p-(t-Bu)C$_6$H$_4$: 99% ee = 92%
R^1 = R^3 = H, R^2 = Me, R^4 = p-(t-Bu)C$_6$H$_4$: 98% ee = 92%
R^1 = H, R^2 = R^3 = Cl, R^4 = p-(t-Bu)C$_6$H$_4$: 95% ee = 92%
R^1 = H, R^2 = R^3 = Cl, R^4 = Bn: 96% ee = 91%

Scheme 1.71 Sulfa-Michael reactions of *tert*-butyl 2-phthalimido acrylates catalysed by bicyclic guanidine.

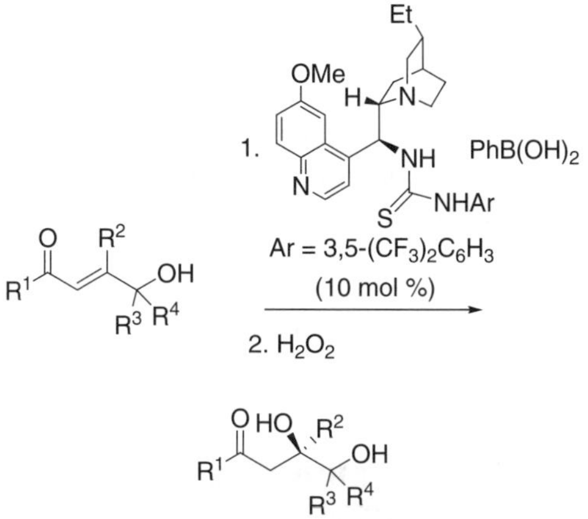

R^1 = Ph, R^2 = R^3 = R^4 = H: 90% ee = 95%
R^1 = p-NO$_2$C$_6$H$_4$, R^2 = R^3 = R^4 = H: 92% ee = 90%
R^1 = 2-Fu, R^2 = R^3 = R^4 = H: 83% ee = 97%
R^1 = Me, R^2 = R^3 = R^4 = H: 78% ee = 89%
R^1 = Ph, R^2 = Me, R^3 = R^4 = H: 78% ee = 98%
R^1 = Ph, R^2 = H, R^3 = R^4 = Me: 71% ee = 99%

Scheme 1.72 Oxa-Michael reactions of γ-hydroxy-α,β-unsaturated ketones catalysed by cinchona alkaloid-derived thiourea.

and gave access to optically active phosphine derivatives in high yields and enantioselectivities of up to 99% ee, as shown in Scheme 1.73. Actually, the formed unstable formylphosphines were *in situ* oxidised into the corresponding phosphine oxides by treatment with iodide. The study revealed that trivalent

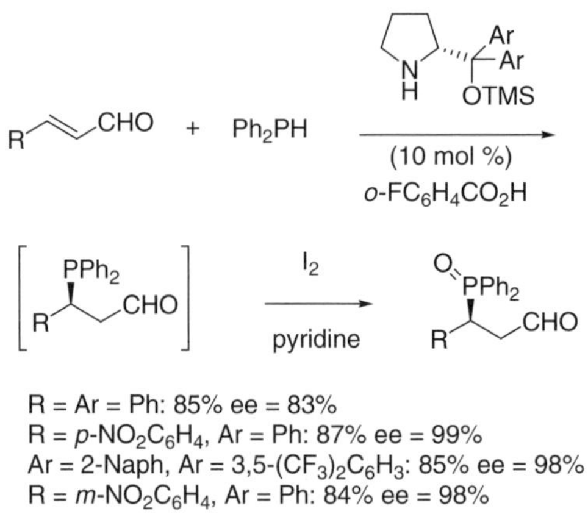

R⌒CHO + Ph₂PH

R = Ar = Ph: 85% ee = 83%
R = *p*-NO₂C₆H₄, Ar = Ph: 87% ee = 99%
Ar = 2-Naph, Ar = 3,5-(CF₃)₂C₆H₃: 85% ee = 98%
R = *m*-NO₂C₆H₄, Ar = Ph: 84% ee = 98%

Scheme 1.73 Hydrophosphinations of α,β-unsaturated aldehydes catalysed by silylated diarylprolinols.

phosphorus nucleophiles were essential in order to achieve the product formation. Moreover, the origin of the high enantioselectivity for the reaction with
diphenylphosphine as the nucleophile was investigated by density functional
theory calculations.

Finally, Kim reported the asymmetric conjugate addition of organoboronic
acids to γ-hydroxy α,β-unsaturated aldehydes catalysed by diarylprolinol silyl
ethers, which afforded the corresponding β-substituted γ-lactols in good
yields.[122] High enantioselectivities of up to 91% ee were obtained in the case of
arylvinylboronic acids, whereas only moderate enantioselectivities (≤22% ee)
were obtained by using arylboronic acids, as shown in Scheme 1.74. The
resulting products were subsequently oxidised into the corresponding β-substituted γ-lactones by treatment with PCC.

1.6   Intermolecular Domino Michael Additions of Other-than-C-Nucleophiles

Chiral pyrrolidines constitute the first class of organocatalysts which have been
successfully used to catalyse various domino Michael addition reactions
involving other-than-C-nucleophiles. As an example, Wang *et al.* have reported
highly enantio- and diastereoselective domino aza-Michael-Michael reactions
of α,β-unsaturated aldehydes with a *trans* 4-amino protected α,β-unsaturated
ester catalysed by diphenylprolinol silyl ether combined with NaOAc as a
base.[123] As summarised in Scheme 1.75, the corresponding trisubstituted synthetically useful, highly functionalised chiral pyrrolidines were produced in
high yields and diastereoselectivities of up to 94% de combined with almost
complete enantioselectivity in all cases of substrates.

This catalyst was also applied by Cordova *et al.* to an enantioselective aminosulfenylation of α,β-unsaturated aldehydes, which gave access

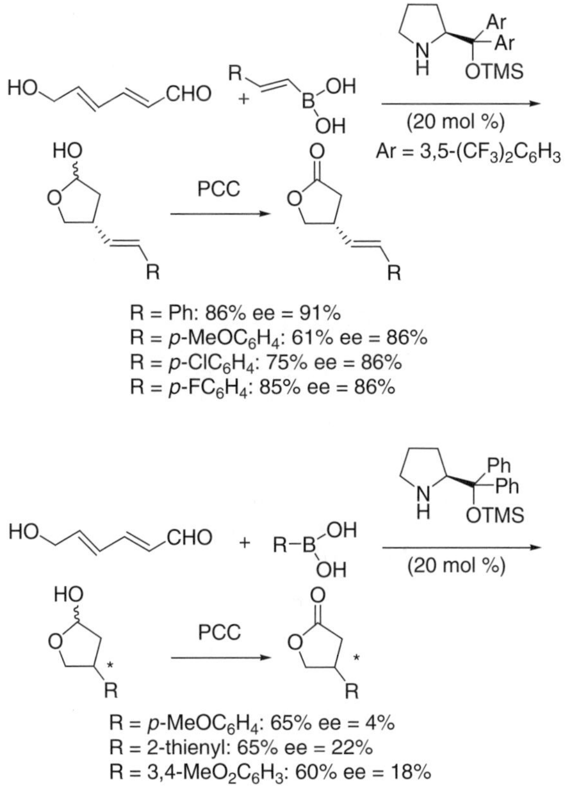
HO
CHO
R
B
OH
OH
Ar
Ar
N
H
OTMS
(20 mol %)
Ar = 3,5-(CF3)2C6H3
PCC
O
R
R = Ph: 86% ee = 91%
R = p-MeOC6H4: 61% ee = 86%
R = p-ClC6H4: 75% ee = 86%
R = p-FC6H4: 85% ee = 86%
Ph
R−B
PCC
R = p-MeOC6H4: 65% ee = 4%
R = 2-thienyl: 65% ee = 22%
R = 3,4-MeO2C6H3: 60% ee = 18%

Scheme 1.74 Michael additions of organoboronic acids catalysed by silylated diarylprolinols.

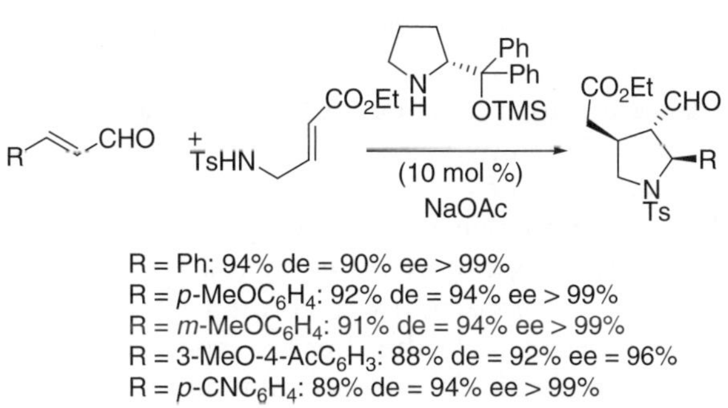
R
CHO
TsHN
CO2Et
Ph
N
H
OTMS
(10 mol %)
NaOAc
Ts
R = Ph: 94% de = 90% ee > 99%
R = p-MeOC6H4: 92% de = 94% ee > 99%
R = m-MeOC6H4: 91% de = 94% ee > 99%
R = 3-MeO-4-AcC6H3: 88% de = 92% ee = 96%
R = p-CNC6H4: 89% de = 94% ee > 99%

Scheme 1.75 Domino aza-Michael–Michael reactions.

to valuable β-amino-α-mercaptoaldehydes in high yields and high enantioselectivities of up to 99% ee, while the diastereoselectivities remained moderate (≤54% de).[124] The best results for this domino reaction are collected in Scheme 1.76.

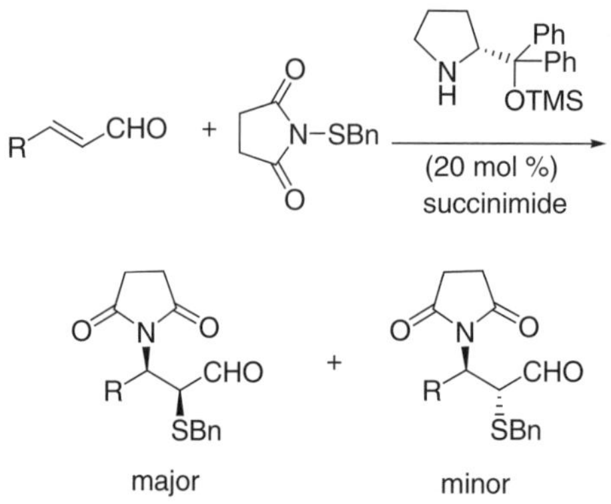

R = Ph: 83% de = 4%
ee (major) = 96% ee (minor) = 99%
R = p-ClC_6H_4: 73% de = 10%
ee (major) = 94% ee (minor) = 99%
R = p-BrC_6H_4: 65% de = 14%
ee (major) = 98% ee (minor) = 99%
R = p-$NO_2C_6H_4$: 79% de = 10%
ee (major) = 93% ee (minor) = 99%
R = p-Tol: 67% de = 42%
ee (major) = ee (minor) = 95%
R = n-Bu: 74% de = 54% ee (major) = 97%

Scheme 1.76 Aminosulfenylation reactions.

A closely related catalyst was found by the same authors to be highly efficient to induce chirality for domino thia-Michael-aldol reactions between 2-mercaptoacetophenone and α,β-unsaturated aldehydes.[125] Indeed, the corresponding chiral thiochromenes bearing three contiguous stereocentres were generated in high yields with good diastereoselectivities of up to 88% de and excellent enantioselectivities of up to 99% ee (Scheme 1.77).This novel domino reaction constituted a simple catalytic highly stereoselective entry to pharmaceutically valuable benzothiopyran derivatives.

An enantioselective domino oxa-Michael–Henry reaction of substituted salicylaldehydes with nitroalkenes, proceeding through aromatic iminium activation has been developed by Xu *et al.* by using a chiral secondary amine organocatalyst and salicylic acid as a co-catalyst.[126] This novel domino reaction served as an efficient method for the preparation of chiral 3-nitro-2*H*-chromenes with moderate to good enantioselectivities of up to 92% ee (Scheme 1.78).

The second class of chiral organocatalysts recently involved in domino Michael reactions of other-than-C-nucleophiles is constituted by the cinchona alkaloid family. In this area, Melchiorre *et al.* have involved a chiral primary amine salt derived from 9-amino-(9-deoxy)-*epi*-hydroquinine to induce chirality in aziridinations of enones.[127] This domino iminium-enamine intramolecular sequence afforded a series of chiral protected aziridines derived from both

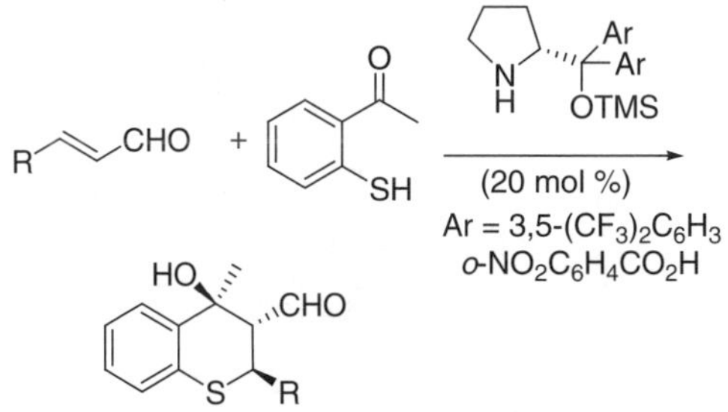

R = Ph: 98% de = 82% ee = 98%
R = *p*-CNC$_6$H$_4$: 63% de = 88% ee = 99%
R = *p*-ClC$_6$H$_4$: 88% de = 82% ee = 99%
R = *p*-BrC$_6$H$_4$: 71% de = 86% ee > 99%
R = 2-Naph: 91% de = 88% ee = 99%
R = CO$_2$Et: 91% de > 88% ee = 96%
R = *n*-Bu: 83% de = 82% ee = 96%
R = *n*-Pr: 68% de = 82% ee = 96%
R = *p*-NO$_2$C$_6$H$_4$: 75% de = 82% ee = 99%

Scheme 1.77 Domino thia-Michael-aldol reactions.

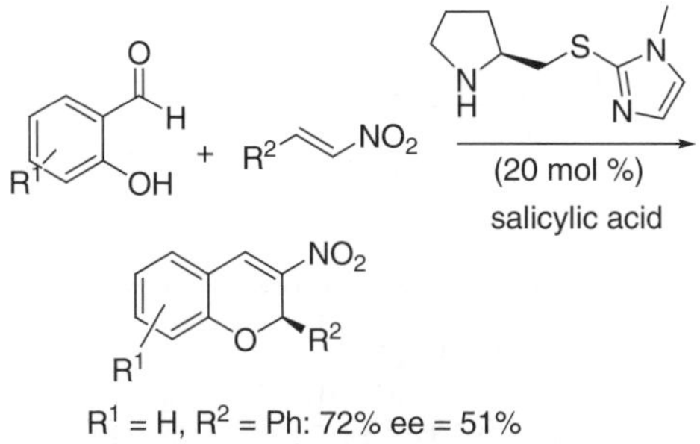

R^1 = H, R^2 = Ph: 72% ee = 51%
R^1 = 3-MeO, R^2 = Ph: 78% ee = 77%
R^1 = 4-MeO, R^2 = Ph: 67% ee = 85%
R^1 = 4-MeO, R^2 = *o*-MeOC$_6$H$_4$: 37% ee = 91%
R^1 = 4-MeO, R^2 = *p*-MeOC$_6$H$_4$: 47% ee = 72%
R^1 = 4-MeO, R^2 = 2-Naph: 35% ee = 70%

Scheme 1.78 Domino oxa-Michael–Henry reactions.

linear and cyclic α,β-unsaturated ketones with excellent yields, diastereo- and enantioselectivities of up to 99% ee (Scheme 1.79).

An unprecedented asymmetric domino thia-Michael–Michael process, involving dynamic kinetic resolution,[128] was developed by Wang *et al.* in the presence of a cinchona alkaloid amine-thiourea at a remarkably low catalyst loading of 2 mol %.[129] This highly diastereo- and enantioselective novel cascade process afforded chiral thiochromanes bearing three contiguous stereogenic

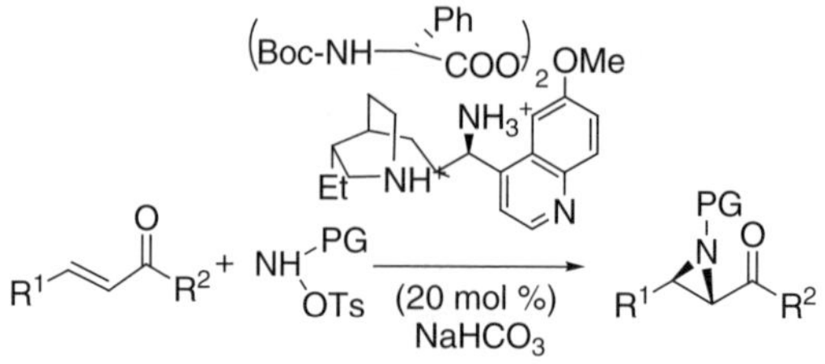

R¹ = *n*-Pent, R² = Me, PG = Cbz: 93% de = 90% ee = 96%
R¹ = *n*-Pent, R² = Me, PG = Boc: 82% de = 90% ee = 99%
R¹ = *p*-NO₂C₆H₄, R² = Me, PG = Cbz: 92% de = 90% ee = 99%
R¹ = CO₂Et, R² = Me, PG = Cbz: 92% de = 90% ee = 99%
R¹ = Me, R² = Et, PG = Cbz: 94% de = 90% ee = 98%
R¹,R² = (CH₂)₃, PG = Cbz: 86% de = 90% ee = 98%

Scheme 1.79   Domino aza-Michael–cyclisation reactions.

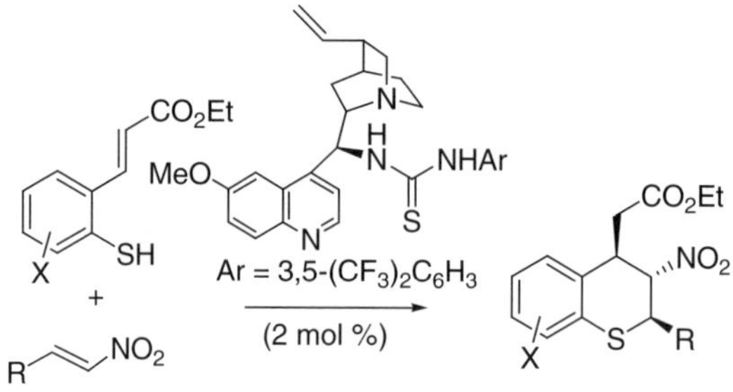

R = Ph, X = H: 90% de = 94% ee = 97%
R = *p*-FC₆H₄, X = H: 92% de = 94% ee = 97%
R = *p*-ClC₆H₄, X = H: 99% de = 94% ee = 97%
R = *p*-BrC₆H₄, X = H: 91% de = 94% ee = 97%
R = *m*-BrC₆H₄, X = H: 83% de = 94% ee = 97%
R = *p*-MeOC₆H₄, X = H: 88% de = 94% ee = 97%
R = *p*-BnOC₆H₄, X = H: 95% de = 94% ee = 97%
R = Ph, X = 5-Me: 93% de = 94% ee = 99%
R = Ph, X = 5-OMe: 42% de = 94% ee = 97%

Scheme 1.80   Domino thia-Michael–Michael reactions.

centres, as shown in Scheme 1.80. Remarkably, it was demonstrated that the reaction also proceeded efficiently with a catalyst loading of only 1 mol %.

A closely related methodology was applied by Zhao *et al.* to the enantioselective synthesis of tetrasubstituted thiochromanes on the basis of a domino thia-Michael–Knoevenagel reaction between 2-mercaptobenzaldehydes and benzylidenemalonates with the same quinine thiourea catalyst.[130] Steric and electron effects were found to affect profoundly the stereoselectivities of the reaction. Thus, it was shown that the diastereoselectivity of the reaction increased if there was a substituent at the *ortho* position of the phenyl ring of the benzylidene moiety, while the enantioselectivity of the reaction decreased if the phenyl ring was substituted with an electron-withdrawing group. The best results with enantioselectivities of up to 96% ee are collected in Scheme 1.81.

In the same context, these authors reported the synthesis of chiral 2,3,4-trisubstituted thiochromanes by using a cupreine-catalysed domino thia-Michael–Henry reaction, occurring between 2-mercaptobenzaldehydes and nitrostyrenes.[131] As shown in Scheme 1.82, the resulting products were

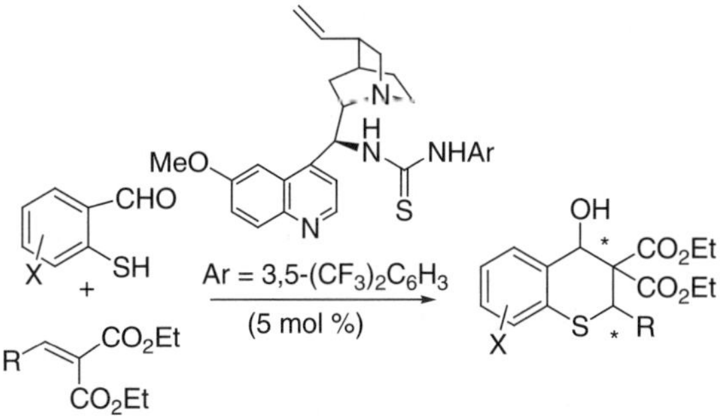

R = *p*-MeOC$_6$H$_4$, X = H: 94% de = 86% ee = 94%
R = *p*-BnOC$_6$H$_4$, X = H: 96% de = 90% ee = 94%
R = *o*-PhC$_6$H$_4$, X = H: 90% de = 80% ee = 93%
R = 2-Naph, X = H: 91% de = 80% ee = 94%
R = *o*-MeOC$_6$H$_4$, X = 4-MeO: 93% de = 80% ee = 91%
R = *o*-MeOC$_6$H$_4$, X = 4-Cl: 90% de = 84% ee = 90%

Scheme 1.81 Domino thia-Michael–Michael reactions.

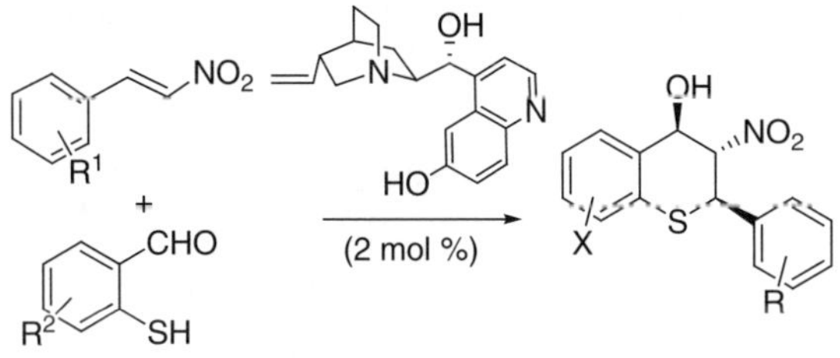

R^1 = R^2 = H: 95% de = 40% ee = 86%
R^1 = 2-Br, R^2 = H: 97% de = 54% ee = 85%
R^1 = 4-Br, R^2 = H: 95% de = 48% ee = 76%
R^1 = 2-NO$_2$, R^2 = H: 95% de = 50% ee = 78%
R^1 = 4-MeO, R^2 = H: 96% de = 36% ee = 85%
R^1 = H, R^2 = 4-MeO: 95% de = 40% ee = 82%
R^1 = 4-Br, R^2 = 4-MeO: 96% de = 44% ee = 80%

Scheme 1.82 Domino thia-Michael–Henry reactions.

obtained with moderate diastereoselectivities of up to 56% de and good enantioselectivities of up to 86% ee.

1.7   Intramolecular Michael Additions

In 2009, a wide range of optically active chromans and tetrahydroquinolines were synthesised by Xiao *et al.* on the basis of the first enantioselective organocatalytic intramolecular hydroarylations of phenol- and aniline-derived enals.[132] Good yields combined with good to high enantioselectivities of up to 96% ee were obtained for the Michael adducts generated by catalysis with a diarylprolinol silyl ether (Scheme 1.83).

On the other hand, several intramolecular aza-Michael additions have been successfully developed. As an example, Bandini *et al.* have reported the first enantioselective synthesis of 3,4-dihydropyrazino[1,2-a]indol-1(2*H*)-ones through an intramolecular phase-transfer-catalysed aza-Michael addition.[133] This highly enantioselective process was performed in the presence of a chiral cinchona ammonium salt and could be successfully applied to a range of substrates, as shown in Scheme 1.84.

In a same area, an efficient and simple method for the enantioselective synthesis of indolines, isoindolines, tetrahydroquinolines and tetraisoquinolines was achieved by means of the organocatalytic intramolecular aza-Michael reaction of the corresponding aniline and benzylamine derivatives.[134] This process was catalysed by a diarylprolinol silyl ether used in the presence of benzoic acid as an additive, which provided the Michael adducts in good yields and excellent enantioselectivities of up to 99% ee (Scheme 1.85). This methodology was applied to the synthesis of the biologically active tetrahydroquinoline alkaloid (+)-angustureine.

Finally, Carter *et al.* have applied a similar protocol to prepare a series of chiral pyrrolidine, piperidine and indoline derivatives.[135] Indeed, these products were generated from the corresponding enals through intramolecular aza-

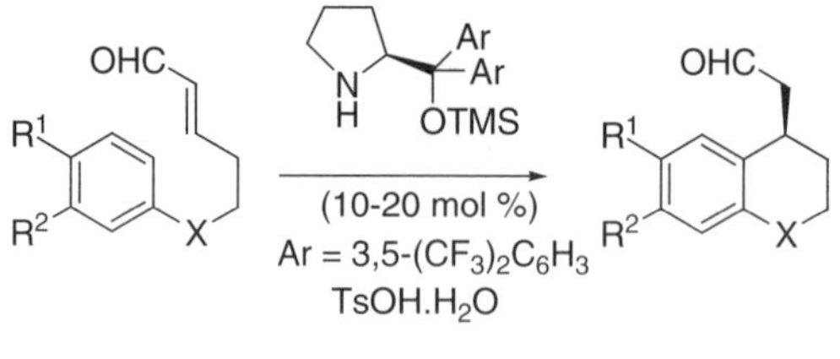

R^1 = H, R^2 = NMe$_2$, X = O: 78% ee = 96%
R^1 = Me, R^2 = NMe$_2$, X = O: 50% ee = 89%
R^1 = OEt, R^2 = H, X = NBn: 80% ee = 74%
R^1 = R^2 = Me, X = NBn: 98% ee = 92%
R^1 = H, R^2 = NMe$_2$, X = NTs: 64% ee = 90%

Scheme 1.83   Intramolecular hydroarylations of phenol- and aniline-derived enals.

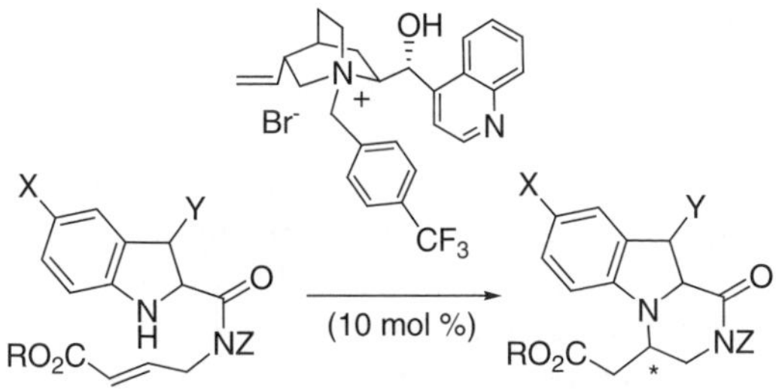

X = F, Y = H, Z = Bn, R = Et: 93% ee = 82%
X = Cl, Y = H, Z = Bn, R = Et: 88% ee = 84%
X = OMe, Y = H, Z = Bn, R = Et: 85% ee = 89%
X = Me, Y = H, Z = Bn, R = Et: 85% ee = 89%
X = Y = H, Z = PMP, R = Et: 75% ee = 82%
X = Y = H, Z = Bn, R = Me: 90% ee = 90%

Scheme 1.84 Phase-transfer-catalysed intramolecular aza-Michael additions.

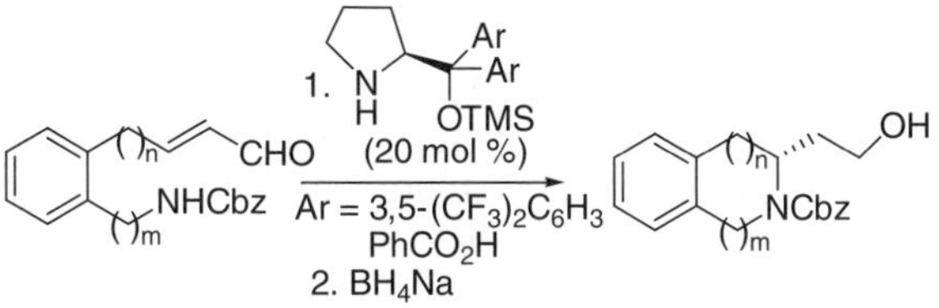

n = 1, m = 0: 70% ee = 93%
n = m = 1: 67% ee = 99%
n = 2, m = 0: 70% ee = 92%

Scheme 1.85 Silylated diarylprolinol-catalysed intramolecular aza-Michael additions.

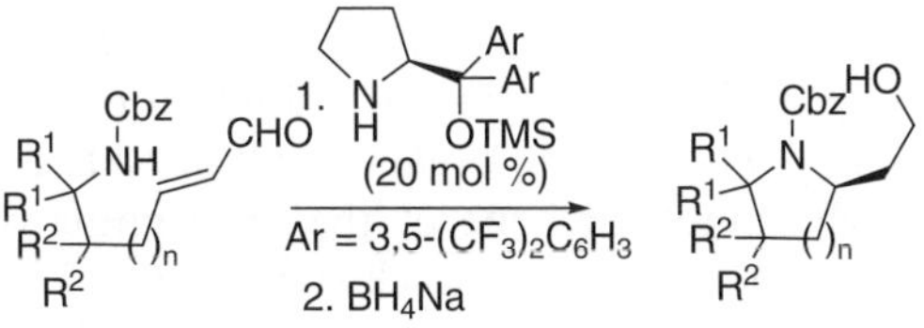

R^1 = R^2 = H, n = 1: 67% ee = 90%
R^1 = Me, R^2 = H, n = 1: 60% ee = 79%
R^1 = H, R^2 = Me, n = 1: 64% ee = 85%
R^1 = H, R^2 = Me, n = 2: 63% ee = 95%
R^1 = R^2 = H, n = 2: 78% ee = 80%

Scheme 1.86 Silylated diarylprolinol-catalysed intramolecular aza-Michael additions.

Michael additions with good yields and high enantioselectivities of up to 95% ee, as shown in Scheme 1.86. This methodology could be applied to the syntheses of several alkaloids such as homopipecolic acid, pelletierine and homoproline.

1.8   Conclusions

The enantioselective Michael reaction is by far the most intensively studied reaction performed in the presence of chiral organocatalysts over the last year and was the subject of a spectacular development especially in 2008. In particular, many successes were achieved by applying a number of novel chiral organocatalysts such as proline derivatives to highly reactive Michael donors or acceptors. Although a wide series of modifications of the structure of proline were accomplished, chiral silylated biarylprolinols remain some of the most used organocatalysts to highly efficiently promote asymmetric Michael additions of C-nucleophiles, and in particular asymmetric domino Michael reactions which uniformly provided spectacular enantioselectivities. On the other hand, excellent results were observed for asymmetric Michael additions of C-nucleophiles by using a number of novel modified cinchona alkaloids. Organocatalytic Michael methodologies involving nitroalkenes as acceptors are among the most widely studied and, in the last year, a large number of results were reported dealing with the asymmetric organocatalysed conjugate addition of C-nucleophiles to nitroolefins. Among them, the Michael addition of aldehydes to nitroalkenes catalysed by chiral silylated biarylprolinols was widely investigated by a number of workers, providing generally excellent enantioselectivities. Excellent results were also reported for these types of reactions by employing several chiral pyrrolidine derivatives as organocatalysts. Furthermore, the asymmetric Michael addition of cyclohexanone to nitroalkenes could be successfully catalysed by supported triazole-pyrrolidine catalysts. In the same area, a new type of polymer-immobilised pyrrolidine-based chiral ionic liquid was proved to be capable of inducing the Michael addition of ketones to nitrostyrenes with exceptional stereoselectivities. The asymmetric Michael addition of activated ketones to nitroolefins catalysed by chiral bifunctional amine-thiourea catalysts was investigated by several authors, providing in each case the corresponding adducts in high yields and stereoselectivities. In addition, a number of highly efficient asymmetric domino nitro-Michael additions of C-nucleophiles catalysed by chiral organocatalysts were described in the last year. Silylated diarylprolinols are one of the most employed chiral organocatalysts for these types of reactions. This class of organocatalysts together with that of cinchona alkaloids was also demonstrated to provide excellent stereoselectivities for asymmetric aza-, oxa- and thia-Michael reactions.

References

1. A. Perlmutter, in *Conjugate Additions in Organic Synthesis*, 1992, Pergamon Press, Oxford.
2. (a) M. Yamaguchi, *Conjugate Addition of Stabilized Carbanions*, in *Comprehensive Asymmetric Catalysis*, E. N. Jacobsen, A. Pfaltz, H. Yamamoto, (ed.), Springer, Berlin, 1999, **Vol. III**, pp. 1121–1139; (b) M. P. Sibi and S. Manyem, *Tetrahedron*, 2000, **56**, 8033–8061; (c) N. Krause and

A. Hoffmann-Röder, *Synthesis*, 2001, 171–196; (d) S. C. Jha and N. N. Joshi, *Arkivoc*, 2002, 167–196.

3. (a) D. Almasi, D. A. Alonso and C. Najera, *Tetrahedron: Asymmetry*, 2007, **18**, 299–365; (b) S. B. Tsogoeva, *Eur. J. Org. Chem.*, 2007, 1701–1716; (c) J. L. Vicario, D. Badia and L. Carrillo, *Synthesis*, 2007, **14**, 2065–2092; (d) S. Sulzer-Mosse and A. Alexakis, *Chem. Commun.*, 2007, **30**, 3123–3135.

4. G. Bartoli and P. Melchiorre, *Synlett*, 2008, **12**, 1759–1772.

5. (a) M. Gruttadauria, F. Giacalone and R. Noto, *Chem. Soc. Rev.*, 2008, **37**, 1666–1688; (b) A. Lattanzi, *Chem. Commun.*, 2009, 1452–1463; (c) A. Mielgo and C. Palomo, *Chem. Asian J.*, 2008, **3**, 922–948.

6. G.-L. Zhao, J. Vesely, J. Sun, K. E. Christensen, C. Bonneau and A. Cordova, *Adv. Synth. Catal.*, 2008, **350**, 657–661.

7. Q Zhu and Y. Lu, *Org. Lett.*, 2008, **10**, 4803–4806.

8. A. Quintard, C. Bournaud and A. Alexakis, *Chem. Eur. J.*, 2008, **14**, 7504–7507.

9. L. Albrecht, B. Richter, H. Krawczyk and K. A. Jorgensen, *J. Org. Chem.*, 2008, **73**, 8337–8343.

10. S. Brandau, A. Landa, J. Franzen, M. Marigo and K. A. Jorgensen, *Ang. Chem., Int. Ed. Engl.*, 2006, **45**, 4305–4309.

11. A. Ma, S. Zhu and D. Ma, *Tetrahedron Lett.*, 2008, **49**, 3075–3077.

12. M. Gruttadauria, F. Giacalone and R. Noto, *Adv. Synth. Catal.*, 2009, **351**, 33–57.

13. D. A. Alonso, S. Kitagaki, N. Utsumi and C. F. Barbas III, *Ang. Chem. Int. Ed. Engl.*, 2008, **47**, 4588–4591.

14. A. Massa, N. Utsumi and C. F. Barbas III, *Tetrahedron Lett.*, 2009, **50**, 145–147.

15. J. Lu, F. Liu and T.-P. Loh, *Adv. Synth. Catal.*, 2008, **350**, 1781–1784.

16. S. Cabrera, E. Reyes, J. Aleman, A. Milelli, S. Kobbelgaard and K. A. Jorgensen, *J. Am. Chem. Soc.*, 2008, **130**, 12031–12037.

17. V. Wascholowski, K. R. Knudsen, C. E. T. Mitchell and S. V. Ley, *Chem. Eur. J.*, 2008, **14**, 6155–6165.

18. R. Ballini, G. Bosica, D. Fiorini, A. Palmieri and M. Petrini, *Chem. Rev.*, 2005, **105**, 933–971.

19. M. Mamgren, J. Granander and M. Amedkouh, *Tetrahedron: Asymmetry*, 2008, **19**, 1934–1940.

20. M. T. Barros and A. M. F. Phillips, *Eur. J. Org. Chem.*, 2008, 2525–2529.

21. H. Miyabe and Y. Takemoto, *Bull. Chem. Soc. Jpn.*, 2008, **81**, 785–795.

22. S. J. Connon, *Chem. Commun.*, 2008, 2499–2510.

23. P. Li, S. Wen, F. Yu, Q. Liu, W. Li, Y. Wang, X. Liang and J. Ye, *Org. Lett.*, 2009, **11**, 753–756.

24. P. Li, Y. Wang, X. Liang and J. Ye, *Chem. Commun.*, 2008, 3302–3304.

25. B. Vakulya, S. Varga and T. Soos, *J. Org. Chem.*, 2008, **73**, 3475–3480.

26. Q. Zhu and Y. Lu, *Org. Lett.*, 2009, **11**, 1721–1724.

27. Q. Zhu, L. Cheng and Y. Lu, *Chem. Commun.*, 2008, 6315–6317.

28. L. Yue, W. Du, Y.-K. Liu and Y.-C. Chen, *Tetrahedron Lett.*, 2008, **49**, 3881–3884.

29. X. Li, L. Cun, C. Lian, L. Zhong, Y. Chen, J. Liao, J. Zhu and J. Deng, *Org. Biomol. Chem.*, 2008, **6**, 349–353.

30. M. Capuzzi, D. Perdicchia and K. A. Jorgensen, *Chem. Eur. J.*, 2008, **14**, 128–135.

31. M. B. Cid, J. Lopez-Cantarero, S. Duce and J. L. Garcia Ruano, *J. Org. Chem.*, 2009, **74**, 431–434.

32. J. Lu, W.-J. Zhou, F. Liu and T.-P. Loh, *Adv. Synth. Catal.*, 2008, **350**, 1796–1800.

33. C. L. Rigby and D. J. Dixon, *Chem. Commun.*, 2008, 3798–3800.

34. J. Aleman, C. B. Jacobsen, K. Frisch, J. Overgaard and K. A. Jorgensen, *Chem. Commun.*, 2008, 632–634.

35. T. Furukawa, N. Shibata, S. Mizuta, S. Nakamura, T. Toru and M. Shiro, *Ang. Chem., Int. Ed. Engl.*, 2008, **47**, 8051–8054.

36. P. Elsner, L. Bernardi, G. Dela Salla, J. Overgaard and K. A. Jorgensen, *J. Am. Chem. Soc.*, 2008, **130**, 4897–4905.

37. S. Shirakawa and S. Shimizu, *Eur. J. Org. Chem.*, 2009, 1916–1924.

38. Z. Jiang, W. Ye, Y. Yang and C.-H. Tan, *Adv. Synth. Catal.*, 2008, **350**, 2345–2351.

39. M. Rueping, B. J. Nachtsheim, S. A. Moreth and M. Bolte, *Ang. Chem., Int. Ed. Engl.*, 2008, **47**, 593–596.

40. (a) H. Pellissier, *Tetrahedron*, 2006, **62**, 1619–1665; (b) H. Pellissier, *Tetrahedron*, 2006, **62**, 2143–2173.

41. (a) T. Arai, H. Sasai, K.-i. Aoe, K. Okamura, T. Date and M. Shibasaki, *Ang. Chem., Int. Ed. Engl.*, 1996, **35**, 104–106; (b) M. Shibasaki, H. Sasai and T. Arai, *Ang. Chem., Int. Ed. Engl.*, 1997, **36**, 1236–1256.

42. (a) D. Enders, C. Grondal and M. R. M. Huettl, *Ang. Chem., Int. Ed. Engl.*, 2007, **46**, 1570–1581; (b) X. Yu and W. Wang, *Org. Biomol. Chem.*, 2008, **6**, 2037–2046.

43. B.-C. Hong, R. Y. Nimje, A. A. Sadani and J.-H. Liao, *Org. Lett.*, 2008, **10**, 2345–2348.

44. J. Wang, F. Yu, X. Zhang and D. Ma, *Org. Lett.*, 2008, **10**, 2561–2564.

45. J. Franzen and A. Fisher, *Ang. Chem., Int. Ed. Engl.*, 2009, **48**, 787–791.

46. S. Bertelsen, R. L. Johansen and K. A. Jorgensen, *Chem. Commun.*, 2008, 3016–3018.

47. J. Vesely, G.-L. Zhao, A. Bartoszewicz and A. Cordova, *Tetrahedron Lett.*, 2008, **49**, 4209–4212.

48. I. Ibrahem, G. L. Zhao, R. Rios, J. Vesely, H. Sunden, P. Dziedzic and A. Cordova, *Chem. Eur. J.*, 2008, **14**, 7867–7879.

49. S. Cabrera, J. Aleman, P. Bolze, S. Bertelsen and K. A. Jorgensen, *Ang. Chem., Int. Ed. Engl.*, 2008, **47**, 121–125.

50. Y. Hayashi, M. Toyoshima, H. Gotoh and H. Ishikawa, *Org. Lett.*, 2009, **11**, 45–48.

51. O. Penon, A. Carlone, A. Mazzanti, M. Locatelli, L. Sambri, G. Bartoli and P. Melchiorre, *Chem. Eur. J.*, 2008, **14**, 4788–4791.

52. P. T. Franke, R. L. Johansen, S. Bertelsen and K. A. Jorgensen, *Chem. Asian J.*, 2008, **3**, 216–224.

53. P. T. Franke, B. Richter and K. A. Jorgensen, *Chem. Eur. J.*, 2008, **14**, 6317–6321.

54. P. Bolze, G. Dickmeiss and K. A. Jorgensen, *Org. Lett.*, 2008, **10**, 3753–3756.

55. (a) M. Rueping, E. Sugiono and E. Merino, *Chem. Eur. J.*, 2008, **14**, 6329–6332; (b) M. Rueping, A. Kuenkel, F. Tato and J. W. Bats, *Ang. Chem., Int. Ed. Engl.*, 2009, **48**, 3699–3702.

56. M. Rueping, E. Sugiono and E. Merino, *Ang. Chem., Int. Ed. Engl.*, 2008, **47**, 3046–3049.

57. J. Jiang, J. Yu, X.-X. Sun, Q.-Q. Rao and L.-Z. Gong, *Ang. Chem., Int. Ed. Engl.*, 2008, **47**, 2458–2462.

58. Y. Chi, S. T. Scroggins and J. M. J. Fréchet, *J. Am. Chem. Soc.*, 2008, **130**, 6322–6323.

59. O. M. Berner, L. Tedeschi and D. Enders, *Eur. J. Org. Chem.*, 2002, 1877–1894.

60. (a) Y. Hayashi, T. Itoh, M. Ohkubo and H. Ishikawa, *Ang. Chem., Int. Ed. Engl.*, 2008, **47**, 4722–4724; (b) H. Ishikawa, T. Suzuki and Y. Hayashi, *Ang. Chem., Int. Ed. Engl.*, 2009, **48**, 1304–1307.

61. P. Garcia-Garcia, A. Ladépçeche, R. Halder and B. List, *Ang. Chem., Int. Ed. Engl.*, 2008, **47**, 4719–4721.

62. N. Ruiz, E. Reyes, J. L. Vicario, D. Badia, L. Carrillo and U. Uria, *Chem. Eur. J.*, 2008, **14**, 9357–9367.

63. S. Zhu, S. Yu and D. Ma, *Ang. Chem., Int. Ed. Engl.*, 2008, **47**, 545–548.

64. Y. Chi, L. Guo, N. A. Kopf and S. H. Gellman, *J. Am. Chem. Soc.*, 2008, **130**, 5608–5609.

65. S. Belot, A. Massaro, A. Tenti, A. Mordini and A. Alexakis, *Org. Lett.*, 2008, **10**, 4557–4560.

66. D. Bonne, L. Salat, J.-P. Dulcère and J. Rodriguez, *Org. Lett.*, 2008, **10**, 5409–5412.

67. S. Mukherjee, J. W. Yang, S. Hoffman and B. List, *Chem. Rev.*, 2007, **107**, 5471–5569.

68. Q. Tao, G. Tang, K. Lin and Y.-F. Zhao, *Chirality*, 2008, **20**, 833–838.

69. Y. Okuyama, H. Nakano, Y. Watanabe, M. Makabe, M. Takeshita, K. Uwai, C. Kabuto and E. Kwon, *Tetrahedron Lett.*, 2009, **50**, 193–197.

70. Q. Zhang, B. Ni and A. D. Headley, *Tetrahedron*, 2008, **64**, 5091–5097.

71. S. Chandrasekhar, T. Bhoopendra, B. B. Parida and C. R. Reddy, *Tetrahedron: Asymmetry*, 2008, **19**, 495–499.

72. Z. Yacob, J. Shah, J. Leistner and J. Liebscher, *Synlett*, 2008, **15**, 2342–2344.

73. T. Karthikeyan and S. Sankararaman, *Tetrahedron: Asymmetry*, 2008, **19**, 2741–2745.

74. D.-Q. Xu, H.-D. Yue, S.-P. Luo, A.-B. Xia, S. Zhang and Z.-Y. Xu, *Org. Biomol. Chem.*, 2008, **6**, 2054–2057.

75. D.-Q. Xu, L.-P. Wang, S.-P. Luo, Y.-F. Wang, S. Zhang and Z.-Y. Xu, *Eur. J. Org. Chem.*, 2008, 1049–1053.

76. C. Wang, C. Yu, C. Liu and Y. Peng, *Tetrahedron Lett.*, 2009, **50**, 2363–2366.

77. E. Tsandi, C. G. Kokotos, S. Kousidou, V. Ragoussis and G. Kokotos, *Tetrahedron*, 2009, **65**, 1444–1449.

78. G. L. Puleo and A. Iuliano, *Tetrahedron: Asymmetry*, 2008, **19**, 2045–2050.

79. G. Chen, Z. Wang and K. Ding, *Chinese J. Chem.*, 2009, **27**, 163–168.

80. B. Tan, X. Zeng, Y. Lu, P. J. Chua and G. Zhong, *Org. Lett.*, 2009, **11**, 1927–1930.

81. Y.-B. Zhao, L.-W. Zhang, L.-Y. Wu, X. Zhong, R. Li and J.-T. Ma, *Tetrahedron: Asymmetry*, 2008, **19**, 1352–1355.

82. G. Lv, R. Jin, W. Mai and L. Gao, *Tetrahedron: Asymmetry*, 2008, **19**, 2568–2572.

83. T. Miao and L. Wang, *Tetrahedron Lett.*, 2008, **49**, 2173–2176.

84. (a) P. Li, L. Wang, M. Wang and Y. Zhang, *Eur. J. Org. Chem.*, 2008, 1157–1160; (b) P. Li, L. Wang, Y. Zhang and G. Wang, *Tetrahedron*, 2008, **64**, 7633–7638.

85. B. Ni, Q. Zhang, K. Dhungana and A. D. Headley, *Org. Lett.*, 2009, **11**, 1037–1040.

86. P. Gao, C. Wang, Y. Wu, Z. Zhou and C. Tang, *Eur. J. Org. Chem.*, 2008, 4563–4566.

87. C.-J. Wang, Z.-H. Zhang, X.-Q. Dong and X.-J. Wu, *Chem. Commun.*, 2008, 1431–1433.

88. F.-Z. Peng, Z.-H. Shao, B.-M. Fan, H. Song, G.-P. Li and H.-B. Zhang, *J. Org. Chem.*, 2008, **73**, 5202–5205.

89. J. M. Andrés, R. Manzano and R. Pedrosa, *Chem. Eur. J.*, 2008, **14**, 5116–5119.

90. P. S. Hynes, P. A. Stupple and D. J. Dixon, *Org. Lett.*, 2008, **10**, 1389–1391.

91. Z.-H. Zhang, X.-Q. Dong, D. Chen and C.-J. Wang, *Chem. Eur. J.*, 2008, **14**, 8780–8783.

92. T. Mandal and C.-G. Zhao, *Ang. Chem., Int. Ed. Engl.*, 2008, **47**, 7714–7717.

93. C. Rabalakos and W. Wulff, *J. Am. Chem. Soc.*, 2008, **130**, 13524–13525.

94. J. Aleman, A. Milelli, S. Cabrera, E. Reyes and K. A. Jorgensen, *Chem. Eur. J.*, 2008, **14**, 10958–10966.

95. X. Han, J. Luo, C. Liu and Y. Lu, *Chem. Commun.*, 2009, 2044–2046.

96. W.-M. Zhou, H. Liu and D.-M. Du, *Org. Lett.*, 2008, **10**, 2817–2820.

97. F. Xue, S. Zhang, W. Duan and W. Wang, *Adv. Synth. Catal.*, 2008, **350**, 2194–2198.

98. Z. Yang, J. Liu, X. Liu, Z. Wang, X. Feng, Z. Su and C. Hu, *Adv. Synth. Catal.*, 2008, **350**, 2001–2006.

99. S. Luo, J. Li, L. Zhang, H. Xu and J.-P. Cheng, *Chem. Eur. J.*, 2008, **14**, 1273–1281.

100. (a) M. Wiesner, J. D. Revell, S. Tonazzi and H. Wennemers, *J. Am. Chem. Soc.*, 2008, **130**, 5610–5611; (b) M. Wiesner, J. D. Revell and H. Wennemers, *Ang. Chem., Int. Ed. Engl.*, 2008, **47**, 1871–1874.
101. Y.-D. Ju, L.-W. Xu, L. Li, G.-Q. Lai, H.-Y. Qiu, J.-X. Jiang and Y. Lu, *Tetrahedron Lett.*, 2008, **49**, 6773–6777.
102. M. Ganesh and D. Seidel, *J. Am. Chem. Soc.*, 2008, **130**, 16464–16465.
103. L. Bernardi, F. Fini, M. Fochi and A. Ricci, *Synlett*, 2008, **12**, 1857–1861.
104. H. Li, S. Zhang, C. Yu, X. Song and W. Wang, *Chem. Commun.*, 2009, 2136–2138.
105. J. Luo, L.-W. Xu, R. A. Siew Hay and Y. Lu, *Org. Lett.*, 2009, **11**, 437–440.
106. B.-C. Hong, R. Y. Nimje, M.-F. Wu and A. A. Sadani, *Eur. J. Org. Chem.*, 2008, 1449–1457.
107. D. Enders, C. Wang and J. W. Bats, *Ang. Chem., Int. Ed. Engl.*, 2008, **47**, 7539–7542.
108. D. Enders, M. R. M. Hüttl, G. Raabe and J. W. Bats, *Adv. Synth. Catal.*, 2008, **350**, 267–279.
109. Y.-K. Liu, H. Liu, W. Du, L. Yue and Y.-C. Chen, *Chem. Eur. J.*, 2008, **14**, 9873–9877.
110. B. Tan, P. J. Chua, Y. Li and G. Zhong, *Org. Lett.*, 2008, **10**, 2437–2440.
111. B. Tan, P. J. Chua, X. LZeng, M. Lu and G. Zhong, *Org. Lett.*, 2008, **10**, 3489–3492.
112. B. Tan, Z. Shi, P. J. Chua and G. Zhong, *Org. Lett.*, 2008, **10**, 3425–3428.
113. X. Lu and L. Deng, *Ang. Chem., Int. Ed. Engl.*, 2008, **47**, 7710–7713.
114. A. Scettri, A. Massa, L. Palombi, R. Villano and M. R. Acocella, *Tetrahedron: Asymmetry*, 2008, **19**, 2149–2152.
115. Q. Lin, D. Meloni, Y. Pan, M. Xia, J. Rodgers, S. Shepard, M. Li, L. Galya, B. Metcalf, T.-Y. Yue, P. Liu and J. Zhou, *Org. Lett.*, 2009, **11**, 1999–2002.
116. P. Ricci, A. Carlone, G. Bartoli, M. Bosco, L. Sambri and P. Melchiorre, *Adv. Synth. Catal.*, 2008, **350**, 49–53.
117. D. Enders and K. Hoffman, *Eur. J. Org. Chem.*, 2009, 1665–1668.
118. D. Leow, S. Lin, S. K. Chittimalla, X. Fu and C.-H. Tan, *Ang. Chem., Int. Ed. Engl.*, 2008, **47**, 5641–5645.
119. (a) S. Shirakawa, T. Kimura, S.-i. Murata and S. Shimizu, *J. Org. Chem.*, 2009, **74**, 1288–1296; (b) S. Shirakawa, A. Moriyama and S. Shimizu, *Eur. J. Org. Chem.*, 2008, 5957–5964.
120. D. R. Li, A. Murugan and J. R. Falck, *J. Am. Chem. Soc.*, 2008, **130**, 46–48.
121. I. Ibrahem, P. Hammar, J. Vesely, R. Rios, L. Eriksson and A. Cordova, *Adv. Synth. Catal.*, 2008, **350**, 1875–1884.
122. S.-G. Kim, *Tetrahedron Lett.*, 2008, **49**, 6148–6151.
123. H. Li, L. Zu, H. Xie, J. Wang and W. Wang, *Chem. Commun.*, 2008, 5636–5638.
124. G.-L. Zhao, R. Rios, J. Vesely, L. Eriksson and A. Cordova, *Ang. Chem., Int. Ed. Engl.*, 2008, **47**, 8468–8472.

125. G.-L. Zhao, J. Vesely, R. Rios, I. Ibrahem, H. Sunden and A. Cordova, *Adv. Synth. Catal.*, 2008, **350**, 237–242.
126. D.-Q. Xu, Y.-F. Wang, S.-P. Luo, S. Zhang, A.-G. Zhong, H. Chen and Z.-Y. Xu, *Adv. Synth. Catal.*, 2008, **350**, 2610–2616.
127. F. Pesciaioli, F. De Vincentiis, P. Galzerano, G. Bencivenni, G. Bartoli, A. Mazzanti and P. Melchiorre, *Ang. Chem., Int. Ed. Engl.*, 2008, **47**, 8703–8706.
128. (a) H. Pellissier, *Tetrahedron*, 2003, **59**, 8291–8327; (b) H. Pellissier, *Tetrahedron*, 2008, **64**, 1563–1601.
129. J. Wang, H. Xie, H. Li, L. Zu and W. Wang, *Ang. Chem., Int. Ed. Engl.*, 2008, **47**, 4177–4179.
130. R. Dodda, T. Mandal and C.-G. Zhao, *Tetrahedron Lett.*, 2008, **49**, 1899–1902.
131. R. Dodda, J. J. Goldman, T. Mandal, C.-G. Zhao, G. A. Broker and E. R. T. Tiekink, *Adv. Synth. Catal.*, 2008, **350**, 537–541.
132. H.-H. Lu, H. Liu, W. Wu, X.-F. Wang, L.-Q. Lu and W.-J. Xiao, *Chem. Eur. J.*, 2009, **15**, 2742–2746.
133. M. Bandini, A. Eichholzer, M. Tragni and A. Umani-Ronchi, *Ang. Chem., Int. Ed. Engl.*, 2008, **47**, 3238–3241.
134. S. Fustero, J. Moscardo, D. Jimenez, M. D. Pérez-Carrion, M. Sanchez-Rosello and C. del Pozo, *Chem. Eur. J.*, 2008, **14**, 9868–9872.
135. E. C. Carlson, L. K. Rathbone, H. Yang, N. D. Collett and R. G. Carter, *J. Org. Chem.*, 2008, **73**, 5155–5158.

CHAPTER 2
Nucleophilic Additions to C=O Double Bonds

2.1 Aldol Reactions

2.1.1 Aldol Reactions Catalysed by Proline Derivatives

The asymmetric aldol reaction is one of the most important topics in modern catalytic synthesis and one of the most advanced types of synthesis in the field of organocatalysis.[1] During the last few years, the organocatalysed intermolecular direct aldol reactions have grown most remarkably, especially those which involve chiral proline derivatives as organocatalysts.[2] Surprisingly, the catalytic potential of proline in asymmetric aldol reactions was not explored further, until recently, in spite of the discovery, in the early 1970s, of the first proline-catalysed intramolecular aldol reaction. The List and Barbas approaches, applying proline as the organocatalyst, are surely among the most efficient and general asymmetric catalytic aldol reactions yet discovered.[3] There are several reasons why proline has become an important molecule in asymmetric catalysis, *e.g.* it is an abundant chiral molecule which is inexpensive and available in both enantiomeric forms. Additionally, there are various chemical reasons that contribute to proline's role in catalysis. Proline is bifunctional, with a carboxylic acid and an amine function, which can both act as an acid or a base, allowing chemical transformations in concert, in a similar manner to enzymatic catalysis. In 2008, a few interesting applications of the enantioselective proline-catalysed aldol methodology were reported, such as that used in the total synthesis of antitumour jaspine B reported by Enders *et al.*[4] Indeed, the key step of this synthesis was a D-proline-catalysed enantioselective aldolisation between pentadecanal and a dioxanone, which afforded the corresponding aldol product with excellent stereoselectivities (Scheme 2.1). Another application of this methodology in total synthesis was reported by MacMillan *et al.* with a total synthesis of callipeltoside C.[5] Indeed, the key step of this synthesis was a D-proline-catalysed enantioselective aldolisation between the Roche

RSC Catalysis Series No. 3
Recent Developments in Asymmetric Organocatalysis
By Hélène Pellissier
© Hélène Pellissier 2010
Published by the Royal Society of Chemistry, www.rsc.org

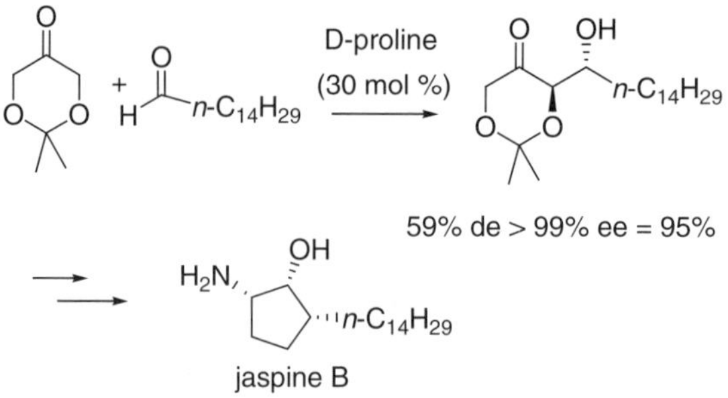
O
O
+
H
n-C14H29
D-proline
(30 mol %)
O
OH
n-C14H29
59% de > 99% ee = 95%
OH
H2N
n-C14H29
jaspine B

Scheme 2.1 Synthesis of jaspine B.

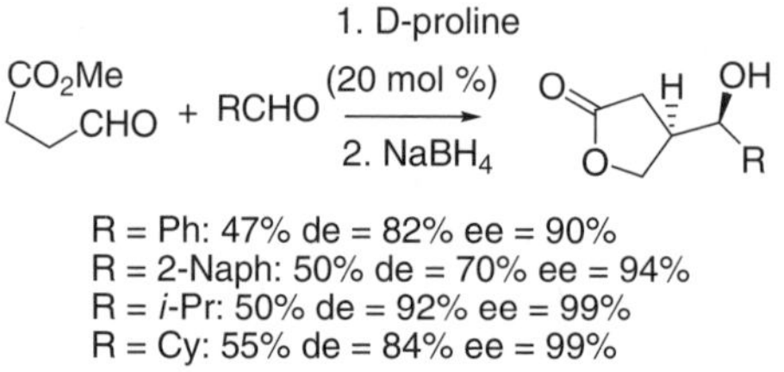
1. D-proline
(20 mol %)
2. NaBH4
CO2Me
CHO
+ RCHO
H
OH
R
R = Ph: 47% de = 82% ee = 90%
R = 2-Naph: 50% de = 70% ee = 94%
R = i-Pr: 50% de = 92% ee = 99%
R = Cy: 55% de = 84% ee = 99%

Scheme 2.2 Synthesis of β-(hydroxyalkyl)-γ-butyrolactones.

ester-derived aldehyde and propionaldehyde, which furnished the corresponding aldehyde in a moderate yield (48%), a good diastereoselectivity (84% de) and an excellent enantioselectivity (99% ee).

In addition, the synthesis of various chiral β-(hydroxyalkyl)-γ-butyrolactones was based on the D-proline-catalysed enantioselective aldolisation between aldehydes and methyl 4-oxobutyrate, followed by a reduction with $NaBH_4$.[6] As shown in Scheme 2.2, a series of lactones could be prepared in moderate yields and diastereoselectivities but with high enantioselectivities of up to 99% ee.

It must be noted that a remarkable improvement of both the chemical yield (from 6% to 82%) and the enantioselectivity (up to >99% ee) of L-proline-catalysed aldol reactions of a wide range of aldehydes with acetone was found by Liebscher *et al.* when hexasubstituted or pentasubstituted guanidinium salts were added as ionic liquids.[7] Furthermore, this study showed that guanidinium salts could be advantageous over imidazolium salt-based ionic liquids.

On the other hand, a number of asymmetric aldol reactions have been performed in the last year in the presence of variously substituted prolines as the organocatalysts. As an example, Zhao *et al.* reported excellent results for the cross-aldol reaction of cyclohexanone with β,γ-unsaturated keto esters catalysed by a *trans*-siloxy-L-proline (Scheme 2.3).[8] This practical and highly efficient protocol could be extended to other ketones, albeit with lower enantioselectivities (≤93% ee).

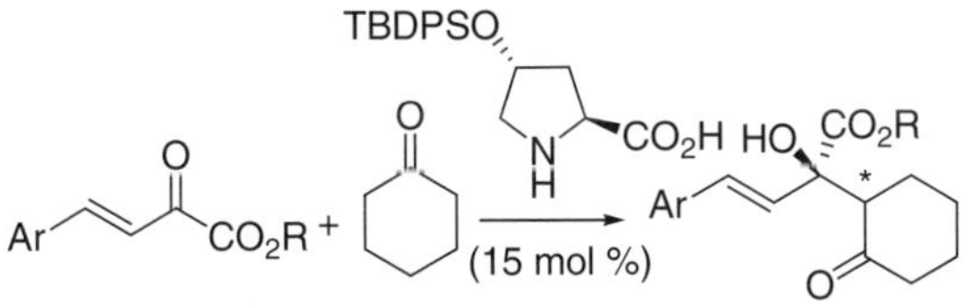

Ar = Ph, R = Et: 99% de = 90% ee > 99%
Ar = Ph, R = CH$_2$CH=CH$_2$: 99% de = 90% ee > 99%
Ar = Ph, R = *i*-Pr: 98% de = 90% ee > 99%
Ar = *p*-FC$_6$H$_4$, R = Me: 99% de = 90% ee > 99%
Ar = *o*-BrC$_6$H$_4$, R = Me: 92% de = 90% ee > 99%
Ar = 2-Fu, R = Me: 72% de = 92% ee > 99%

Scheme 2.3 Cross-aldol reactions of cyclohexanone with β,γ-unsaturated keto esters catalysed by *trans*-siloxy-L-proline.

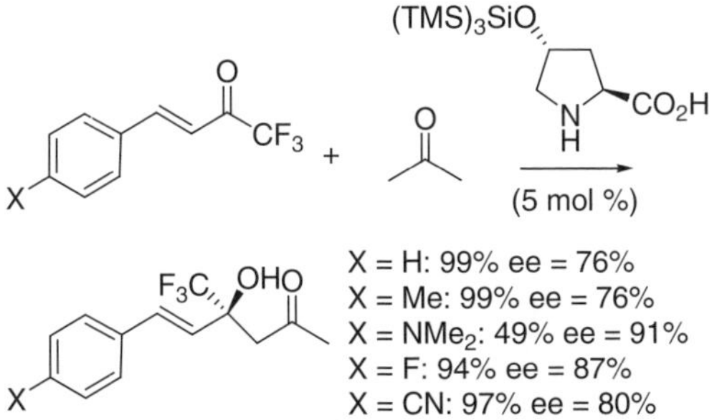

X = H: 99% ee = 76%
X = Me: 99% ee = 76%
X = NMe$_2$: 49% ee = 91%
X = F: 94% ee = 87%
X = CN: 97% ee = 80%
X = Br: 90% ee = 91%

Scheme 2.4 1,2-Aldol reactions of acetone with α,β-unsaturated trifluoromethyl-ketones catalysed by *trans*-siloxy-L-proline.

A closely related but more sterically hindered organocatalyst was used by Yuan and Zhang to promote an unexpected enantioselective 1,2-aldol reaction of acetone with α,β-unsaturated trifluoromethylketones.[9] Indeed, the corresponding tertiary alcohols were obtained in high yields and enantioselectivities, as summarised in Scheme 2.4.

From a green chemistry perspective, novel 4-substituted acyloxyproline derivatives were investigated as organocatalysts by Gruttadauria *et al.* for the enantioselective aldolisation of cyclic ketones with substituted benzaldehydes in water.[10] The best results were obtained with catalysts bearing the most hydrophobic acyl chains, such as 4-phenylbutanoate or 4-(pyren-1-yl)butan-oate, which provided excellent stereoselectivities in general, as shown in Scheme 2.5. It was shown that these catalysts could be used down to 0.5 mol % as catalyst loading with an excellent stereoselectivity, albeit a moderate yield.

Highly diastereo- and enantioselective aldol reactions between aromatic aldehydes and cyclohexanone performed in water were developed by Tao *et al.* by using a novel simple amphiphilic proline-derived organocatalyst bearing a

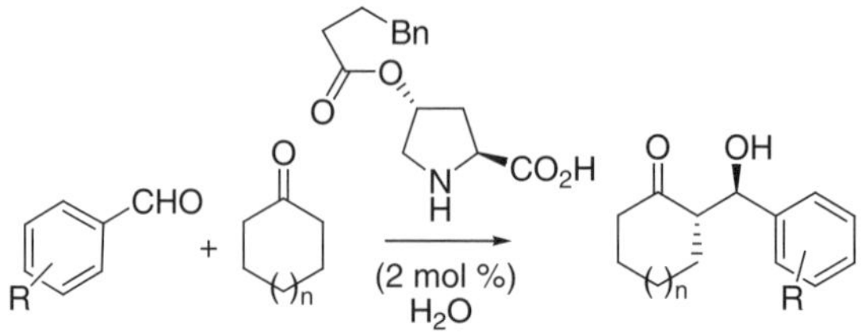

R = H, n = 1: 55% de = 92% ee = 96%
R = 4-NO$_2$, n = 1: 99% de = 92% ee = 99%
R = 2-Naph, n = 1: 45% de = 84% ee = 99%
R = 4-CN, n = 1: 99% de = 94% ee = 98%
R = 2-Cl, n = 1: 99% de = 94% ee = 98%
R = 2-NO$_2$, n = 1: 97% de = 98% ee = 99%
R = H, n = 0: 83% de = 78% ee = 97%
R = 4-NO$_2$, n = 0: 99% de = 74% ee = 98%
R = 2-Naph, n = 0: 84% de = 78% ee = 97%
R = 4-CN, n = 0: 99% de = 70% ee = 99%
R = 2-Cl, n = 0: 99% de = 80% ee = 98%
R = 2-NO$_2$, n = 0: 99% de = 84% ee = 99%

Scheme 2.5 Aldolisations of cyclic ketones with substituted benzaldehydes catalysed by 4-phenylbutanoate proline.

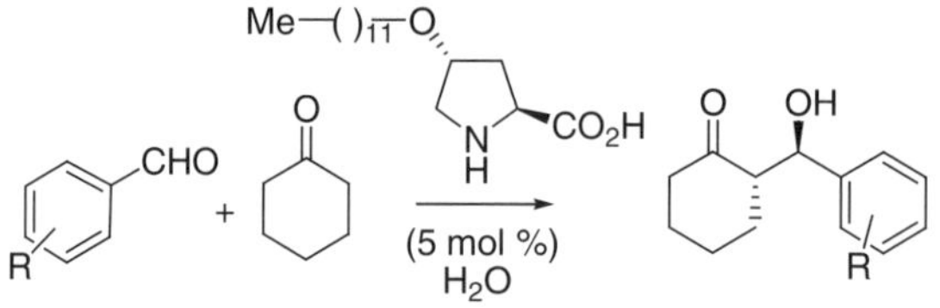

R = 2-NO$_2$: 98% de = 90% ee = 99%
R = 3-NO$_2$: 93% de = 94% ee = 98%
R = 4-NO$_2$: 92% de = 94% ee = 98%
R = 2-Cl: 93% de = 88% ee = 99%
R = 3-Cl: 97% de = 92% ee = 98%
R = 4-Cl: 96% de = 88% ee = 98%
R = 4-F: 85% de = 64% ee = 98%
R = 2-MeO: 67% de = 96% ee = 98%

Scheme 2.6 Aldolisations of cyclic ketones with benzaldehydes catalysed by amphiphilic proline.

long alkyl chain on the 4-position *via* a stable ether bond.[11] The catalytic activity obtained with 5 mol % of catalyst loading was better than that obtained at 30 mol % of proline itself. Especially, high yields of up to 99%, excellent enantioselectivities of up to 99% ee, combined with excellent diastereoselectivities of up to 98% de were achieved in the reactions of aromatic aldehydes with cyclic ketones with 5 mol % of catalyst loading in water, as shown in Scheme 2.6. It must be noted that this catalyst could also be used in organic solvents, such as THF or DMF, providing the corresponding aldol products albeit in lower yields and stereoselectivities. The catalytic role of this novel amphiphilic chiral molecule was more like an enzyme process than that of proline itself. Amazingly, the authors have observed that the configuration of the aldol product of acetone with *ortho*-nitrobenzaldehyde was related to the amount of water.

Zlotin *et al.* demonstrated that the asymmetric aldol reaction between aldehydes and ketones could be catalysed by novel chiral ionic liquids bearing a 4-hydroxyproline moiety and pyridinium cation in water.[12] The use of the catalyst with the readily regenerated anion PF_6^- allowed the aldol products to be obtained in excellent stereoselectivities and yields, as shown in Scheme 2.7.

In the same context, Trombini *et al.* developed other organocatalysts of this type bearing an imidazolium.[13] Under solvent-free conditions, these catalysts were shown to be highly efficient at a remarkably low catalyst loading of 0.1 mol % to induce excellent stereoselectivities in the aldol reaction of cyclohexanone with aldehydes in the presence of water (Scheme 2.8). Moreover, exceptionally high values of TON (up to 930) were achieved in the case of the most reactive aromatic aldehydes.

In addition, a new functional polymer in which proline was bonded to polystyrene through a 1,2,3-triazole linker was applied to the asymmetric aldolisation of cyclohexanone with benzaldehydes in water.[14] The use of this resin allowed the combination of excellent yields, diastereo- and enantioselectivities to be recorded in all the studied cases, as shown in Scheme 2.9.

These reactions were also successfully performed in the presence of another efficient polystyrene anchored L-proline developed by Tao *et al.*[15] Therefore, by using 5 mol % of this linear catalyst in the presence of water in DMF, the reaction provided the corresponding aldol products with good yields of up to 91%, good enantioselectivities of up to 98% ee and good diastereoselectivities of up to 86% de, as shown in Scheme 2.10. It was noted that the yields of these reactions performed in a ketone/water mixture were lower than those

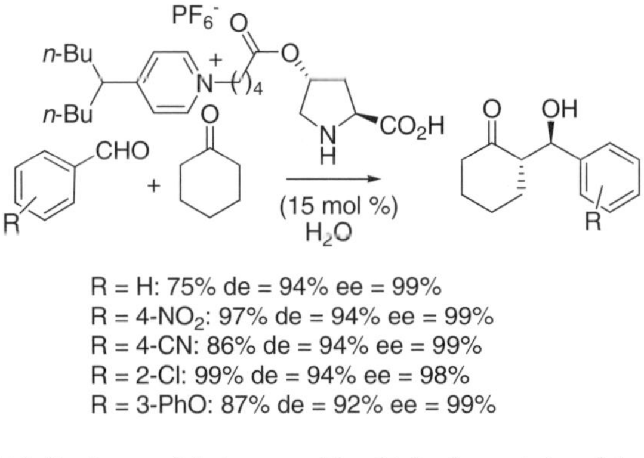

R = H: 75% de = 94% ee = 99%
R = 4-NO$_2$: 97% de = 94% ee = 99%
R = 4-CN: 86% de = 94% ee = 99%
R = 2-Cl: 99% de = 94% ee = 98%
R = 3-PhO: 87% de = 92% ee = 99%

Scheme 2.7 Aldolisations of ketones with aldehydes catalysed by proline-derived ionic liquids.

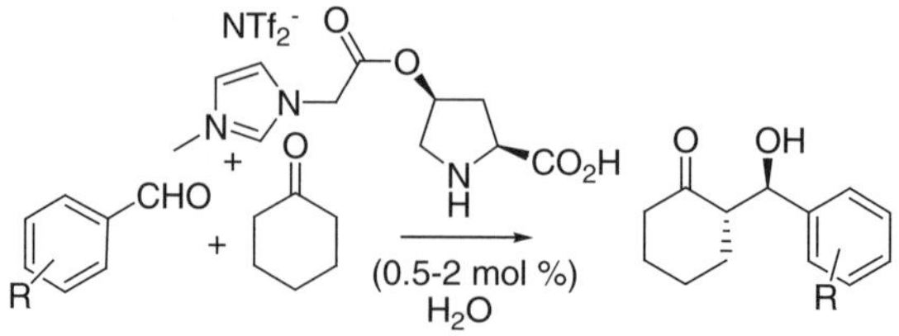

R = 4-NO$_2$: 99% de = 90% ee > 99%
R = 4-CN: 99% de = 84% ee > 99%
R = 2-NO$_2$: 95% de = 84% ee > 99%
R = 4-Br: 95% de = 90% ee > 99%
R = 2-Naph: 90% de = 76% ee = 98%

Scheme 2.8   Aldolisations of cyclohexanone with aldehydes catalysed by proline-derived ionic liquids.

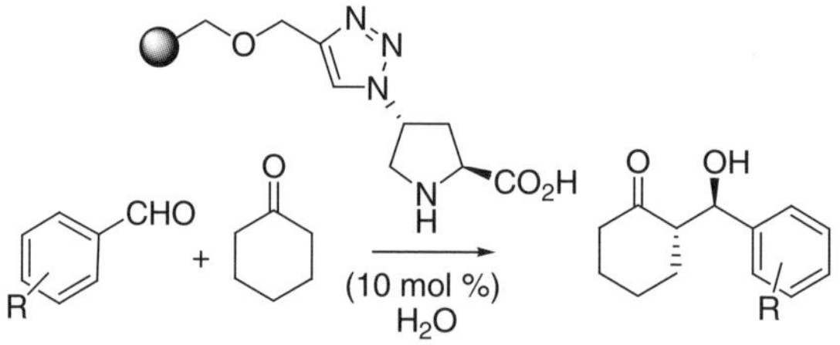

R = H: 74% de = 92% ee = 98%
R = 4-NO$_2$: 80% de = 90% ee = 96%
R = 2-Naph: 82% de = 94% ee > 99%
R = 4-Br: 90% de = 94% ee = 96%
R = 4-CF$_3$: 98% de = 94% ee = 96%

Scheme 2.9   Aldolisations of cyclohexanone with aldehydes catalysed by proline-derived resin.

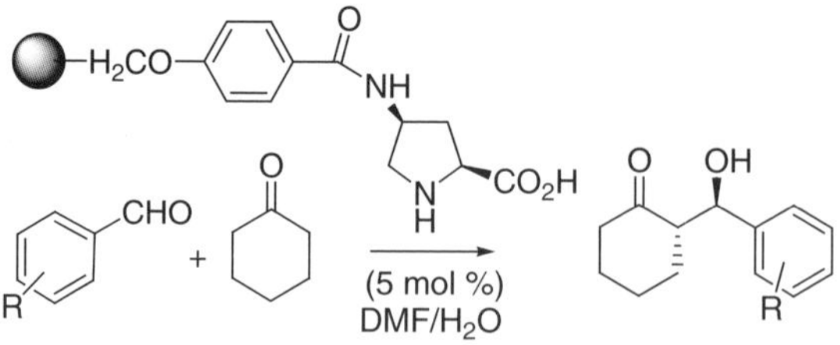

R = 4-NO$_2$: 91% de = 84% ee = 98%
R = 2-NO$_2$: 63% de = 82% ee = 93%
R = 2-Cl: 72% de = 82% ee = 89%
R = 3-Br: 65% de = 84% ee = 96%

Scheme 2.10   Aldolisations of cyclohexanone with aldehydes catalysed by polystyrene anchored L-proline.

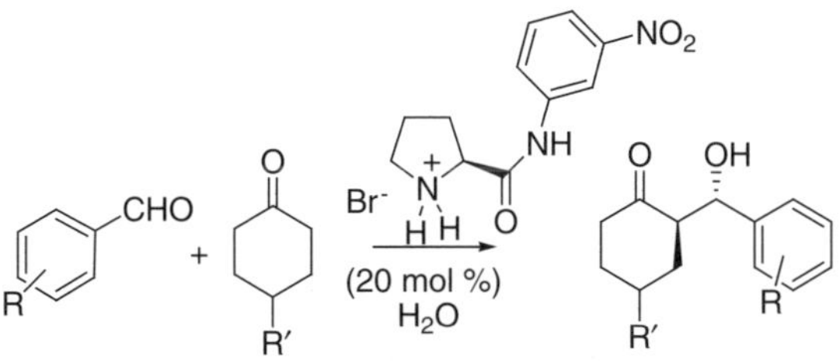

R = 4-NO$_2$, R′ = H: 92% de = 88% ee = 95%
R = 2-NO$_2$, R′ = H: 86% de = 84% ee = 94%
R = 4-Cl, R′ = H: 73% de = 84% ee = 95%
R = 2,6-Cl$_2$, R′ = H: 89% de > 98% ee = 98%
R = 4-NO$_2$, R′ = Me: 88% ee = 95%
R = 2-NO$_2$, R′ = Me: 85% ee = 95%
R = 4-Cl, R′ = Me: 80% ee = 97%
R = 4-NO$_2$, R′ = t-Bu: 90% ee = 92%

Scheme 2.11 Aldolisations of cyclohexanone with aldehydes catalysed by protonated simple prolinamide.

performed in wet DMF (up to 84%). However, the stereoselectivity was comparable with values of up to 84% de and 96% ee. In addition, this catalyst could be recovered by a simple precipitation and filtration process, allowing its reuse for at least five times without obvious loss of the catalytic efficiency.

In the last year, an impressive number of highly efficient enantioselective aldolisations have been catalysed by variously substituted prolinamides. As an example, Chimni *et al.* have demonstrated that very simple protonated (*S*)-prolinamide derivatives efficiently catalysed aldol reactions of ketones with aromatic aldehydes in water.[16] The best results concerning a range of cyclohexanones are collected in Scheme 2.11. When applied to aliphatic ketones, the protocol provided the aldol adducts in good yields, albeit with lower diastereo- and enantioselectivities (≤63% de and ≤48% ee, respectively).

In 2009, Fu *et al.* developed highly 4-phenoxy-substituted prolinamide phenols, which could promote the asymmetric aldolisation of cyclohexanone with a range of aldehydes with a high degree of diastereo- and enantioselectivity in a large amount of water (Scheme 2.12).[17] The best enantioselectivities of up to 97% ee were obtained with the most steric hindered catalyst that bore a *tert*-butyl group on the phenol moiety. The scope of the reaction could be extended to aliphatic ketones with enantioselectivities of up to 94% ee, albeit with low to moderate diastereoselectivities of up to 50% de.

When the aldolisation of cyclohexanone with aldehydes was performed in the presence of 4-hydroxy-prolinamide alcohols as the organocatalysts, the aldol adduct was obtained in comparable high yields and excellent enantioselectivities of up to 99% ee.[18] Interestingly, the extension of this methodology to acetone provided by reaction with various benzaldehydes the corresponding aldol adducts in still excellent enantioselectivities in a range of 97–99% ee, as shown in Scheme 2.13.

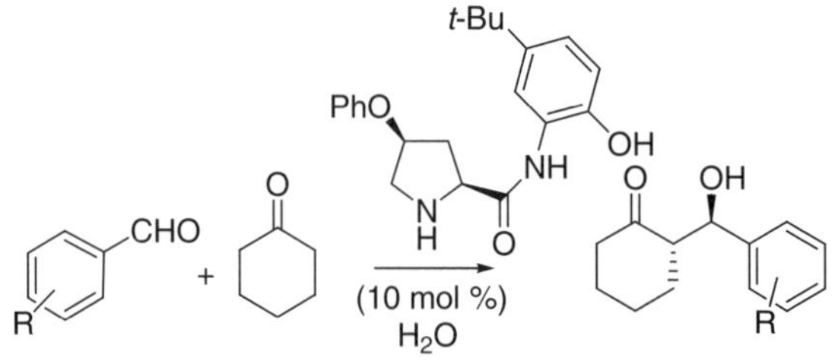

R = 4-NO$_2$: 99% de = 98% ee = 94%
R = 3-NO$_2$: 97% de = 94% ee = 95%
R = 2-NO$_2$: 99% de = 98% ee = 97%
R = 4-CF$_3$: 98% de = 90% ee = 93%
R = 4-CN: 93% de = 88% ee = 93%
R = H: 72% de = 85% ee = 84%
R = 4-Me: 65% de = 80% ee = 97%

Scheme 2.12   Aldolisations of cyclohexanone with aldehydes catalysed by 4-phenoxy prolinamide phenol.

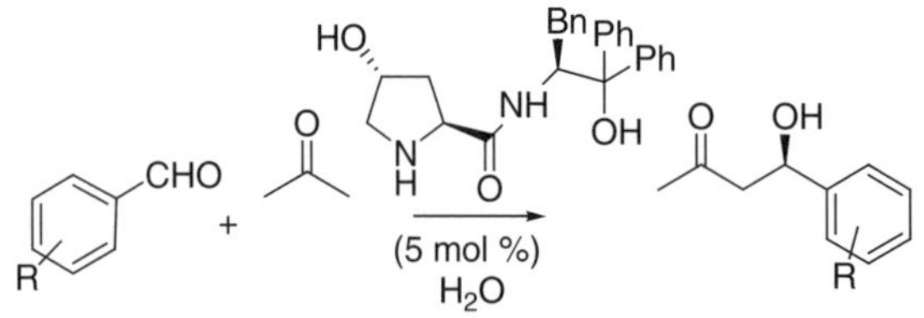

R = 4-MeO: 95% ee = 98%
R = 4-F: 82% ee = 99%
R = 2-Cl: 95% ee = 97%
R = 4-Cl: 92% ee = 97%
R = H: 48% ee = 99%

Scheme 2.13   Aldolisations of acetone with aldehydes catalysed by 4-hydroxy-prolinamide alcohol.

The silyl ether of a closely related 4-hydroxy-prolinamide alcohol was also successfully applied as an organocatalyst to the asymmetric aldolisation of either cyclic or acyclic ketones with a range of aromatic aldehydes in water.[19] This catalyst was highly efficient and as low as 1 mol % of catalyst loading allowed high yields and excellent diastereo- and enantioselectivities to be obtained, as shown in Scheme 2.14.

Ma *et al.* have found that the presence of molecular sieves increased the diastereo- and enantioselectivity in the aldolisation of cyclic ketones with various aromatic aldehydes using binaphthyl-based axially chiral prolinamides as the organocatalysts.[20] The beneficial influence of the molecular sieves could be due to their water-trapping properties, preventing water-mediated background reactions. The best results are collected in Scheme 2.15.

In a similar area, Lüdtke *et al.* have employed a robust cysteine-derived prolinamide as an organocatalyst to promote the asymmetric aldolisation of

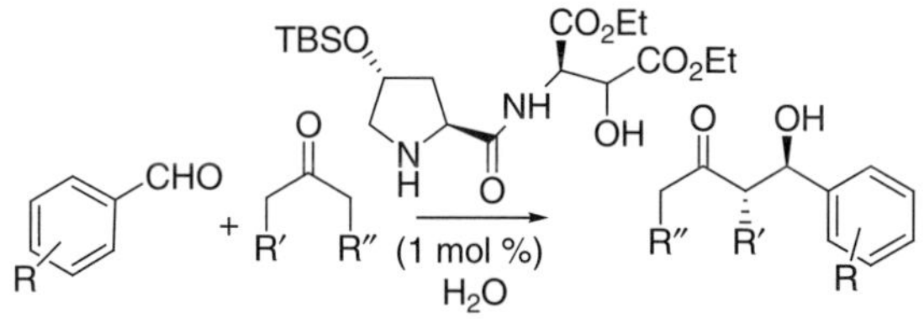

R = 4-NO$_2$, R', R" = (CH$_2$)$_3$: 99% de > 98% ee = 94%
R = 2-NO$_2$, R', R" = (CH$_2$)$_3$: 85% de > 98% ee = 95%
R = 3-NO$_2$, R', R" = (CH$_2$)$_3$: 95% de = 96% ee = 92%
R = 4-Br, R', R" = (CH$_2$)$_3$: 94% de > 98% ee = 93%
R = 4-NO$_2$, R' = R" = H: 85% ee = 71%
R = 4-NO$_2$, R' = H, R" = *n*-Pent: 80% ee = 85%
R = 4-NO$_2$, R' = H, R" = CH$_2$Bn: 88% ee = 89%
R = 4-NO$_2$, R' = H, R" = *n*-Pr: 75% ee = 85%
R = 2,6-Cl$_2$, R', R" = (CH$_2$)$_3$: 95% de > 98% ee = 98%

Scheme 2.14 Aldolisations of ketones with aldehydes catalysed by silylated 4-hydroxy-prolinamide alcohol.

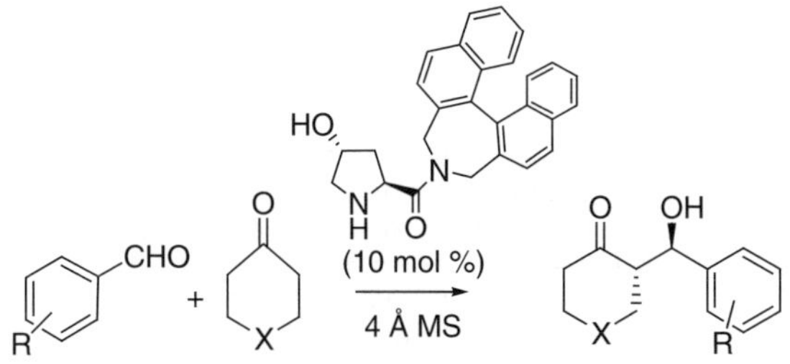

R = 4-NO$_2$, X = CH$_2$: 99% de = 84% ee > 99%
R = 4-NO$_2$, X = O: 78% de = 30% ee = 92%
R = 3-NO$_2$, X = S: 43% de = 94% ee = 96%
R = 4-NO$_2$, X = *t*-BuCH$_2$: 99% de = 96% ee = 95%
R = 2-NO$_2$, X = CH$_2$: 79% de = 82% ee = 87%

Scheme 2.15 Molecular sieves-controlled aldolisations of cyclic ketones with aldehydes catalysed by silylated 4-hydroxy-prolinamide.

acetone with aromatic aldehydes, providing the aldol adducts in high enantioselectivities of up to 94% ee (Scheme 2.16),[21] while similar reactions gave moderate enantioselectivities (7–69% ee) and low to moderate yields when catalysed by electronic-tuned *N*-(2-hydroxyphenyl)-(*S*)-prolinamides.[22]

A series of new robust proline-based dipeptides with two amide units were evaluated by Peng *et al.* as organocatalysts for the asymmetric aldolisation of cyclohexanone in organic solvents.[23] The reactions proceeded smoothly in the presence of an additive, such as AcOH or *o*-TolCO$_2$H, providing high yields and enantio- and diastereoselectivities of up to 98% ee and 98% de, respectively (Scheme 2.17). In order to explain the enantioselectivity of the reaction,

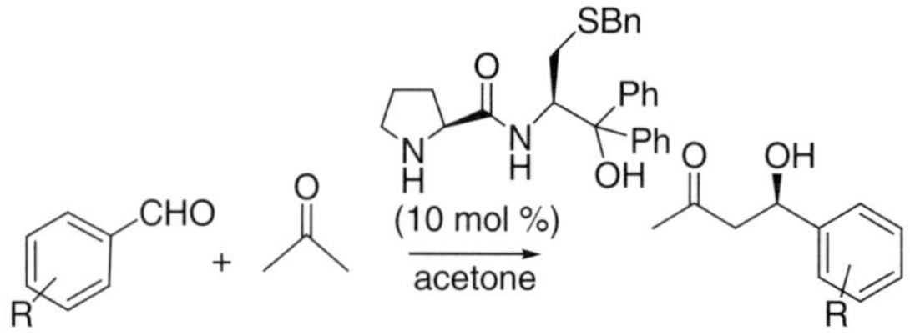

R = H: 83% ee = 94%
R = 4-NO$_2$: 40% ee = 64%
R = 4-Cl: 57% ee = 87%
R = 4-Me: 56% ee = 85%
R = 2-Me: 56% ee = 81%

Scheme 2.16   Aldolisations of acetone with aldehydes catalysed by cysteine-derived prolinamide.

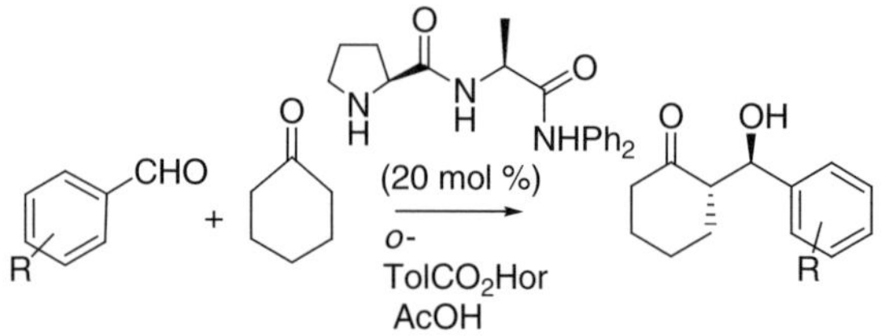

R = 4-NO$_2$: 98% de = 96% ee = 93%
R = 3-NO$_2$: 94% de = 74% ee = 96%
R = 2-NO$_2$: 60% de = 64% ee = 94%
R = 4-Br: 75% de = 80% ee = 98%
R = 2,4-Cl$_2$: 93% de = 98% ee = 97%
R = 2-Br: 73% de = 90% ee = 94%

Scheme 2.17   Aldolisations of cyclohexanone with aldehydes catalysed by proline-based dipeptide.

the authors proposed that the two amide units played a key role in stabilising the transition state by forming hydrogen bonds between the catalyst and the aldehyde, and these two tunable functionalities not only activated the aldol acceptor but also favoured one face of the acceptor to the attack of the enamine.

The synthesis of a novel Merrifield resin-supported dipeptide Pro-Ala-O-P, derived from proline and alanine, was reported by Wang and Yan with the aim of being used as an organocatalyst in asymmetric aldol reactions of ketones with aldehydes.[24] Indeed, this supported dipeptide was found to be an efficient catalyst to promote the asymmetric aldol reaction under neat conditions between aromatic aldehydes and cyclic ketones, generating the corresponding aldol products with moderate to high yields and diastereoselectivities of up to 80% de combined with good enantioselectivities of up to 95% ee, as shown in Scheme 2.18. Moreover, this catalyst could be used for seven times with only a minor decrease in product yields, but maintained stereoselectivities.

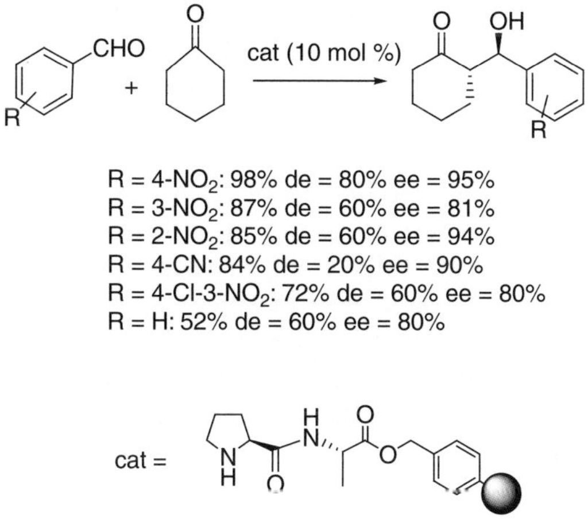
CHO
O
+
cat (10 mol %)
O OH
R
R
R = 4-NO2: 98% de = 80% ee = 95%
R = 3-NO2: 87% de = 60% ee = 81%
R = 2-NO2: 85% de = 60% ee = 94%
R = 4-CN: 84% de = 20% ee = 90%
R = 4-Cl-3-NO2: 72% de = 60% ee = 80%
R = H: 52% de = 60% ee = 80%
cat =
N
H
H
N
O
O
O

Scheme 2.18 Aldolisations of ketones with aldehydes catalysed by Merrifield resin-supported dipeptide.

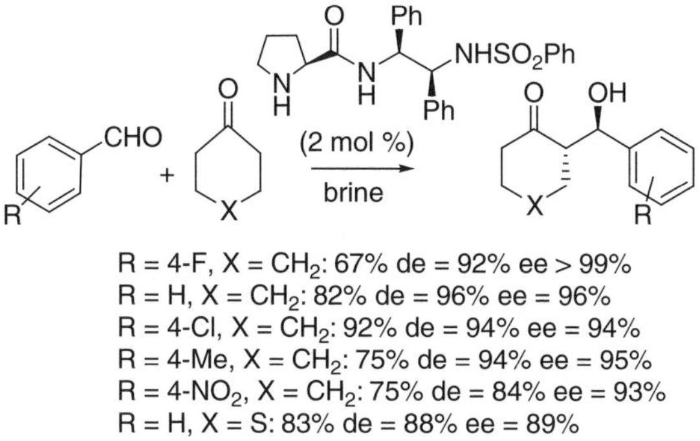
O
Ph
NHSO2Ph
N
H
N
H
Ph
CHO
O
+
(2 mol %)
brine
X
O OH
X
R
R
R = 4-F, X = CH2: 67% de = 92% ee > 99%
R = H, X = CH2: 82% de = 96% ee = 96%
R = 4-Cl, X = CH2: 92% de = 94% ee = 94%
R = 4-Me, X = CH2: 75% de = 94% ee = 95%
R = 4-NO2, X = CH2: 75% de = 84% ee = 93%
R = H, X = S: 83% de = 88% ee = 89%

Scheme 2.19 Aldolisations of ketones with aldehydes catalysed by β-amino sulfonamide.

Although chiral sulfur ligands have been widely employed in a number of transition-metal-catalysed asymmetric transformations,[25] the incorporation of sulfur in organocatalysts is still rather rare in the literature. In this context, Singh and Gandhi disclosed the synthesis of an organocatalyst derived from L-proline and a β-amino sulfonamide, and its application in aldolisation of ketones with aromatic aldehydes.[26] This organocatalyst has proved to be very efficient for this reaction, with both cyclic as well as acyclic ketones in brine with 2 mol% of catalyst loading, affording the aldol products in excellent yields, diastereo- and enantioselectivities, as shown in Scheme 2.19.

Some of the highest levels of both diastereo- and enantioselectivity reported for many of these reactions have been recorded by Carter and Yang by involving a simple sulfonamide derived from proline as the organocatalyst.[27]

Indeed, the application of this catalyst to the aldolisation of a range of ketones with aromatic aldehydes provided remarkable results collected in Scheme 2.20. Moreover, it was demonstrated that a catalyst loading of only 2 mol % could be employed in the absence of any organic solvent with continued high levels of selectivity.

On the other hand, the introduction of a phenoxy group at the hydroxyl function of *N*-arylsulfonyl derivatives of *trans*-4-hydroxy-L-proline resulted in more hydrophobic catalysts, which could be used to promote aldolisations of cyclic ketones with aromatic aldehydes in water with excellent yields, diastereo- and enantioselectivities, as shown in Scheme 2.21.[28]

In the same context, Wang *et al.* reported similar highly enantioselective reactions performed in water by using another proline-derived sulfonamide, such as fluorous (*S*)-pyrrolidine sulfonamide.[29] A notable feature of this simple

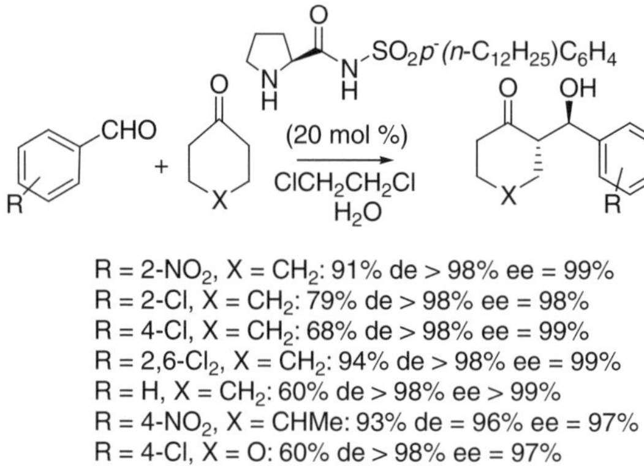
SO2p-(n-C12H25)C6H4
CHO
(20 mol %)
ClCH2CH2Cl
H2O
O
OH
R
X
R = 2-NO2, X = CH2: 91% de > 98% ee = 99%
R = 2-Cl, X = CH2: 79% de > 98% ee = 98%
R = 4-Cl, X = CH2: 68% de > 98% ee = 99%
R = 2,6-Cl2, X = CH2: 94% de > 98% ee = 99%
R = H, X = CH2: 60% de > 98% ee > 99%
R = 4-NO2, X = CHMe: 93% de = 96% ee = 97%
R = 4-Cl, X = O: 60% de > 98% ee = 97%

Scheme 2.20 Aldolisations of ketones with aldehydes catalysed by sulfonamide.

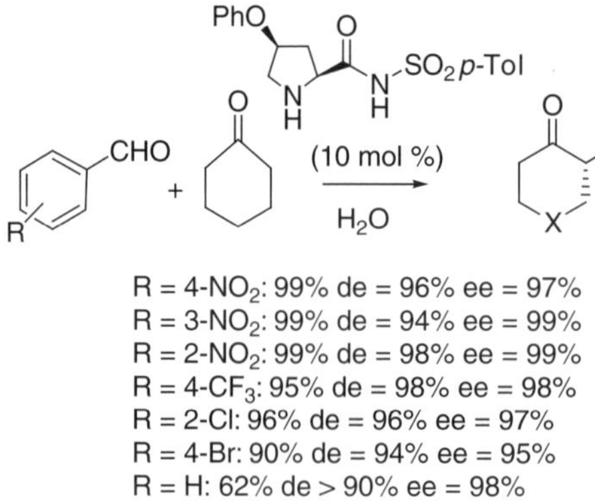
PhO
SO2p-Tol
CHO
(10 mol %)
H2O
O
OH
R
X
R = 4-NO2: 99% de = 96% ee = 97%
R = 3-NO2: 99% de = 94% ee = 99%
R = 2-NO2: 99% de = 98% ee = 99%
R = 4-CF3: 95% de = 98% ee = 98%
R = 2-Cl: 96% de = 96% ee = 97%
R = 4-Br: 90% de = 94% ee = 95%
R = H: 62% de > 90% ee = 98%

Scheme 2.21 Aldolisations of ketones with aldehydes catalysed by sulfonamide in water.

organocatalyst was that it could be recovered from the reaction mixtures by simple fluorous solid-phase extraction and subsequently reused up to seven times without loss of either the catalytic activity or the enantioselectivity. Employed at a 10 mol % of catalyst loading, the aldol products, generated from a range of cyclic as well as acyclic ketones and aromatic aldehydes, were formed in high yields, excellent enantioselectivities of up to 97% ee combined with high diastereoselectivities of up to 90% de. In addition, Kokotos *et al.* investigated other sulfonamides derived from homoproline and dipeptides as organocatalysts for the asymmetric aldolisation of acetone with *para*-nitrobenzaldehyde.[30] The best enantioselectivity of 84% ee combined with 79% yield was obtained with the methanesulfonamide of dipeptide Pro-Phe used at 20 mol % of catalyst loading in the presence of TEA in CH_2Cl_2. On the other hand, an *N*-(2-thienylsulfonyl)prolinamide was found by Toru *et al.* to work as an efficient organocatalyst in the aldol reaction of acetone with substituted isatins.[31] Therefore, the reaction of acetone with various substituted isatins occurred with moderate to excellent yields and enantioselectivities of up to 97% ee in the presence of a low catalyst loading of 5 mol %, providing the expected convolutamydine A derivatives, as shown in Scheme 2.22. Surprisingly, the reaction of unsubstituted isatin with acetone gave the corresponding product in an excellent yield but a low enantioselectivity (3% ee). The authors proposed that the hydrogen bonding between the amide proton and the 2-thienyl sulfur atom in the *N*-(2-thienylsulfonyl)prolinamide catalyst played an important role in the enantioselectivity since this catalyst allowed a higher enantioselectivity to be obtained than that obtained by using the corresponding *p*-tolylsulfonimide catalyst.

A rational combination of two privileged chiral backbones, such as those of cinchonidine and proline, was efficiently applied by Xiao *et al.* to promote the aldolisation of aliphatic ketones, such as acetone and 2-butanone, with aromatic aldehydes, providing the corresponding aldol adducts in good yields and high enantioselectivities of up to 97% ee (Scheme 2.23).[32] The presence of the

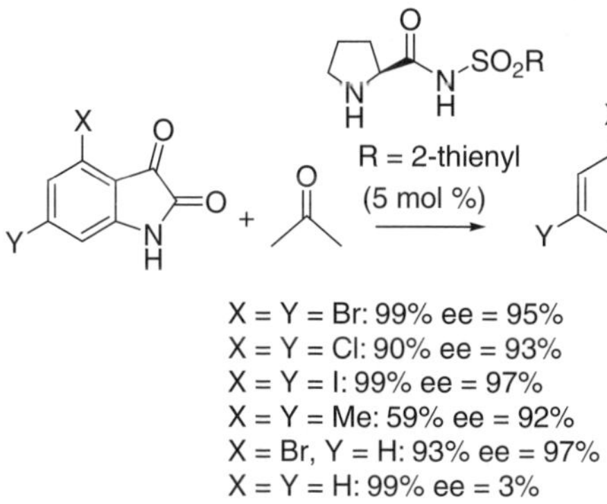

R = 2-thienyl
(5 mol %)

X = Y = Br: 99% ee = 95%
X = Y = Cl: 90% ee = 93%
X = Y = I: 99% ee = 97%
X = Y = Me: 59% ee = 92%
X = Br, Y = H: 93% ee = 97%
X = Y = H: 99% ee = 3%

Scheme 2.22 Aldolisations of acetone with isatins catalysed by sulfonamide.

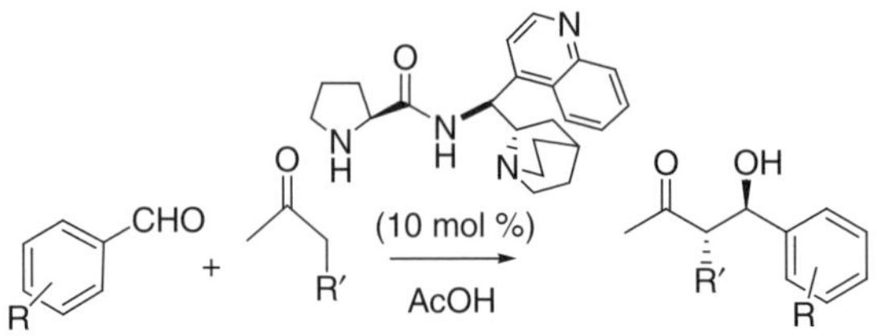

R = 4-NO$_2$, R' = H: 97% ee = 92%
R = 3-NO$_2$, R' = H: 94% ee = 93%
R = 2-NO$_2$, R' = H: 80% ee = 97%
R = 4-CN, R' = H: 60% ee = 93%
R = 4-NO$_2$, R' = Me: 53% de = 86% ee = 98%
R = 3-NO$_2$, R' = Me: 71% de = 86% ee = 92%

Scheme 2.23 Aldolisations of aliphatic ketones with aldehydes catalysed by proline-derived cinchonidine.

cinchonidine backbone in the catalyst structure was demonstrated to be essential to both the reaction efficiency and the enantioselectivity.

Recoverable (*S*)-binam-L-prolinamide in combination with benzoic acid was investigated by Guillena *et al.* as a novel organocatalyst for the enantioselective aldolisation of a range of both cyclic and acyclic ketones with aromatic aldehydes under solvent-free conditions.[33] The reactions were performed in the presence of PhCO$_2$H as an additive and led to the formation of the corresponding *anti*-aldol products with enantioselectivities in a range of 16 to 97% ee, which were highly dependent on the ketones. Moreover, both yields and diastereoselectivities were very dependent on the substrates with values of 34–98% and 0–98% de, respectively. Under the same conditions, the aldol reaction between aldehydes was also possible, yielding after an *in situ* reduction the corresponding chiral 1,3-diols with moderate to high enantioselectivities (40–85% ee) combined with diastereoselectivities of 18–78% de and 45–96% yields.

Najera *et al.* have demonstrated that L-prolinethioamides were more effective catalysts for the asymmetric aldolisation of ketones with aldehydes than the corresponding L-prolinamides under solvent-free conditions.[34] Thus, the use of a novel recyclable L-prolinethioamide, derived from L-proline and (*R*)-1-aminoindane, allowed a wide range of chiral aldol adducts to be obtained with a combination of excellent yields and stereoselectivities in general, and at a low catalyst loading of 5 mol % (Scheme 2.24). Furthermore, an intramolecular version of this process could be developed, providing chiral bicyclic diketones with enantioselectivities of up to 88% ee.

Moreover, a water-compatible proline-derived thiourea-amine bearing a hydrophobic camphor scaffold was evaluated as an organocatalyst for the aldolisation of cyclohexanone with *para*-nitrobenzaldehyde, affording the aldol product with an excellent yield, an enantioselectivity of 99% ee and an excellent *anti*-diastereoselectivity (Scheme 2.25).[35] The reaction was carried out in the

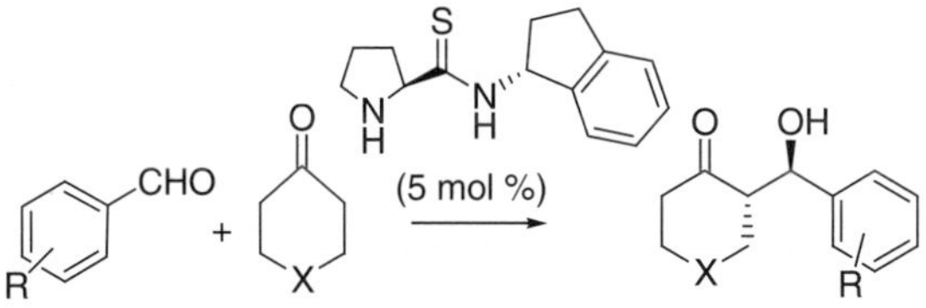

R = 4-Cl, X = CH$_2$: 70% de = 94% ee = 94%
R = 2-NO$_2$, X = CH$_2$: 82% de = 92% ee = 96%
R = 3-NO$_2$, X = CH$_2$: 92% de = 92% ee = 95%
R = 3,5-Cl$_2$, X = CH$_2$: 91% de > 96% ee = 98%
R = 4-NO$_2$, X = O: 99% de > 96% ee = 96%
R = 4-NO$_2$, X = S: 82% de = 94% ee = 88%

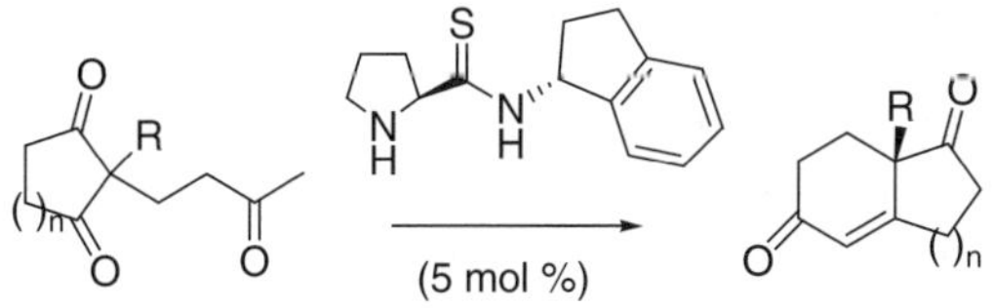

R = Me, n = 2: 99% ee = 86%
R = Et, n = 1: 99% ee = 84%
R = Me, n = 1: 71% ee = 88%

Scheme 2.24 Inter- and intramolecular aldolisations of ketones catalysed by L-prolinethioamide.

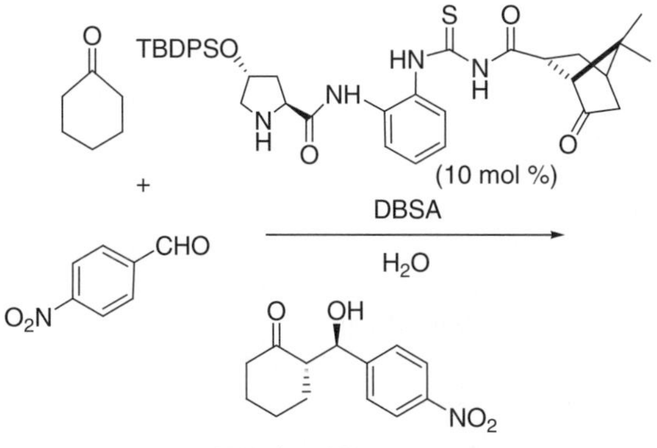

95% de = 92% ee = 99%

Scheme 2.25 Aldolisation of cyclohexanone catalysed by pyrrolidinyl-camphor thiourea-amine.

presence of DBSA as an additive, and its scope could be extended to 2-butanone, providing the corresponding product in 58% yield, 86% de and 98% ee.

A novel class of chiral organocatalysts derived from proline, such as L-proline-based phosphamides, was studied by Li *et al.*[36] These authors demonstrated that these catalysts could promote the aldolisation of cyclohex-anone and acetone with aromatic aldehydes with moderate to good yields (37–81%), low to high diastereoselectivities (16–98% de) and moderate to high enantioselectivities (48–99% ee). In addition, chiral pyrrolidine-based phos-phinyl oxides were synthesised by Liu *et al.* and then evaluated as organoca-talysts in the presence of AcOH as an additive in DMSO for the same reactions.[37] As an example, the products derived from cyclohexanone and various aromatic aldehydes were achieved in moderate to good yields (44–76%), high enantioselectivities in general (80–93% ee), albeit in moderate diastereoselectivities in general (38–48% de) when the reaction was catalysed with 20 mol% of diphenylphosphinyl oxide pyrrolidine, which was revealed to be the most efficient catalyst of the series studied. On the other hand, another type of proline-derived organocatalysts, prolinamide-based oxazolines, was investigated by Doherty *et al.* for the aldolisation of cyclohexanone with aro-matic aldehydes in a mixture of DMSO and water at 30 mol % of catalyst loading.[38] In the case of *para*-nitrobenzaldehyde, an enantioselectivity of 84% ee was obtained in combination with a good yield of 88% but with a moderate diastereoselectivity of 50% de. In general, these reactions were found to be highly substrate-specific with electron-deficient aldehydes giving the highest yields and enantioselectivities, whereas their less electrophilic counterparts gave poor conversions and low ee values. More successfully, Gruttadauria *et al.* introduced a novel prolinamide derivative anchored to a polystyrene support as a heterogeneous catalyst for the aldolisation of a wide range of ketones with aromatic aldehydes.[39] The optimal reaction conditions were found when a 1:2 water/chloroform mixture was used. Noticeably, the enantioselectivities obtained employing acetone as the ketone were probably the highest achieved with a supported proline derivative (Scheme 2.26). Furthermore, this catalyst could be easily recovered, regenerated and recycled, without loss of the activity, at least for 12 cycles.

On the other hand, low enantioselectivities (≤28% ee) were reported by Chandrasekhar *et al.* for the aldolisation of cyclohexanone with various aro-matic aldehydes by using pyrrolidine-triazole as a chiral organocatalyst, in spite of high yields (76–95%) and moderate to high diastereoselectivities (40–96% de).[40] A library of bifunctionalised chiral ionic liquids was synthesised by Luo *et al.* through a facile combinatorial strategy and then evaluated as organo-catalysts for the aldolisation of cyclohexanone with *para*-nitrobenzaldehyde.[41] A moderate result was obtained with a proline-derived catalyst bearing an imidazole moiety and a phthalic mono-anion, with 94% yield combined with 40% de and 73% ee.

Even among organocatalysed-mediated aldol reactions, there is only one successful reaction of acetaldehyde in spite of extensive research on this class of reactions in recent years.[42] Indeed, this unique example, reported by Hayashi *et al.*, described the highly enantioselective aldolisation of acetal-dehyde with various aldehydes catalysed by a diarylprolinol as the organoca-talyst.[43] This reaction was performed in DMF and provided the corresponding β-hydroxy α-unsubstituted aldehydes, which were subsequently reduced into

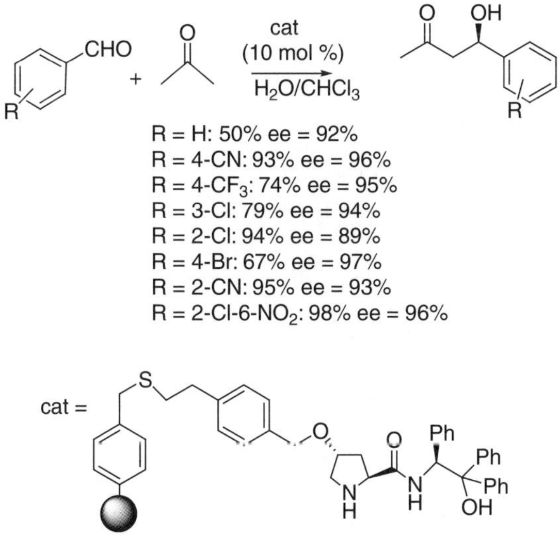
CHO
O
cat
(10 mol %)
H2O/CHCl3
O OH
R
R = H: 50% ee = 92%
R = 4-CN: 93% ee = 96%
R = 4-CF3: 74% ee = 95%
R = 3-Cl: 79% ee = 94%
R = 2-Cl: 94% ee = 89%
R = 4-Br: 67% ee = 97%
R = 2-CN: 95% ee = 93%
R = 2-Cl-6-NO2: 98% ee = 96%
cat =
S
O
O
Ph
Ph
Ph
N H
N H
OH

Scheme 2.26 Prolinamide-supported-aldolisations of acetone.

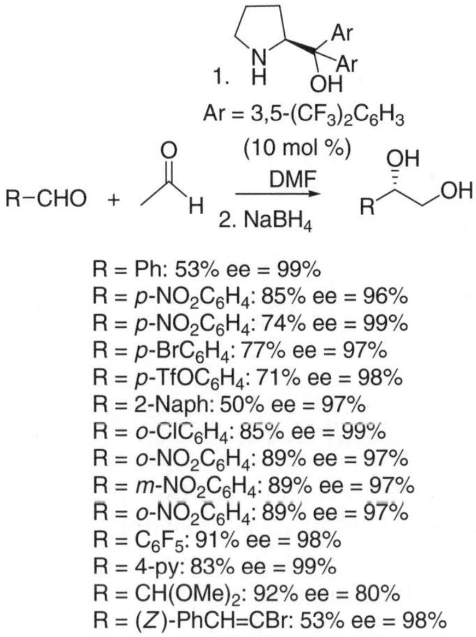
Ar
Ar
1. N H
OH
Ar = 3,5-(CF3)2C6H3
(10 mol %)
DMF
2. NaBH4
R–CHO +
O
H
OH
R
OH
R = Ph: 53% ee = 99%
R = p-NO2C6H4: 85% ee = 96%
R = p-NO2C6H4: 74% ee = 99%
R = p-BrC6H4: 77% ee = 97%
R = p-TfOC6H4: 71% ee = 98%
R = 2-Naph: 50% ee = 97%
R = o-ClC6H4: 85% ee = 99%
R = o-NO2C6H4: 89% ee = 97%
R = m-NO2C6H4: 89% ee = 97%
R = o-NO2C6H4: 89% ee = 97%
R = C6F5: 91% ee = 98%
R = 4-py: 83% ee = 99%
R = CH(OMe)2: 92% ee = 80%
R = (Z)-PhCH=CBr: 53% ee = 98%

Scheme 2.27 Aldolisations of acetaldehyde with aldehydes.

the corresponding 1,3-diols by treatment with $NaBH_4$. Exceptional enantioselectivities of up to 99% ee combined with high yields are shown in Scheme 2.27. Moreover, a self-aldol reaction of acetaldehyde was catalysed by the same catalyst, affording the corresponding trimer acetal, which was

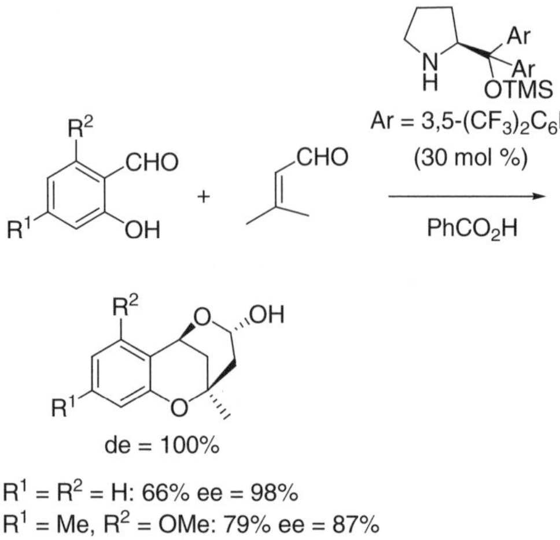

Scheme 2.28   Domino aldol-oxa-Michael reactions.

generated by the reaction of the self-aldol product with another acetaldehyde molecule in a moderate yield (56%) and a good enantioselectivity (82% ee).[44] Though diaryl prolinol silyl ether is widely used as an effective organocatalyst, the present reaction is a rare example of its precursor, diarylprolinol, acting as an outstanding catalyst, and the first example of the use of diarylprolinol in an enantioselective aldol reaction.

In 2009, a diarylprolinol silyl ether was involved as an organocatalyst in an asymmetric domino aldol-oxa-Michael reaction occurring between salicylaldehyde and senecialdehyde.[45] As shown in Scheme 2.28, this process generated the corresponding chiral tricyclic systems in good yields and high enantioselectivities of up to 98% ee. One of these products could be further converted into 4-dehydroxydiversonol.

2.1.2   Aldol Reactions Catalysed by Non-proline Derivatives

In addition to proline-derived organocatalysts, a large number of chiral primary amines have been successfully used to catalyse asymmetric aldolisations. Thus, several primary amine catalysts derived from chiral 1,2-cyclohexyldiamine have been applied as organocatalysts to promote the aldolisation of ketones. Remarkable combinations of both excellent yields and enantioselectivities associated to high diastereoselectivities of up to 94% de were obtained by Maruoka and Nakayama for the asymmetric aldolisation of various cyclic ketones with benzaldehydes by using a common substituted *cis*-cyclohexyldiamine in a 1:1 THF/H$_2$O mixture as the solvent (Scheme 2.29).[46]

Chiral *trans-N,N*-dialkylated diaminocyclohexanes have been used as organocatalysts for the asymmetric aldolisation of dihydroxyacetone with benzaldehydes in the presence of TfOH as an additive.[47] In this case, the authors

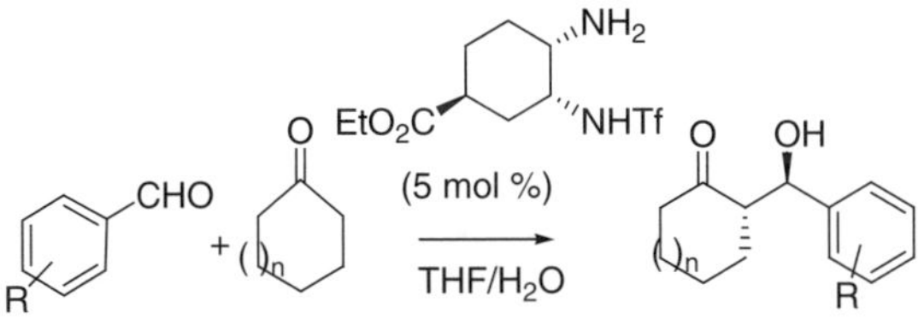

R = 4-NO_2, n = 1: 90% de = 94% ee = 98%
R = 3-NO_2, n = 1: 99% de = 86% ee = 97%
R = 4-CN, n = 1: 98% de = 84% ee = 97%
R = 4-CF_3, n = 1: 99% de = 84% ee = 96%
R = 4-Br, n = 1: 96% de = 86% ee = 98%
R = 4-CO_2Me, n = 1: 99% de = 78% ee = 93%
R = H, n = 1: 53% de = 92% ee = 98%
R = 4-NO_2, n = 0: 99% de = 84% ee = 93%

Scheme 2.29 Aldolisations of cyclic ketones catalysed by *cis*-cyclohexyldiamine.

assumed that the reaction occurred through a *Z*-enamine intermediate, leading to *syn* stereoselectivity. Therefore, the corresponding *syn*-aldol products were formed in excellent yields, *syn* diastereoselectivities and enantioselectivities, as shown in Scheme 2.30. The scope of this reaction could be extended to hydroxyacetone, which gave similar results (up to 96% ee) to those obtained with dihydroxyacetone, and to cyclic dihydroxyacetone, which afforded the corresponding *anti*-aldol products through *E*-enamine intermediates. These products were obtained with high yields, excellent enantioselectivities of up to 97% ee and good diastereoselectivities of up to 82% de (Scheme 2.30). In addition, aliphatic aldehydes have also been investigated for both the *syn*- and *anti*-aldol reactions but gave lower yields due to side reactions.

Trans-*N*,*N*-didecyl diaminocyclohexane was also applied by Cheng *et al.* to the asymmetric aldolisation of pyruvic derivatives with benzaldehydes, providing the corresponding aldol products with excellent *syn* diastereoselectivities and enantioselectivities of up to 99% ee (Scheme 2.31).[48] It is important to note that enantioselectivities of up to 99% ee combined with *syn* diastereoselectivities of up to 90% de were also observed in the case of using aliphatic aldehydes, albeit with low yields (≤32%). A synthetic application of this methodology was found in the synthesis of enantiopure furanoses, which are important precursors for the synthesis of bioactive compounds.

Corresponding *trans*-*N*,*N*-dipropyl diaminocyclohexane catalyst was successfully introduced by these authors as the first magnetic nanoparticle-supported (MNP) chiral primary amine catalyst.[49] Indeed, magnetic nanoparticles have recently appeared as a new type of catalyst support because of their easy preparation and functionalisation, large surface area ratio, facile separation *via* magnetic force as well as low toxicity and price. These fascinating features have made MNPs a promising alternative to porous/mesoporous catalyst supports. Applied to the asymmetric aldolisation of cyclohexanone with benzaldehydes, they led to the expected aldol products in excellent yields and enantioselectivities

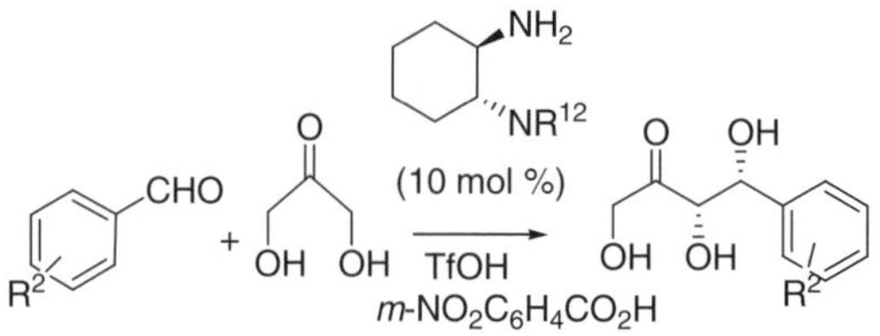

R^1 = n-Dec, R^2 = 4-NO$_2$: 97% de = 94% ee = 99%
R^1 = n-Dec, R^2 = 2-NO$_2$: 93% de = 94% ee = 98%
R^1 = n-Dec, R^2 = 3-NO$_2$: 91% de = 94% ee = 98%
R^1 = n-Dec, R^2 = 4-CF$_3$: 60% de = 94% ee = 98%
R^1 = n-Dec, R^2 = 4-Cl: 61% de = 92% ee = 98%
R^1 = n-Dec, R^2 = 4-CN: 74% de = 94% ee = 96%
R^1 = n-Dec, R^2 = 3-Br: 50% de = 98% ee = 84%
R^1 = n-Dec, R^2 = H: 43% de = 84% ee = 85%

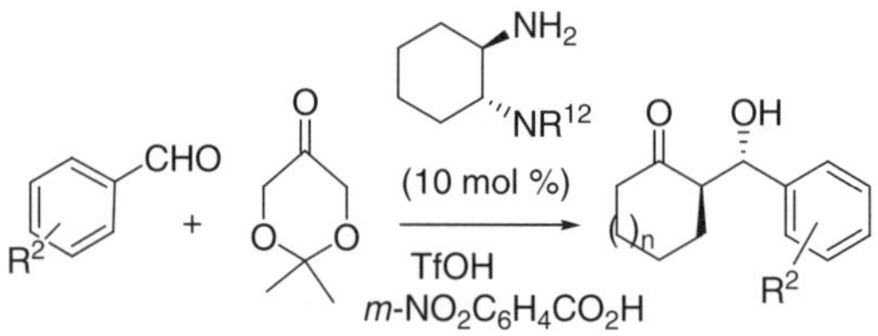

R^1 = Et, R^2 = 4-NO$_2$: 90% de = 72% ee = 94%
R^1 = Et, R^2 = 2-NO$_2$: 99% de = 60% ee = 95%
R^1 = Et, R^2 = 3-NO$_2$: 95% de = 66% ee = 95%
R^1 = Et, R^2 = 4-CF$_3$: 83% de = 60% ee = 94%
R^1 = Et, R^2 = 4-CN: 78% de = 60% ee = 97%
R^1 = Et, R^2 = 4-Ph: 85% de = 66% ee = 85%

Scheme 2.30 Aldolisations of dihydroxyacetones catalysed by *trans*-cyclo-hexyldiamines.

combined with moderate to good *anti*-diastereoselectivities of up to 84% de, as shown in Scheme 2.32. Moreover, these catalysts could be easily recycled *via* magnetic force and reused for up to 11 times with no loss of either the activity or the enantioselectivity.

A non-covalent immobilisation of chiral *trans-N,N*-dipropyl diaminocyclo-hexane was also investigated by the same authors on the basis of acid–base interactions.[50] This immobilisation strategy employed solid acids, such as polystyrene (PS)/sulfonic acids. The involvement of these heterogeneous cata-lysts in the aldolisation of cyclohexanone with benzaldehydes allowed the aldol products to be synthesised in excellent yields and enantioselectivities of up to 99% ee combined with moderate to high *anti*-diastereoselectivities of up to 92% de, as shown in Scheme 2.33. Furthermore, these polymeric sulfonic acids could be easily recovered by filtration and reused for six cycles with a similar stereoselectivity but a slightly decreased activity.

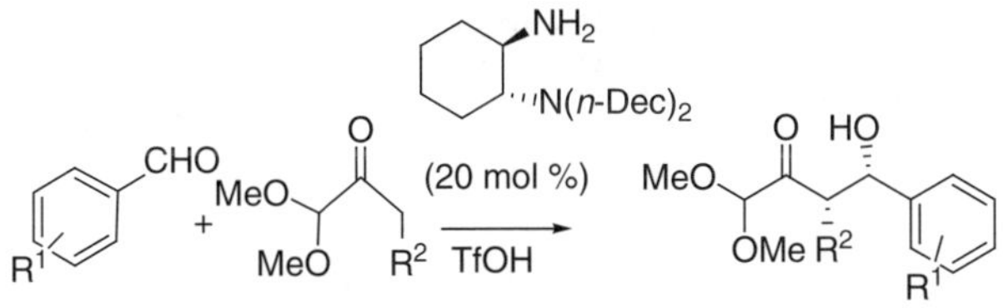

R^1 = 4-NO$_2$, R^2 = H: 90% ee = 93%
R^1 = 3-NO$_2$, R^2 = H: 94% ee = 94%
R^1 = 2-NO$_2$, R^2 = H: 98% ee = 94%
R^1 = 4-CF$_3$, R^2 = H: 63% ee = 91%
R^1 = 4-CN, R^2 = H: 76% ee = 93%
R^1 = 4-NO$_2$, R^2 = Me: 65% de = 92% ee = 97%
R^1 = 3-NO$_2$, R^2 = Me: 58% de = 94% ee > 99%
R^1 = 2-NO$_2$, R^2 = Me: 41% de = 96% ee > 99%
R^1 = 4-CF$_3$, R^2 = Me: 52% de = 90% ee = 99%
R^1 = 4-NO$_2$, R^2 = Et: 32% de = 90% ee = 99%
R^1 = 4-NO$_2$, R^2 = n-Pr: 30% de = 80% ee > 99%

Scheme 2.31 Aldolisations of pyruvic derivatives catalysed by *trans*-cyclohexyldiamine.

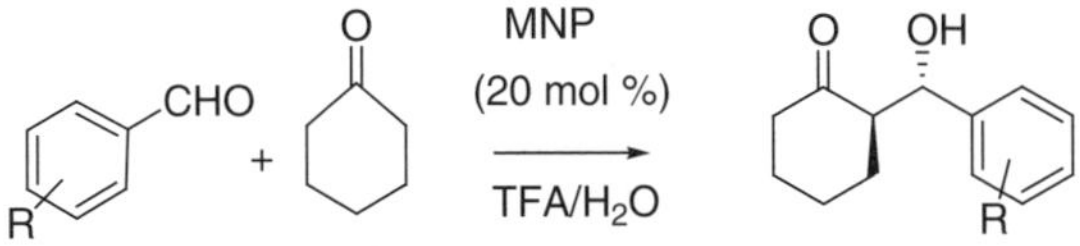

R = 4-NO$_2$: 98% de = 84% ee = 98%
R = 3-NO$_2$: 89% de = 82% ee = 92%
R = 2-NO$_2$: 92% de = 78% ee = 87%
R = 4-CF$_3$: 81% de = 72% ee = 88%

MNP =

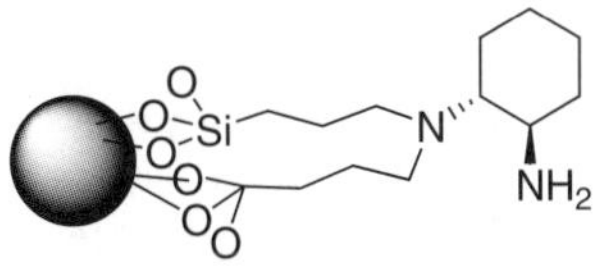

Scheme 2.32 Aldolisations of cyclohexanone catalysed by MNP-supported primary amine catalyst.

The aldolisation of cyclic ketones with benzaldehydes performed in water was carried out by using a novel bifunctional *trans*-cyclohexyldiamine-derived primary amine, which bore both central and axial chiral elements.[51] The aldol products were isolated with excellent levels of diastereo- and enantioselectivities and high yields, as shown in Scheme 2.34. Moreover, the high efficiency of this binaphthyl catalyst could be extended to aliphatic ketones such as acetone, which led to the corresponding aldol adduct in 71% yield and 87% ee in similar conditions.

On the other hand, Lai *et al.* have explored the potential of using hydrophobic polysiloxane in asymmetric aldolisation reactions.[52] Therefore, these

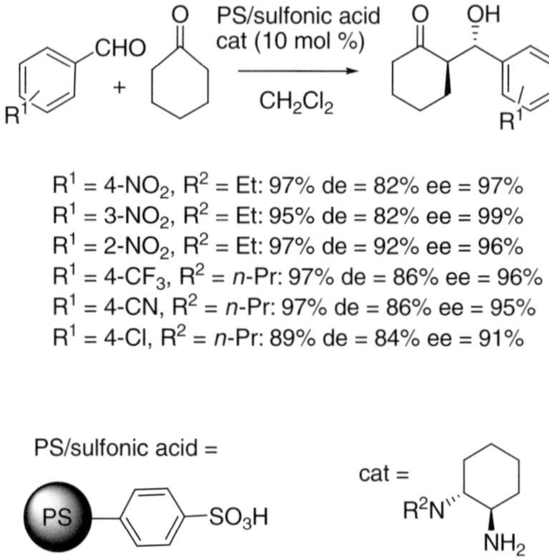
PS/sulfonic acid
cat (10 mol %)
CH2Cl2
CHO
R1
O
OH
R1 = 4-NO2, R2 = Et: 97% de = 82% ee = 97%
R1 = 3-NO2, R2 = Et: 95% de = 82% ee = 99%
R1 = 2-NO2, R2 = Et: 97% de = 92% ee = 96%
R1 = 4-CF3, R2 = n-Pr: 97% de = 86% ee = 96%
R1 = 4-CN, R2 = n-Pr: 97% de = 86% ee = 95%
R1 = 4-Cl, R2 = n-Pr: 89% de = 84% ee = 91%
PS/sulfonic acid =
PS
SO3H
cat =
R2N
NH2

Scheme 2.33 Aldolisations of cyclohexanone catalysed by polymeric sulfonic acids.

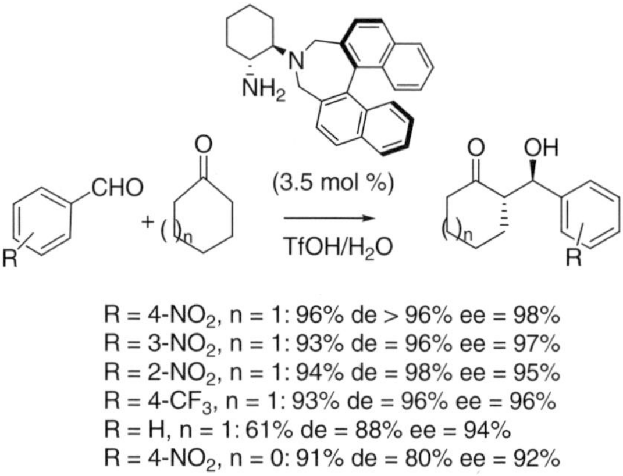
N
NH2
(3.5 mol %)
TfOH/H2O
CHO
R
O
OH
R = 4-NO2, n = 1: 96% de > 96% ee = 98%
R = 3-NO2, n = 1: 93% de = 96% ee = 97%
R = 2-NO2, n = 1: 94% de = 98% ee = 95%
R = 4-CF3, n = 1: 93% de = 96% ee = 96%
R = H, n = 1: 61% de = 88% ee = 94%
R = 4-NO2, n = 0: 91% de = 80% ee = 92%

Scheme 2.34 Aldolisations of cyclic ketones catalysed by binaphthyl *trans*-cyclohexyldiamine.

authors have prepared nitrogen amounts of polysiloxane with a chiral primary amine backbone and used these homogeneous novel organocatalysts for the aldolisation of cyclic ketones with benzaldehydes. Under solvent-free conditions, moderate to high yields and *anti*-diastereoselectivities were observed associated with good to excellent enantioselectivities, as shown in Scheme 2.35.

On the other hand, asymmetric aldolisations of functionalised ketones have been studied by Hu *et al.*, using a primary amine of an amino acid introduced into the bispidine framework.[53] In the presence of a weak acidic additive such

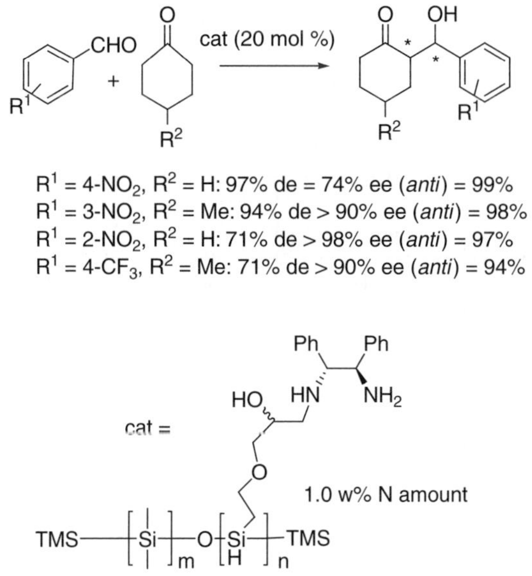
CHO
+
cat (20 mol %)
O
OH
R1
R2
R1 = 4-NO2, R2 = H: 97% de = 74% ee (anti) = 99%
R1 = 3-NO2, R2 = Me: 94% de > 90% ee (anti) = 98%
R1 = 2-NO2, R2 = H: 71% de > 98% ee (anti) = 97%
R1 = 4-CF3, R2 = Me: 71% de > 90% ee (anti) = 94%
cat =
Ph
Ph
HO
HN
NH2
O
1.0 w% N amount
TMS
Si
O
Si
H
TMS
m
n

Scheme 2.35 Aldolisations of cyclohexanones catalysed by polysiloxane-modified primary amine.

as HCO_2H, these reactions afforded the corresponding aldol products in high yields and enantioselectivities for a wide substrate scope including α-keto phosphonates, α-keto esters and α,α-dialkoxyketones (Scheme 2.36).

Highly enantioselective aldolisations of hydroxyketones with aliphatic aldehydes have been achieved by using bifunctional primary amino acid catalysts, although simple amino acids and their siloxyl derivatives are known to be unsuitable organocatalysts for aldolisation due to their poor catalytic activities. In this context, Zhao *et al.* have developed a novel L-threonine-derived catalyst for the aldolisation of hydroxacetone with unactivated aliphatic aldehydes that had proved to be difficult in the presence of other catalysts. These reactions provided the corresponding aldol products with excellent yields, enantioselectivities of up to 98% ee and good to high *syn* diastereoselectivities of up to 94% de (Scheme 2.37).[54] It must be noted that this catalyst, which was used at a catalyst loading of only 2 mol %, proved to be highly specific for aliphatic aldehydes, while for aromatic aldehydes only moderate enantioselectivities (76–82% ee) could be obtained.

A closely related catalyst was also employed by Barbas *et al.* for the aldolisation of dihydroxyacetone with both aliphatic and aromatic aldehydes in DMF.[55] The *syn* aldol products were formed in good yields and enantioselectivities of up to 99% ee, whereas the diastereoselectivities were moderate (0–78% de). Furthermore, the reaction of TBS-protected dihydroxyacetone with both aliphatic and aromatic aldehydes proceeded in aqueous media with high yields, enantioselectivities of up to 98% ee and diastereoselectivities of up to 84% de (Scheme 2.38).

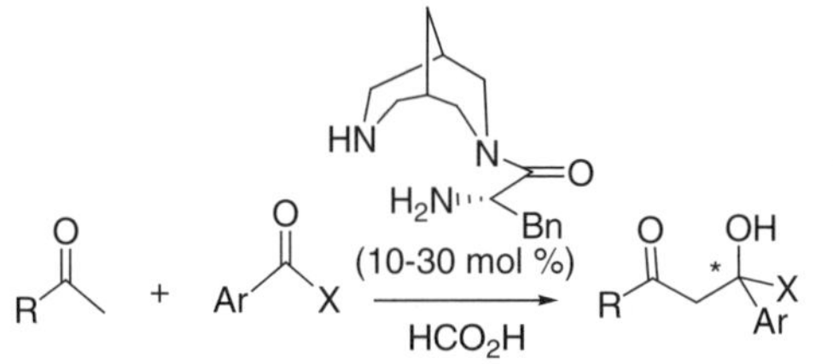

Ar = Ph, R = Me, X = PO(OEt)$_2$: 94% ee = 96%
Ar = Ph, R = Me, X = PO(OMe)$_2$: 97% ee = 96%
Ar = p-NO$_2$C$_6$H$_4$, R = Me, X = (p-NO$_2$C$_6$H$_4$CH$_2$O)$_2$CH:
90% ee = 98%
Ar = p-Tol, R = Me, X = PO(OEt)$_2$: 72% ee = 96%
Ar = p-MeOC$_6$H$_4$, R = Me, X = PO(OMe)$_2$: 75% ee = 94%
Ar = p-ClC$_6$H$_4$, R = Me, X = PO(OEt)$_2$: 94% ee = 96%
Ar = m-Tol, R = Me, X = PO(OEt)$_2$: 90% ee = 95%
Ar = p-FC$_6$H$_4$, R = Me, X = CO$_2$Me: 96% ee = 93%
Ar = p-NO$_2$C$_6$H$_4$, R = Me, X = H: 97% ee = 96%
Ar = Ph, R = Et, X = PO(OEt)$_2$: 65% ee = 93%
Ar = Ph, R = Et, X = CO$_2$Me: 41% ee = 90%
Ar = p-NO$_2$C$_6$H$_4$, R = Et, X = (p-NO$_2$C$_6$H$_4$CH$_2$O)$_2$CH:
40% ee = 96%

Scheme 2.36 Aldolisations of functionalised ketones catalysed by bispidine-derived primary amine.

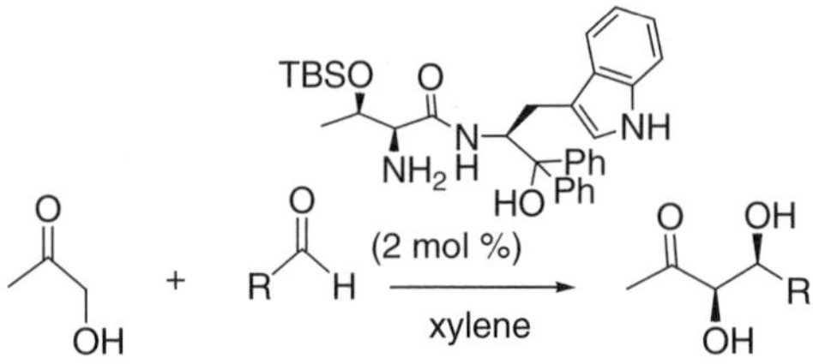

R = n-Pr: 92% de > 88% ee = 98%
R = n-Bu: 93% de = 88% ee = 97%
R = n-Pent: 87% de > 82% ee = 97%
R = i-Bu: 98% de > 94% ee = 96%

Scheme 2.37 Aldolisations of ketones with aliphatic aldehydes catalysed by primary amino acid catalyst.

While primary amine organocatalysts derived from amino acids have shown very low enantioselectivities and activities in the typical aldol reactions of acetone with aldehydes, Da *et al.* demonstrated that the introduction of the optimal co-catalyst DNP (2,4-dinitrophenol) dramatically elevated both the activities and the enantioselectivities of these catalysts.[56] As shown in Scheme 2.39, the combination of a primary amino acid with DNP allowed the aldol products to be obtained in moderate to high yields and excellent enantio-selectivities of up to 99% ee. In addition, the scope of this methodology could

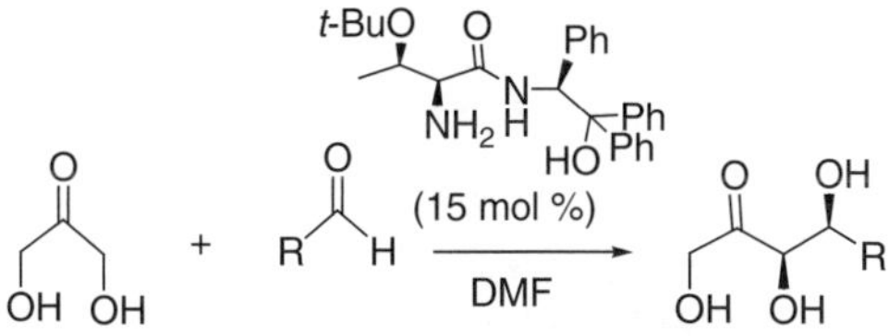

R = *n*-Pent: 70% de = 50% ee > 99%
R = Cy: 74% de = 78% ee = 50% ee = 94%
R = *i*-Bu: 98% de > 94% ee = 96%
R = *p*-NO$_2$C$_6$H$_4$: 78% de = 72% ee = 92%
R = CH(MeO)$_2$: 62% de = 50% ee = 94%

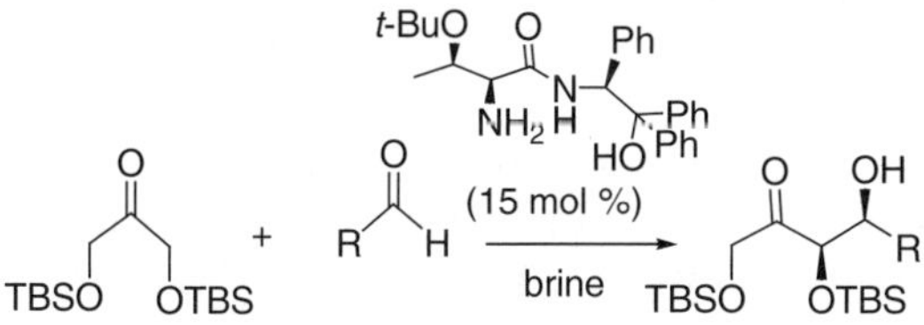

R = *n*-Pent: 77% de = 84% ee = 95%
R = CH$_2$Bn: 83% de = 74% ee = 96%
R = *p*-NO$_2$C$_6$H$_4$: 81% de = 66% ee = 87%
R = CH(MeO)$_2$: 62% de = 50% ee = 94%

Scheme 2.38   Aldolisations of ketones with aliphatic aldehydes catalysed by primary amino acid catalyst.

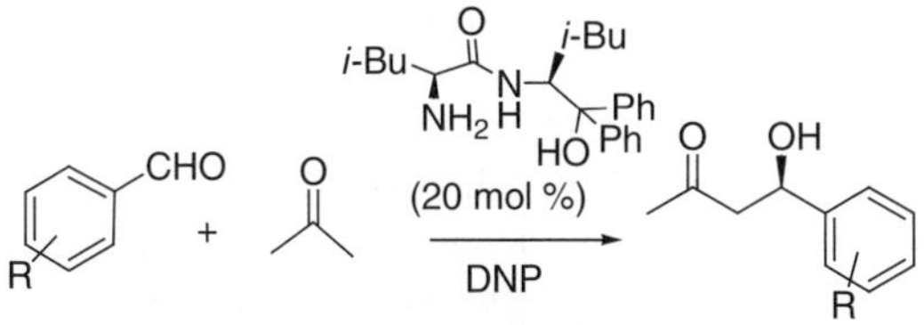

R = 4-NO$_2$: 82% ee = 96%
R = 2-NO$_2$: 69% ee = 96%
R = 3-NO$_2$: 69% ee = 97%
R = 2-Cl: 60% ee = 97%
R = 2-Br: 43% ee > 99%
R = 4-Br: 50% ee = 95%

Scheme 2.39   Aldolisations of acetone with aromatic aldehydes catalysed by primary amino acid catalyst and DNP.

be successfully extended to other methyl alkyl ketones with good enantio-selectivities (88–92% ee). It must be noted that the same catalyst was also used by Gong *et al.* in water for the aldolisation of dihydroxyacetone with aromatic aldehydes, which gave enantioselectivities of up to 99% ee and diastereo-selectivities of up to 90% de.[57]

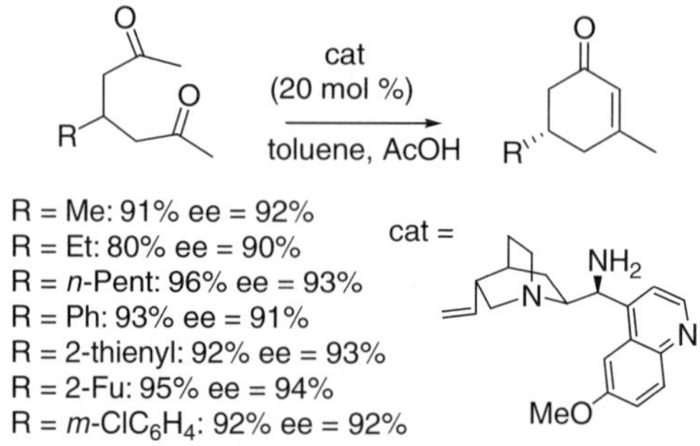

R = Me: 91% ee = 92%
R = Et: 80% ee = 90%
R = n-Pent: 96% ee = 93%
R = Ph: 93% ee = 91%
R = 2-thienyl: 92% ee = 93%
R = 2-Fu: 95% ee = 94%
R = m-ClC$_6$H$_4$: 92% ee = 92%

Scheme 2.40   Intramolecular aldolisations of 4-substituted-2,6-heptanediones catalysed by quinine-derived primary amine.

In addition, a recyclable non-immobilised siloxy-L-serine organocatalyst was
applied to promote the aldolisation of cyclohexanone with aromatic aldehydes
in the ionic liquid [bmim][BF$_4$].[58] At 10 mol % of catalyst loading, the aldol
products were produced in moderate to high yields (40–86%) and enantioselectivities (41–92% ee) combined with low to good *anti*-diastereoselectivities
(4–77% de).

Finally, a quinine-derived primary amine was successfully applied by List *et al.*
to catalyse the intramolecular aldolisation of 4-substituted-2,6-heptanediones to
highly valuable synthetic targets, chiral 5-substituted-3-methyl-2-cyclohexene-1-
ones, which has been a long-term challenge in asymmetric catalysis.[59] Therefore,
these reactions performed in the presence of 9-amino-9-deoxyepiquinine combined with acetic acid as an additive afforded the desired products with excellent
yields and enantioselectivities, as shown in Scheme 2.40.

Few chiral secondary amine organocatalysts have been applied to the asymmetric aldol reaction, in 2008. As an example, Maruoka and Kano have designed
a binaphthyl-based amino acid, which was applied to induce the asymmetric
aldolisation of both acyclic and cyclic ketones with both aliphatic and aromatic
aldehydes performed in DMF.[60] Remarkable results were obtained in general
with enantioselectivities of up to 99% ee for either cyclic or acyclic ketones,
which led to the *anti*-products (Scheme 2.41). Furthermore, these authors have
used a closely related binapthyl-based amino sulfonamide to promote the cross-
aldol reaction between aldehydes, which yielded the corresponding *syn* products
in moderate to high yields (71–91%) and diastereoselectivities (72–90% de)
combined with excellent enantioselectivities (94–99% ee).

In addition, Chen *et al.* developed the aldolisation of cyclohexanone with
benzaldehydes in water by using a chiral bifunctional thiourea-secondary
amine catalyst bearing a hydrophobic camphor scaffold.[61] In the presence of
DBSA as an additive, these reactions afforded the *anti*-aldol products in
moderate to high yields, and high to excellent diastereo- and enantioselectivities, as shown in Scheme 2.42.

Although chiral primary and secondary amines have been explored in
asymmetric aldolisations, tertiary amines are mostly overlooked for this type of

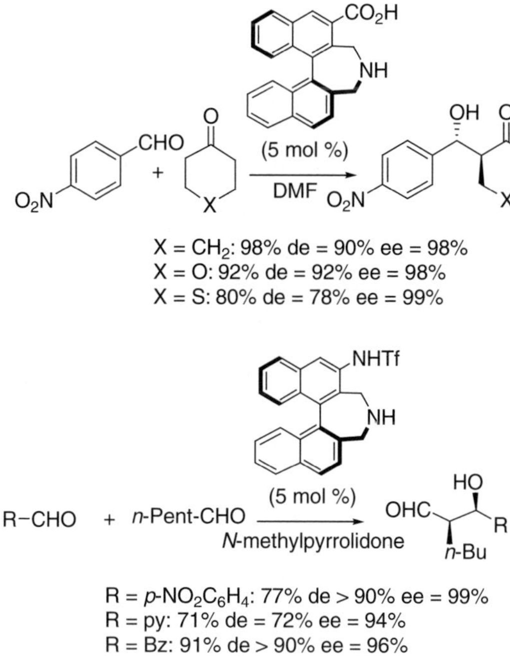
CO_2H
NH
CHO
O
+
X
O_2N
(5 mol %)
DMF
OH O
X = CH_2: 98% de = 90% ee = 98%
X = O: 92% de = 92% ee = 98%
X = S: 80% de = 78% ee = 99%
NHTf
NH
R−CHO + *n*-Pent-CHO
(5 mol %)
N-methylpyrrolidone
OHC
HO
R
n-Bu
R = *p*-$NO_2C_6H_4$: 77% de > 90% ee = 99%
R = py: 71% de = 72% ee = 94%
R = Bz: 91% de > 90% ee = 96%

Scheme 2.41 Aldolisations catalysed by binapthyl-based secondary amines.

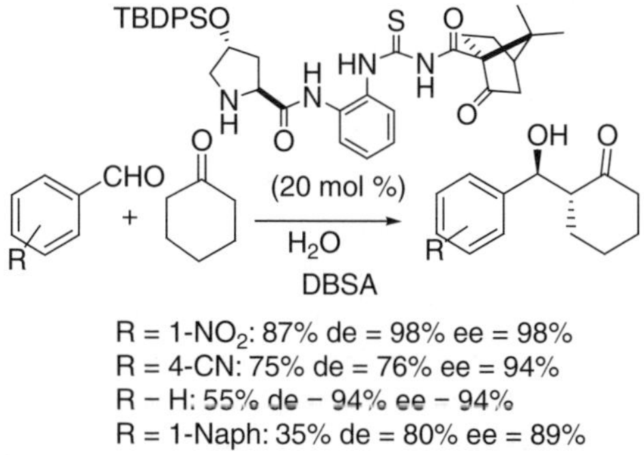
TBDPSO
S
O
HN
N
H
N
H
H
N
O
CHO
+
R
(20 mol %)
H_2O
DBSA
OH O
R = 1-NO_2: 87% de = 98% ee = 98%
R = 4-CN: 75% de = 76% ee = 94%
R − H: 55% de − 94% ee − 94%
R = 1-Naph: 35% de = 80% ee = 89%

Scheme 2.42 Aldolisations of cyclohexanone with aromatic aldehydes catalysed by camphor-derived thiourea-secondary amine.

reaction. This is certainly because of mechanistic considerations where iminium ions or enamines are likely to be active intermediates in the catalytic cycle. Among the rare examples of asymmetric tertiary amine-catalysed aldolisations are those reported by Vilarrasa *et al.*, who dealt with the aldolisation of both acyclic and cyclic ketones with both aromatic and aliphatic aldehydes.[62] These reactions, which afforded a single *anti*-aldol product in each case of substrate,

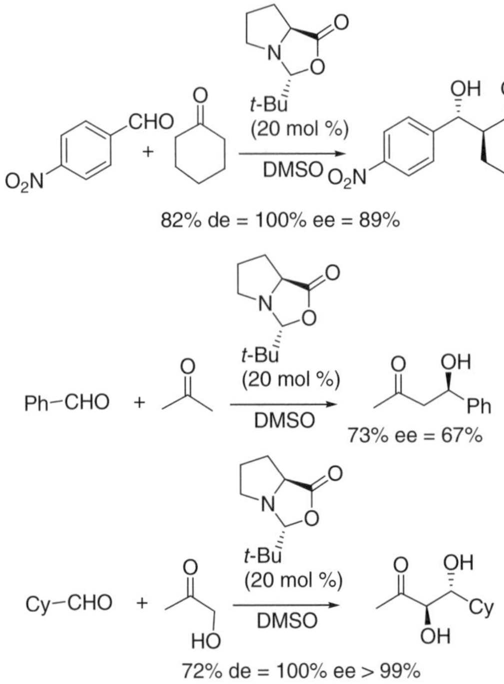

Scheme 2.43 Aldolisations of ketones with aldehydes catalysed by Seebach's proline-derived oxazolidinone.

were catalysed by bicyclic Seebach's proline-derived oxazolidinone, which represents the most efficient organocatalyst amongst those of the bicyclic oxazolidinone type to date (Scheme 2.43).

A comparative study of the efficiency of chiral bimorpholine- and bipiperidine-type organocatalysts for asymmetric aldolisations was reported by Kanger *et al.*[63] These chiral tertiary amines were investigated for both the aldolisation of acetone with *para*-nitrobenzaldehyde and the intramolecular aldolisation of triketones. In both cases of aldolisations, the bimorpholine derivatives were found to be more reactive as well as more selective than the corresponding bipiperidine derivatives, as summarised in Scheme 2.44. The intramolecular aldolisations yielded the Wieland–Miescher ketone and its analogue in enantioselectivities of up to 95% ee by using the bimorpholine-type catalyst combined with an additive such as TfOH, while the aldol product from acetone was isolated in the same conditions with enantioselectivities of up to 88% ee.

In addition, the tertiary amine-catalysed asymmetric aldolisation of hydroxyacetone with a variety of aromatic aldehydes was investigated by Mlynarski *et al.* in the presence of quinidine as the organocatalyst.[64] Although the corresponding products were formed in reasonable yields (10–96%), the moderate enantioselectivities (≤47% ee) remained to be improved. Finally, Huang *et al.*

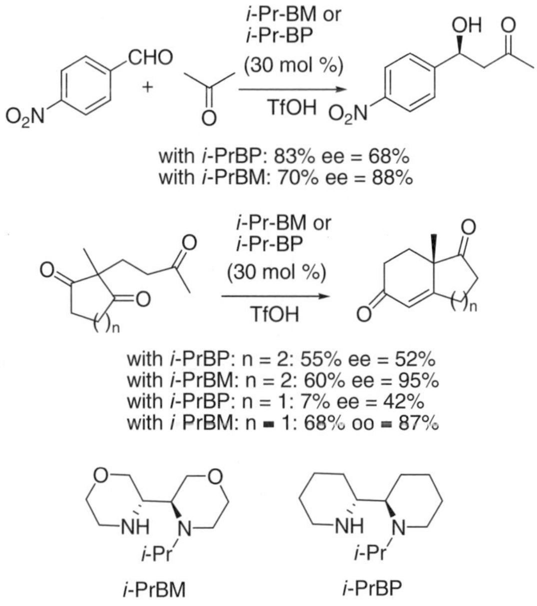

Scheme 2.44 Inter- and intramolecular aldolisations of ketones catalysed by bimorpholine- and bipiperidine derivatives.

have synthesised *meta*-substituted inherently chiral calyx[4]arene catalysts, and further evaluated their abilities as bifunctional organocatalysts for the enantioselective aldolisation of cyclic ketones with aromatic aldehydes in the presence of AcOH.[65] Especially, the reaction between *para*-nitrobenzaldehyde and cyclopentanone led quantitatively to the corresponding *anti*-aldol product with an enantioselectivity of 94% ee, albeit with a low diastereoselectivity of 12% de, while the *anti*-aldol product derived from the reaction between cyclohexanone and *para*-cyanobenzaldehyde was isolated with a diastereoselectivity of 88% de, along with 93% yield and an enantioselectivity of 79% ee. In 2009, Yashima *et al.* have introduced chiral functional poly(phenyl isocyanide)s with macromolecular helicity memory.[66] These novel polymer-based chiral organocatalysts, incorporating amines such as piperazine, were studied for their ability to promote the asymmetric aldolisation of ketones with *ortho*- and *para*-nitrobenzaldehydes but only low enantioselectivities ($\leq$12% ee) were obtained for the aldol adducts combined with moderate yields (0–56%).

2.2 Miscellaneous Reactions

In addition to the classic aldol reaction, several modified versions have been reported. These methods are based on the use of nucleophiles related to the standard ketones. In particular, nitromethane is an interesting carbon nucleophile in aldol reactions and the use of this type of substrate has been investigated in aldol reactions catalysed by organocatalysts. The asymmetric catalytic nitroaldol reaction, also known as the asymmetric Henry reaction, is another example of an aldol-related synthesis of considerable general interest. In this reaction, nitromethane (or a related nitroalkane) reacts in the presence of a chiral catalyst with an aldehyde, yielding the corresponding optically active β-nitro alcohol. The β-nitro alcohols are valuable intermediates in the synthesis of a broad variety of chiral building blocks such as β-amino alcohols. Actually, there are only a few examples of catalytic asymmetric metal-free Henry reactions. In 2008, Feng *et al.* developed an efficient secondary amine amide catalyst for the asymmetric nitroaldol reaction of α-ketophosphonates under mild conditions.[67] Therefore, in the presence of 5 mol% of catalyst loading combined with 2,4-dinitrophenol as an additive, the corresponding nitroaldol products were obtained in moderate to high yields and excellent enantioselectivities for most substrates, as shown in Scheme 2.45.

On the other hand, the asymmetric nitroaldol reaction of α-keto esters was explored by Sohtome *et al.*, using a guanidine-thiourea bifunctional organocatalyst at temperatures below the freezing point of water.[68] This novel methodology could be extended to various aliphatic α-keto esters and nitroalkanes, providing the corresponding chiral *tert*-nitroaldols in moderate to high yields (35–90%), low to high enantioselectivities (5–93% ee) and moderate to high diastereoselectivities (58–94% de), as shown in Scheme 2.46.

In addition, an unprecedented organocatalysed nitroaldol condensation of fluoromethyl ketones in the presence of cinchona alkaloids was reported by Umani-Ronchi *et al.*[69] The aldol products derived from both tri- and difluoromethyl ketones were obtained with high levels of stereoinduction (76–99% ee) under mild conditions and a low loading of catalyst (1–5 mol %). As shown in

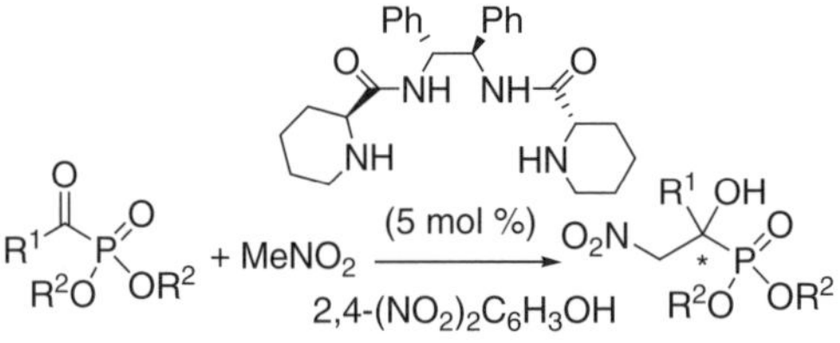

$R^1 = Ph, R^2 = Me: 86\%\ ee = 94\%$
$R^1 = p\text{-}FC_6H_4, R^2 = Et: 88\%\ ee = 99\%$
$R^1 = p\text{-}FC_6H_4, R^2 = Me: 90\%\ ee = 99\%$
$R^1 = p\text{-}ClC_6H_4, R^2 = Et: 93\%\ ee = 96\%$
$R^1 = p\text{-}ClC_6H_4, R^2 = Me: 90\%\ ee = 97\%$
$R^1 = p\text{-}Tol, R^2 = Et: 73\%\ ee = 98\%$
$R^1 = p\text{-}(t\text{-}Bu)C_6H_4, R^2 = Et: 85\%\ ee = 97\%$
$R^1 = Bn, R^2 = Et: 95\%\ ee = 99\%$
$R^1 = Me, R^2 = Et: 90\%\ ee = 91\%$

Scheme 2.45 Nitroaldolisations of α-ketophosphonates catalysed by secondary amine amide.

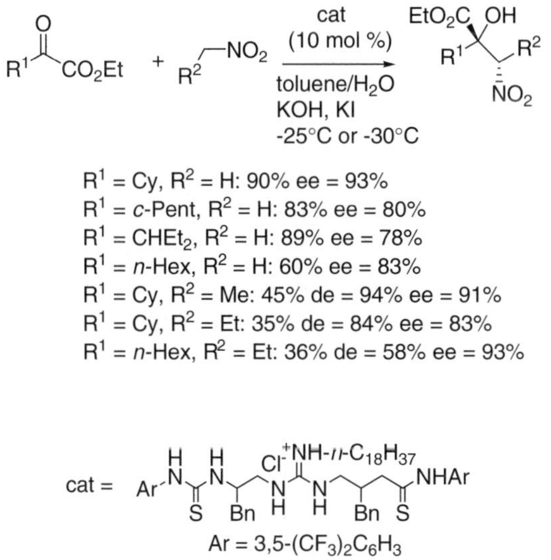

$$R^1 = Cy, R^2 = H: 90\%\ ee = 93\%$$
$$R^1 = c\text{-Pent}, R^2 = H: 83\%\ ee = 80\%$$
$$R^1 = CHEt_2, R^2 = H: 89\%\ ee = 78\%$$
$$R^1 = n\text{-Hex}, R^2 = H: 60\%\ ee = 83\%$$
$$R^1 = Cy, R^2 = Me: 45\%\ de = 94\%\ ee = 91\%$$
$$R^1 = Cy, R^2 = Et: 35\%\ de = 84\%\ ee = 83\%$$
$$R^1 = n\text{-Hex}, R^2 = Et: 36\%\ de = 58\%\ ee = 93\%$$

Scheme 2.46 Nitroaldolisations of α-keto esters catalysed by guanidine-thiourea.

Scheme 2.47, the best results were observed by using a C9-benzoylcupreine bearing an electron-withdrawing substituent.

An indirect aldol condensation between trichlorosilyloxicyclohexene and benzaldehyde was reported by Juaristi *et al.* in the presence of novel chiral thioureas, incorporating chiral moieties derived from (*R*)- or (*S*)-α-phenyl-ethylamine, (*R*)-phenylglycine or (1*R*,2*S*)-ephedrine.[70] However, the enantios-electivities of the formed aldol products remained low ($\leq 10\%$ ee).

The enantioselective allylation of aldehydes is another C–C bond-forming reaction of wide interest. The resulting unsaturated alcohols are used as ver-satile intermediates in the construction of many interesting molecules. A broad variety of organocatalysts have been found to catalyse the enantioselective allylation of aldehydes with allyltrialkylsilancs,[71] such as optically active urea derivatives, phosphoramides, *P*-oxides, *N*-oxides and bis-sulfoxides. Amine *N*-oxides are good electron-pair donors, and this property has been exploited in organocatalytic reactions in a chiral environment. In particular, chiral pyridine *N*-oxide catalysts have been studied by several groups to promote the asym-metric allylation of aldehydes with allyltrichlorosilane. In 2008, Chelucchi *et al.* reported the synthesis of *N*,*N*-dioxides of C_2-symmetric dipyridylmethane ligands derived from natural monoterpenes, which were further investigated as organocatalysts for the asymmetric allylation of aromatic aldehydes with allyltrichlorosilane.[72] The most efficient catalyst derived from (+)-3-carene allowed the corresponding allylic alcohols to be obtained in generally good yields and enantioselectivities of up to 83% ee, as summarised in Scheme 2.48.

In the same context, other novel chiral bipyridine *N*,*N'*-dioxides derived from terpenes were employed by Malkov *et al.* to catalyse similar reactions.[73] A

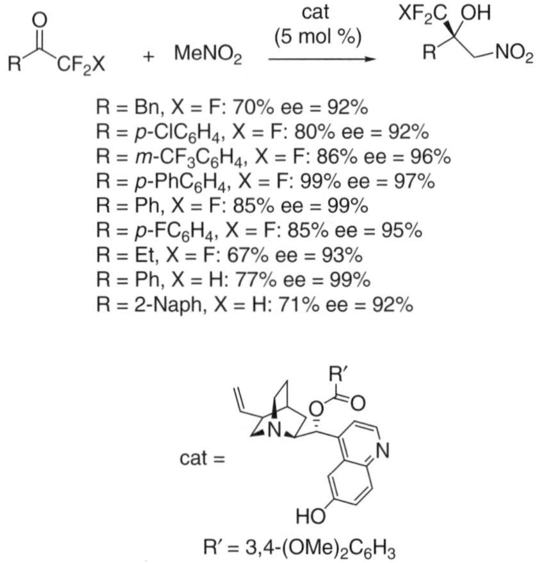

R = Bn, X = F: 70% ee = 92%
R = p-ClC₆H₄, X = F: 80% ee = 92%
R = m-CF₃C₆H₄, X = F: 86% ee = 96%
R = p-PhC₆H₄, X = F: 99% ee = 97%
R = Ph, X = F: 85% ee = 99%
R = p-FC₆H₄, X = F: 85% ee = 95%
R = Et, X = F: 67% ee = 93%
R = Ph, X = H: 77% ee = 99%
R = 2-Naph, X = H: 71% ee = 92%

cat =

R′ = 3,4-(OMe)₂C₆H₃

Scheme 2.47 Nitroaldolisations of fluoromethyl ketones catalysed by cinchona alkaloid.

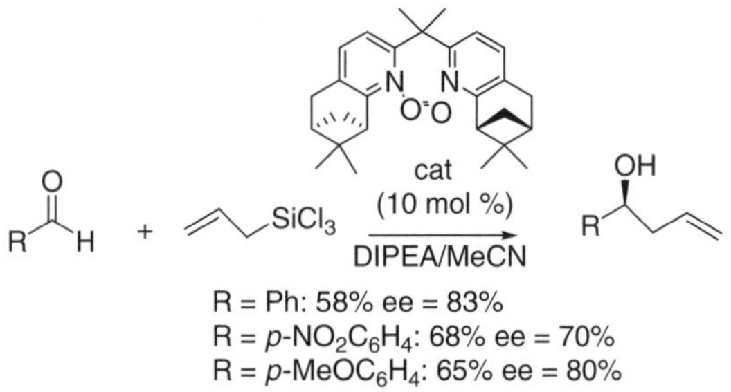

R = Ph: 58% ee = 83%
R = p-NO₂C₆H₄: 68% ee = 70%
R = p-MeOC₆H₄: 65% ee = 80%

Scheme 2.48 Allylations of aldehydes catalysed by bipyridine *N,N′*-dioxide.

dioxide, whose chirality originated from the annulated terpene unit, reached an enantioselectivity of up to 56% ee, whereas a C_2-symmetric *N,N′*-dioxide exhibited a good enantioselectivity (81% ee) and a high reactivity even at −80°C with a catalyst loading as low as 1 mol % (Scheme 2.49).

On the other hand, the asymmetric allylation of aldehydes was also successfully performed in the presence of chiral easily available biheteroaromatic diphosphine oxides, such as tetraMe-BITIOPO, which is the precursor of the industrially produced tetraMe-BITIOP.[74] Using this organocatalyst, the reaction afforded homoallyllic alcohols in fair-to-good yields and with enantioselectivities of up to 95% ee, as shown in Scheme 2.50.

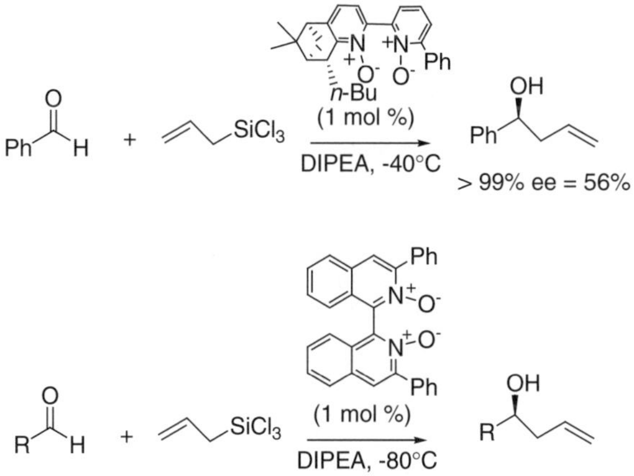

R = Ph: > 99% ee = 81%
R = *p*-Tol: > 99% ee = 69%
R = *p*-ClC_6H_4: > 99% ee = 76%
R = 3,5-$Me_2C_6H_3$: > 99% ee = 71%

Scheme 2.49 Allylations of aldehydes catalysed by bipyridine *N*,*N*′-dioxides.

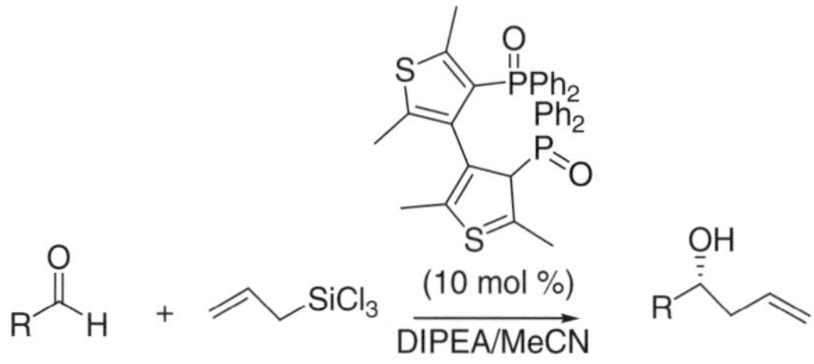

R = Ph: 85% ee = 93%
R = *p*-$NO_2C_6H_4$: 51% ee = 93%
R = *p*-$MeOC_6H_4$: 95% ee = 91%
R = *p*-ClC_6H_4: 55% ee = 95%
R = *o*-$MeOC_6H_4$: 95% ee = 83%
R = *m*-$MeOC_6H_4$: 83% ee = 87%
R = 1-Naph: 98% ee = 81%

Scheme 2.50 Allylations of aldehydes catalysed by tetraMe-BITIOPO.

More recently, a chiral tetradentate bis-sulfoxide was applied as an organocatalyst to these reactions, providing the allylic alcohols in moderate to good yields (43–69%) and enantioselectivities (47–70% ee).[75] The results are collected in Scheme 2.51.

A series of other nucleophiles have been condensed onto C=O double bonds in the presence of chiral organocatalysts. As an example, Rueping *et al.* described the use of chiral silylated *N*-triflylphosphoramides derived from

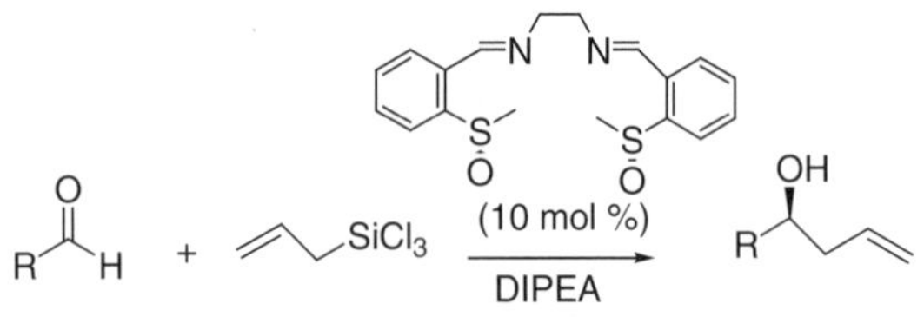

R = Ph: 57% ee = 53%
R = 5-NO$_2$-2-thienyl: 65% ee = 70%
R = 5-NO$_2$-2-Fu: 69% ee = 66%
R = BnCH$_2$: 60% ee = 62%

Scheme 2.51   Allylations of aldehydes catalysed by bis-sulfoxide.

BINOL to catalyse the enantioselective 1,2-addition of *N*-methylindole to β,γ-unsaturated α-keto esters.[76] The corresponding bis-indoles were isolated as the sole products of the reactions in low to moderate enantioselectivities (3–62% ee). The benzoin condensation is a nucleophilic acylation of great synthetic utility, involving the addition of an acyl anion equivalent to an aldehyde, resulting in the formation of a 2-hydroxyketone.[77] Heteroazolium salts in the presence of a base are the most frequently used catalysts for the umpolung of an aldehyde for an asymmetric addition to another aldehyde or an imine. The asymmetric cyanation of aldehydes is an important and well-documented reaction, due to the fact that the product cyanohydrins are important intermediates for the synthesis of α-hydroxy acids, α-amino acids and β-amino alcohols. Hydrocyanation is actually one of the first examples of asymmetric organocatalysis in general. Indeed, as early as 1912, Bredig reported that the addition of HCN to benzaldehyde was accelerated by the alkaloids, quinine and quinidine, and that the resulting cyanohydrins were optically active.[78] There are, however, very few examples of organocatalysed cyanide addition to aldehydes. As an example, an enantioselective organocatalytic version of the benzoin condensation was developed by Enders and Han, in 2008.[79] In this context, enantiopure 1,2,4-triazolium salts were prepared starting from the inexpensive (*S*)-pyroglutamic acid, which gave by subsequent treatment with base, the corresponding *N*-heterocyclic carbenes. When these carbenes were applied as organocatalysts for the benzoin condensation, they led to the acyloin products in low to good yields and enantioselectivities of up to 95% ee (Scheme 2.52).

   In 2009, Feng *et al.* reported the asymmetric hydrophosphonylation of α-keto esters catalysed for the first time by cinchona-derived thiourea organocatalysts.[80] Thus, a series of aromatic and heteroaromatic α-keto esters reacted with dimethyl phosphite to afford the corresponding α-hydroxy phosphonates in high yields and enantioselectivities of up to 91% ee (Scheme 2.53).

   The Morita–Baylis–Hillman reaction is a powerful transformation in organic synthesis, consisting of the formation of α-methylene-β-hydroxy-carbonyl compounds by the addition of aldehydes to α,β-unsaturated carbonyl compounds, such as vinyl ketones, acrylonitriles or acrylic esters.[81] For the

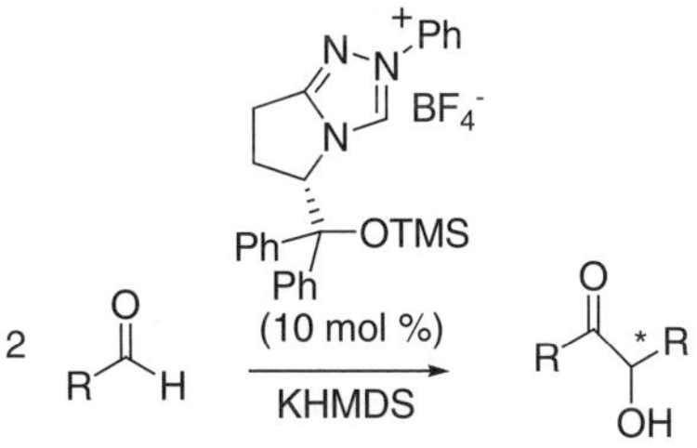

R = Ph: 66% ee = 95%
R = 2-Naph: 71% ee = 50%
R = *p*-ClC$_6$H$_4$: 65% ee = 77%
R = *m*-ClC$_6$H$_4$: 50% ee = 84%

Scheme 2.52 Benzoin condensations catalysed by 1,2,4-triazolium salt.

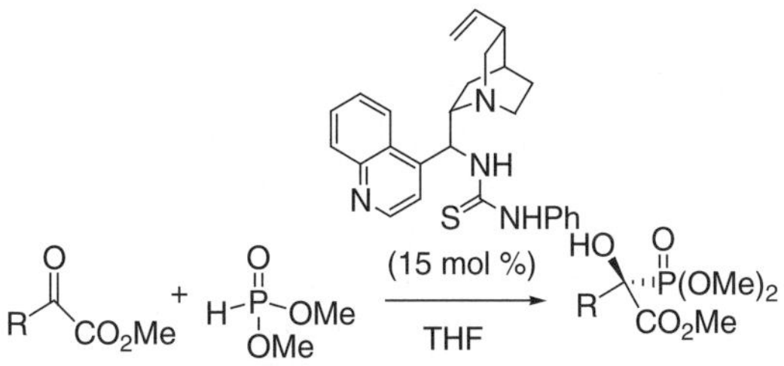

R = Ph: 92% ee = 90%
R = *m*-Tol: 92% ee = 90%
R = *p*-Tol: 90% ee = 90%
R = *p*-MeOC$_6$H$_4$: 90% ee = 91%
R = *p*-CF$_3$C$_6$H$_4$: 94% ee = 90%
R = 2-Naph: 86% ee = 88%
R = 2-Thio: 91% ee = 91%

Scheme 2.53 Hydrophosphonylations of α-keto esters catalysed by cinchona-derived thiourea.

reaction to occur, the presence of catalytically active nucleophiles is required. The reaction is initiated by the addition of the catalytically active nucleophile to the enone. The resulting enolate adds to the aldehyde, establishing the new stereogenic centre by proton transfer from the α-position of the carbonyl moiety to the alcoholate oxygen atom, with concomitant elimination of the nucleophile, which is available for the next catalytic cycle. A considerable amount of effort has been devoted to the development of catalytic, enantioselective versions of the processes. Discovering catalytic systems for asymmetric Morita–Baylis–Hillman reactions has proved to be a synthetic challenge and, to date, only a few successful chiral organocatalysts, such as chiral nucleophilic amines or phosphines, have been demonstrated for this process. Among these, simple L-proline in the presence of NaHCO$_3$ was found by Gruttadauria *et al.*

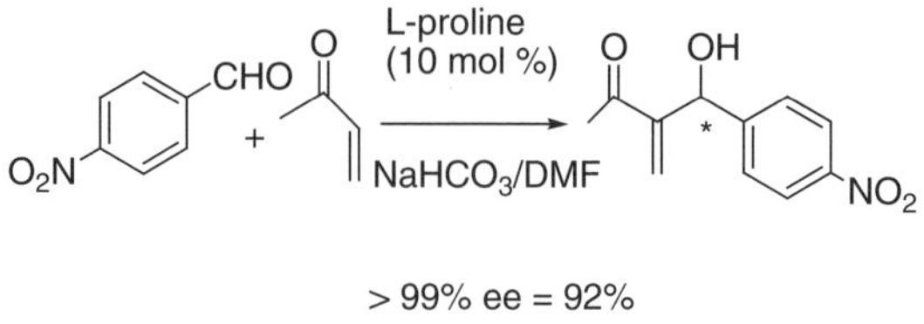

> 99% ee = 92%

Scheme 2.54   L-Proline-catalysed Morita–Baylis–Hillman reaction.

to be an efficient organocatalyst for the asymmetric Morita–Baylis–Hillman reaction between alkyl vinyl ketones and aryl aldehydes.[82] As shown in Scheme 2.54, the reaction of methyl vinyl ketone with *para*-nitrobenzaldehyde led to the corresponding Morita–Baylis–Hillman product in an almost quantitative yield and a high enantioselectivity of 92% ee. According to quantum mechanical calculations, the amino group of proline attacked the methyl vinyl ketone to give the zwitterionic iminium species, which underwent intramolecular nucleophilic attack by the carboxylate group to give the bicyclic enaminolactone species as an intermediate. The hydrogen carbonate ion seemed to provide hydrogen-bond assistance in the C–C bond formation step. Moreover, other secondary amino acids, such as sarcosine, pipecolinic acid and homoproline in the presence of $NaHCO_3$ were also found to be capable of catalysing these reactions, albeit in lower enantioselectivities ($\leq 80\%$ ee).

As an extension of this work, the same authors have used polystyrene-supported proline as a recyclable catalyst in the Morita–Baylis–Hillman reaction of a range of aryl aldehydes with methyl or ethyl vinyl ketone.[83] These reactions were performed in the presence of imidazole and provided a series of Morita–Baylis–Hillman adducts in moderate to high yields (17–88%) combined with high enantioselectivities of up to 95% ee (Scheme 2.55). This study represented the first example of supported proline as heterogeneous catalyst for the Morita–Baylis–Hillman reaction. In addition, Zhou *et al.* reported that these reactions could be catalysed by combinations of L-proline with chiral tertiary amines derived from various readily available chiral sources, such as L-proline or (*S*)-α-phenylethylamine.[84] In these conditions, the Morita–Baylis–Hillman adducts were obtained in reasonable chemical yields (34–97%) and low to good enantioselectivities (12–83% ee). In this study, it was demonstrated that the proline stereochemistry was the sole factor to determine the configuration of the newly formed chiral centre.

On the other hand, the Morita–Baylis–Hillman reaction has been investigated in the presence of various chiral thioureas as organocatalysts. In particular, a chiral bis-thiourea prepared from *trans*-1,2-diaminocyclohexane was found to promote, in the presence of DMAP, the Morita–Baylis–Hillman reaction of cyclic enones with a wide range of aldehydes.[85] Indeed, aromatic as well as aliphatic aldehydes led to the corresponding Morita–Baylis–Hillman products in moderate to excellent yields and ee values reached up to 90% ee, as summarised in Scheme 2.56.

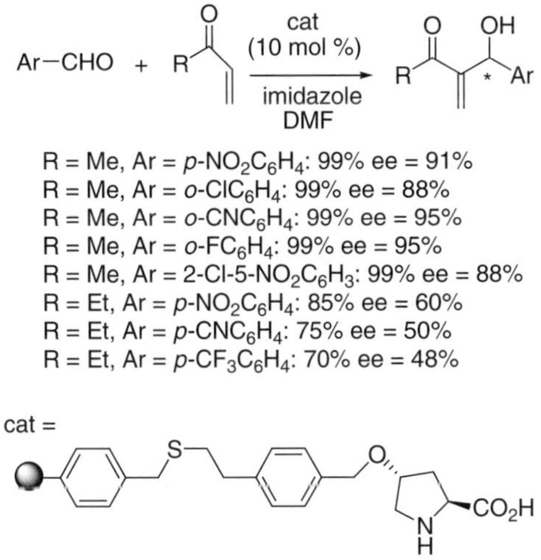

Scheme 2.55 Polystyrene-supported proline-catalysed Morita–Baylis–Hillman reactions.

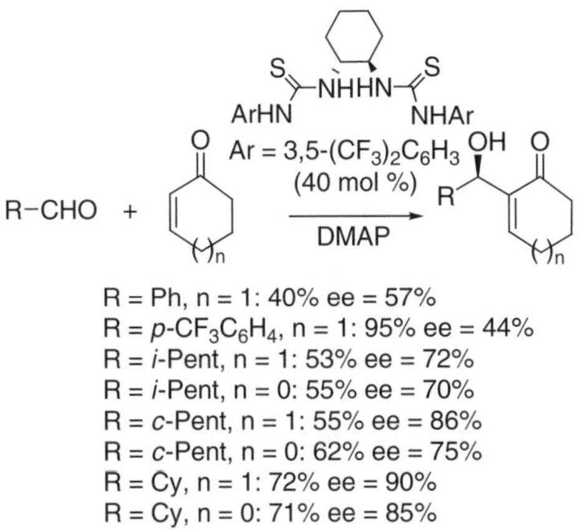

Scheme 2.56 Bis-thiourea-catalysed Morita–Baylis–Hillman reactions.

The Morita–Baylis–Hillman reaction of cyclic enones with aromatic aldehydes was also studied by Shi and Liu by involving bis-thiourea organocatalysts derived from BINAM.[86] The reactions were performed in the presence of DABCO and afforded the desired products in good yields and enantioselectivities of up to 88% ee, as shown in Scheme 2.57.

Good to excellent yields combined with excellent enantioselectivities of up to 94% ee were obtained for the Morita–Baylis–Hillman reaction of methyl vinyl ketone with various aromatic aldehydes by using a new class of chiral phosphino-

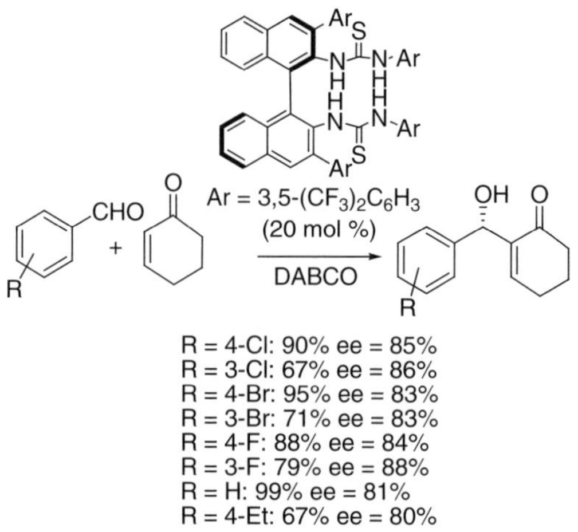

Scheme 2.57   Bis-thiourea-catalysed Morita–Baylis–Hillman reactions.

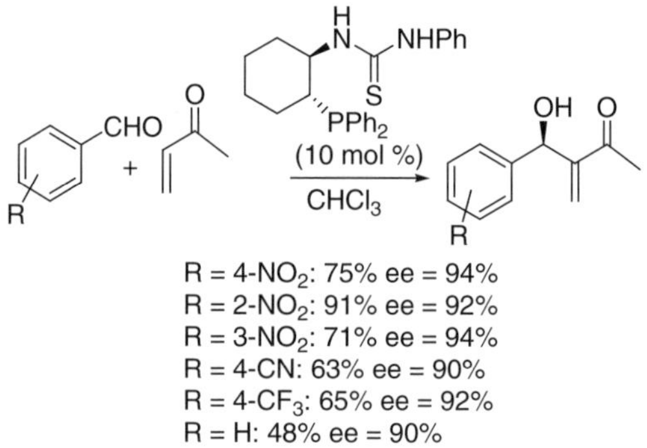

Scheme 2.58   Phosphinothiourea-catalysed Morita–Baylis–Hillman reactions.

thioureas derived from *trans*-2-amino-1-(diphenylphosphino)cyclohexane.[87] The best results for this reaction performed under very mild conditions are collected in Scheme 2.58.

The Petasis reaction is a multicomponent condensation occurring between boronic acids, amines and aldehydes. The asymmetric version of this reaction is very attractive for the synthesis of chiral α-amino acids.[88] In this context, Schaus and Lou reported the use of chiral biphenols as organocatalysts for the asymmetric Petasis reaction of (*E*)-diethyl styrylboronate with secondary amines and ethyl glyoxylate.[89] The corresponding α-amino esters were obtained in high yields and enantioselectivities of up to 97% ee by using a vaulted biaryl phenol such as (*S*)-VAPOL as the organocatalyst in the presence of 3-Å molecular sieves (Scheme 2.59).

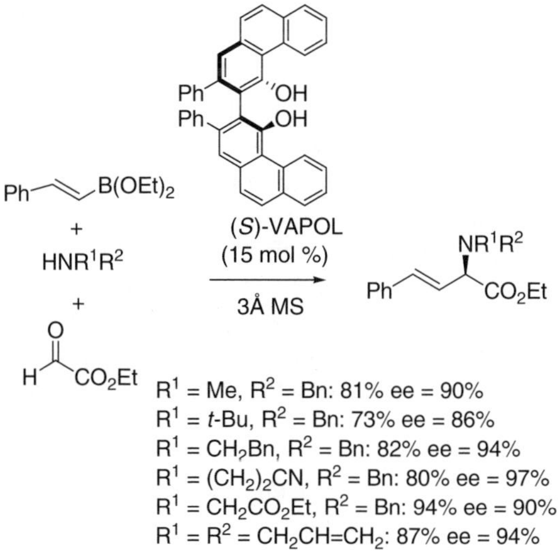

Ph$\diagdown$$\diagup$B(OEt)$_2$

+

HNR^{1}R^2

+

H$\diagup$$\overset{\displaystyle O}{\diagdown}CO_2$Et

(S)-VAPOL
(15 mol %)

$\xrightarrow{\text{3Å MS}}$

Ph$\diagdown$$\diagup$$\overset{\displaystyle NR^1R^2}{\diagup}CO_2$Et

R^1 = Me, R^2 = Bn: 81% ee = 90%
R^1 = t-Bu, R^2 = Bn: 73% ee = 86%
R^1 = CH$_2$Bn, R^2 = Bn: 82% ee = 94%
R^1 = (CH$_2$)$_2$CN, R^2 = Bn: 80% ee = 97%
R^1 = CH$_2$CO$_2$Et, R^2 = Bn: 94% ee = 90%
R^1 = R^2 = CH$_2$CH=CH$_2$: 87% ee = 94%

Scheme 2.59   (S)-VAPOL-catalysed Petasis reactions.

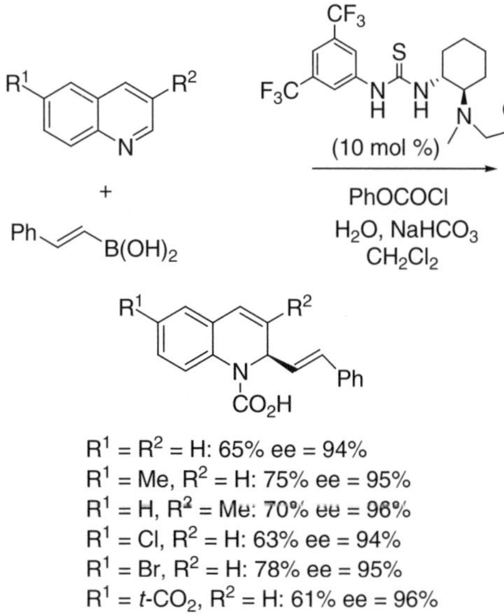

R^1$\diagup$quinoline$\diagdown$R^2

+

Ph$\diagdown$$\diagup$B(OH)$_2$

(10 mol %)

$\xrightarrow{\substack{\text{PhOCOCl} \\ \text{H}_2\text{O, NaHCO}_3 \\ \text{CH}_2\text{Cl}_2}}$

R^1$\diagdown$$\diagup$$\overset{\displaystyle R^2}{}$$\diagdown$$\underset{\displaystyle CO_2H}{N}$$\diagup$$\diagdown$Ph

R^1 = R^2 = H: 65% ee = 94%
R^1 = Me, R^2 = H: 75% ee = 95%
R^1 = H, R^2 = Me: 70% ee = 96%
R^1 = Cl, R^2 = H: 63% ee = 94%
R^1 = Br, R^2 = H: 78% ee = 95%
R^1 = t-CO$_2$, R^2 = H: 61% ee = 96%

Scheme 2.60   Petasis-type reactions of quinolines catalysed by amine-thiourea.

In addition, an asymmetric Petasis-type transformation of quinolines with vinylboronic acid was developed by Miyabe and Takemoto by using a chiral bifunctional tertiary amine-thiourea, providing the corresponding 1,2-adducts in the presence of phenyl chloroformate as an activator.[90] As shown in Scheme 2.60, the products were obtained in good yields and excellent enantioselectivities of up to 96% ee in general, when H_2O and $NaHCO_3$ were added to the reaction mixture.

The Biginelli reaction offers an efficient way to obtain pharmacologically and biologically important 3,4-dihydropyrimidin-2(1*H*)-ones through the simple multicomponent reaction of an aldehyde, a urea or a thiourea and an easily enolisable carbonyl compound. In spite of their usefulness, the enantioselective versions of this reaction are rather limited. In 2009, Zhao *et al.* reported one of these rare versions using novel chiral substituted 5-(pyrrolidin-2-yl)tetrazoles as organocatalysts.[91] As shown in Scheme 2.61, the reaction between an aromatic aldehyde, a urea or a thiourea and an α-keto ester led to the corresponding 3,4-dihydropyrimidin-2(1*H*)-one in good yields (63–88%) combined with good enantioselectivities (68–80% ee) at room temperature. Similar reactions were

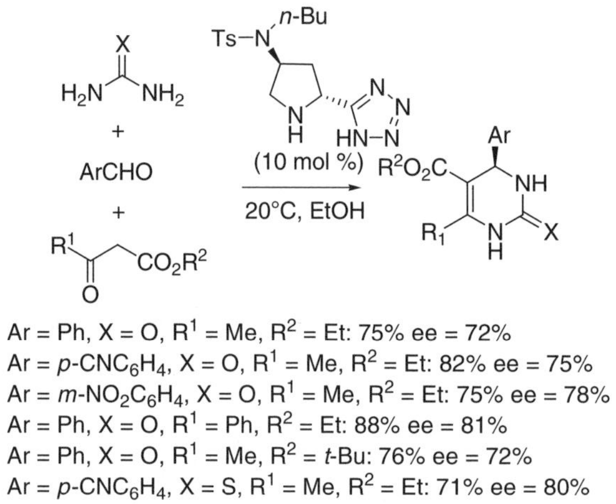
n-Bu
Ts–N
N
H
N
N
HN–N
(10 mol %)
20°C, EtOH
X
H2N
NH2
+
ArCHO
+
R1
CO2R2
O
Ar
R2O2C
NH
R1
N
H
X
Ar = Ph, X = O, R1 = Me, R2 = Et: 75% ee = 72%
Ar = p-CNC6H4, X = O, R1 = Me, R2 = Et: 82% ee = 75%
Ar = m-NO2C6H4, X = O, R1 = Me, R2 = Et: 75% ee = 78%
Ar = Ph, X = O, R1 = Ph, R2 = Et: 88% ee = 81%
Ar = Ph, X = O, R1 = Me, R2 = t-Bu: 76% ee = 72%
Ar = p-CNC6H4, X = S, R1 = Me, R2 = Et: 71% ee = 80%

Scheme 2.61 Biginelli reactions catalysed by substituted 5-(pyrrolidin-2-yl)tetrazole.

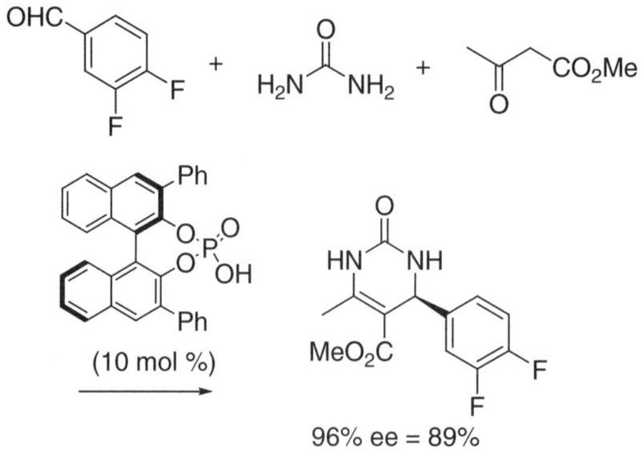
OHC
F
F
+
O
H2N
NH2
+
CO2Me
O
Ph
O
P
O
O
OH
Ph
(10 mol %)
O
HN
NH
MeO2C
F
F
96% ee = 89%

Scheme 2.62 Biginelli reactions catalysed by BINOL-derived phosphoric acid.

also investigated by Juaristi *et al.* by employing (1*S*,4*S*)-2,5-diazadi-cyclo[2.2.1]heptane derivatives as the organocatalysts.[92] Therefore, the enantio-selective Biginelli reaction of methyl and ethyl acetoacetate with aromatic aldehydes and urea afforded the corresponding 3,4-dihydropyrimidin-2(1*H*)-ones in moderate to good yields combined with moderate enantioselectivities (≤46% ee).

An enantioselective synthesis of SNAP-7941, a potent melanin concentrating hormone receptor antagonist, was achieved by Schauss and Goss by using an asymmetric organocatalytic Biginelli reaction.[93] This reaction involving a β-keto ester, urea and an aromatic aldehyde was catalysed by a chiral BINOL-derived phosphoric acid and led to the expected Biginelli product in excellent yield and enantioselectivity, as shown in Scheme 2.62.

2.3   Conclusions

The asymmetric aldol reaction is the most advanced type of synthesis in the field of organocatalysis. During the last year, the organocatalysed aldol reactions have grown most remarkably, especially those which involve chiral proline derivatives as organocatalysts, which provided uniformly spectacular stereoselectivities combined with excellent yields. From a green chemistry perspective, in comparison to the reasonable catalyst loading of 10 mol% generally applied to the asymmetric organocatalytic Michael reactions, there are a number of asymmetric organocatalytic aldol reactions which could provide the corresponding aldol products in excellent stereoselectivities and yields at lower catalyst loadings down to 0.1 mol%. Thus, a proline derivative bearing an imidazolium was shown to be capable at this remarkably low catalyst loading to induce excellent stereoselectivities in the aldol reaction of cyclohexanone with aldehydes under solvent-free conditions and, moreover, exceptionally high values of TON (up to 930) were achieved. In addition, several other proline derivatives, such as 4-substituted acyloxyproline deriva-tives, were applied to the enantioselective aldolisation of cyclic ketones with substituted benzaldehydes in water at 0.5 mol % of catalyst loading, and also provided excellent stereoselectivities. Furthermore, highly enantioselective aldol reactions could be performed in the presence of a polystyrene anchored L-proline catalyst employed at 5 mol % of catalyst loading. In addition to the highly efficient proline-derived organocatalysts, a wide number of chiral pri-mary amines have been successfully applied as organocatalysts to promote asymmetric aldolisations, giving in most cases stereoselectivities as high as those obtained with the proline-derived catalysts. For example, a series of primary amine catalysts derived from chiral 1,2-cyclohexyldiamine provided excellent stereoselectivities for the asymmetric aldolisation of ketones. Fur-thermore, this type of catalyst could be immobilised on polystyrene/sulfonic acids and further successfully employed in similar reactions. Moreover, *trans*-*N*,*N*-dipropyl diaminocyclohexane catalyst was successfully introduced as the first magnetic nanoparticle-supported chiral primary amine catalyst and then applied to the asymmetric aldolisation of cyclohexanone with benzaldehydes, leading to the expected aldol products in excellent yields and enantioselectivities. In conclusion, the results obtained for the asymmetric organocatalytic aldol reactions developed in the last year are unexpectedly and uniformly excellent.

References

1. (a) G. Guillena, C. Nàjera and D. J. Ramon, *Tetrahedron: Asymmetry*, 2007, **18**, 2249–2293; (b) L. M. Geary and P. G. Hultin, *Tetrahedron: Asymmetry*, 2009, **20**, 131–173; (c) J. Mlynarski and J. Paradowska, *Chem. Soc. Rev.*, 2008, **37**, 1502–1511.
2. B. List, *Tetrahedron*, 2002, **58**, 5573–5590.
3. (a) B. List, R. A. Lerner and C. F. Barbas III, *J. Am. Chem. Soc.*, 2000, **122**, 2395–2396; (b) K. Sakthivel, W. Notz, T. Bui and C. F. Barbas, *J. Am. Chem. Soc.*, 2001, **123**, 5260–5267.
4. (a) D. Enders, V. Terteryan and J. Palecek, *Synthesis*, 2008, **14**, 2278–2282; (b) D. Enders and A. A. Narine, *J. Org. Chem.*, 2008, **73**, 7857–7870.
5. J. Carpenter, A. B. Northrup, D. Chung, J. J. M. Wiener, S.-G. Kim and D. W. C. MacMillan, *Ang. Chem., Int. Ed. Engl.*, 2008, **47**, 3568–3572.
6. S. Hajra and A. K. Giri, *J. Org. Chem.*, 2008, **73**, 3935–3937.
7. J. Shah, H. Blumenthal, Z. Yacob and J. Liebscher, *Adv. Synth. Catal.*, 2008, **350**, 1267–1270.
8. C. Zheng, Y. Wu, X. Wang and G. Zhao, *Adv. Synth. Catal.*, 2008, **350**, 2690–2694.
9. D. Zhang and C. Yuan, *Tetrahedron*, 2008, **64**, 2480–2488.
10. F. Giacalone, M. Gruttadauria, P. L. Meo, S. Riela and R. Noto, *Adv. Synth. Catal.*, 2008, **350**, 2747–2760.
11. Y.-Q. Fu, Y.-J. An, W.-M. Liu, Z.-C. Li, G. Zhang and J.-C. Tao, *Catal. Lett.*, 2008, **124**, 397–404.
12. D. E. Siyutkin, A. S. Kucherenko and S. G. Zlotin, *Tetrahedron*, 2009, **65**, 1366–1372.
13. (a) M. Lombardo, S. Easwar, A. De Marco, F. Pasi and C. Trombini, *Org. Biomol. Chem.*, 2008, **6**, 4224–4229; (b) M. Lombardo, S. Easwar, F. Pasi and C. Trombini, *Adv. Synth. Catal.*, 2009, **351**, 276–282.
14. D. Font, S. Sayalero, A. Bastero, C. Jimeno and M. A. Pericas, *Org. Lett.*, 2008, **10**, 337–340.
15. Y.-X. Liu, Y.-N. Sun, H.-H. Tan and J.-C. Tao, *Catal. Lett.*, 2008, **120**, 281–287.
16. S. S. Chimni, S. Singh and D. Mahajan, *Tetrahedron: Asymmetry*, 2008, **19**, 2276–2284.
17. S.-p. Zhang, X.-k. Fu and S.-d. Fu, *Tetrahedron Lett.*, 2009, **50**, 1173–1176.

18. Y. Okuyama, H. Nakano, Y. Watanabe, M. Makabe, M. Takeshita, K. Uwai, C. Kabuto and E. Kwon, *Tetrahedron Lett.*, 2009, **50**, 193–197.
19. J.-F. Zhao, L. He, J. Jiang, Z. Tang, L.-F. Cun and L.-Z. Gong, *Tetrahedron Lett.*, 2008, **49**, 3372–3375.
20. X.-J. Li, G.-W. Zhang, L. Wang, M.-Q. Hua and J.-A. Ma, *Synlett*, 2008, **8**, 1255–1259.
21. R. S. Schwab, F. Z. Galetto, J. B. Azeredo, A. L. Braga, D. S. Lüdtke and M. W. Paixao, *Tetrahedron Lett.*, 2008, **49**, 5094–5097.
22. J. Du, Z. Li, D.-M. Du and J. Xu, *Arkivoc*, 2008, **xvii**, 145–146.
23. F. Chen, S. Huang, H. Zhang, F. Liu and Y. Peng, *Tetrahedron*, 2008, **64**, 9585–9591.
24. J. Yan and L. Wang, *Synthesis*, 2008, **13**, 2065–2072.
25. (a) H. Pellissier, *Tetrahedron*, 2007, **63**, 1297–1330; (b) M. Mellah, A. Voituricz and E. Schulz, *Chem. Rev.*, 2007, **107**, 5133–5209.
26. S. Gandhi and V. Singh, *J. Org. Chem.*, 2008, **73**, 9411–9416.
27. H. Yang and R. G. Carter, *Org. Lett.*, 2008, **10**, 4649–4652.
28. S.-p. Zhang, X.-k. Fu, S.-d. Fu and J.-f. Pan, *Catal. Commun.*, 2009, **10**, 401–405.
29. L. Zu, H. Xie, H. Li, J. Wang and W. Wang, *Org. Lett.*, 2008, **10**, 1211–1214.
30. E. Tsandi, C. G. Kokotos, S. Kousidou, V. Ragoussis and G. Kokotos, *Tetrahedron*, 2009, **65**, 1444–1449.
31. S. Nakamura, N. Hara, H. Nakashima, K. Kubo, N. Shibata and T. Toru, *Chem. Eur. J.*, 2008, **14**, 8079–8081.
32. J.-R. Chen, X.-L. An, X.-Y. Zhu, X.-F. Wang and W.-J. Xiao, *J. Org. Chem.*, 2008, **73**, 6006–6009.
33. G. Guillena, M. d. C. Hita, C. Najera and S. F. Viozquez, *J. Org. Chem.*, 2008, **73**, 5933–5943.
34. D. Almasi, D. A. Alonso and C. Najera, *Adv. Synth. Catal.*, 2008, **350**, 2467–2472.
35. Z.-H. Tzeng, H.-Y. Chen, R. J. Reddy, C.-T. Huang and K. Chen, *Tetrahedron*, 2009, **65**, 2879–2888.
36. G. Yu, Z.-M. Ge, T.-M. Cheng and R.-T. Li, *Chinese J. Chem.*, 2008, **26**, 911–915.
37. X.-W. Liu, T. N. Le, Y. Lu, Y. Xiao, J. Ma and X. Li, *Org. Biomol. Chem.*, 2008, **6**, 3997–4003.
38. S. Doherty, J. G. Knight, A. McRae, R. W. Harrington and W. Clegg, *Eur. J. Org. Chem.*, 2008, 1759–1766.
39. (a) M. Gruttadauria, F. Giacalone, A. M. Marculescu, A. M. P. Salvo and R. Noto, *Arkivoc*, 2009, **viii**, 5–15; (b) M. Gruttadauria, F. Giacalone, A. M. Marculescu and R. Noto, *Adv. Synth. Catal.*, 2008, **350**, 1397–1405.
40. S. Chandrasekhar, T. Bhoopendra, B. B. Parida and C. R. Reddy, *Tetrahedron: Asymmetry*, 2008, **19**, 495–499.
41. L. Zhang, S. Luo, X. Mi, S. Liu, Y. Qiao, H. Xu and J.-P. Cheng, *Org. Biomol. Chem.*, 2008, **6**, 567–576.

42. B. Alcaide and P. Almendros, *Ang. Chem., Int. Ed. Engl.*, 2008, **47**, 4632–4634.
43. Y. Hayashi, T. Itoh, S. Aratake and H. Ishikawa, *Ang. Chem., Int. Ed. Engl.*, 2008, **47**, 2082–2084.
44. Y. Hayashi, S. Samanta, T. Itoh and H. Ishikawa, *Org. Lett.*, 2008, **10**, 5581–5583.
45. N. Volz, M. C. Bröhmer, M. Nieger and S. Bräse, *Synlett*, 2009, **4**, 550–553.
46. K. Nakayama and K. Maruoka, *J. Am. Chem. Soc.*, 2008, **130**, 17666–17667.
47. S. Luo, H. Xu, L. Zhang, J. Li and J.-P. Cheng, *Org. Lett.*, 2008, **10**, 653–656.
48. S. Luo, H. Xu, L. Chen and J.-P. Cheng, *Org. Lett.*, 2008, **10**, 1775–1778.
49. S. Luo, X. Zheng and J.-P. Cheng, *Chem. Commun.*, 2008, 5719–5721.
50. S. Luo, J. Li, L. Zhang, H. Xu and J.-P. Cheng, *Chem. Eur. J.*, 2008, **14**, 1273–1281.
51. F.-Z. Peng, Z.-H. Shao, X.-W. Pu and H.-B. Zhang, *Adv. Synth. Catal.*, 2008, **350**, 2199–2204.
52. L.-W. Xu, Y.-D. Ju, L. Li, H.-Y. Qiu, J.-X. Jiang and G.-Q. Lai, *Tetrahedron Lett.*, 2008, **49**, 7037–7041.
53. J. Liu, Z. Yang, Z. Wang, F. Wang, X. Chen, X. Liu, X. Feng, Z. Su and C. Hu, *J. Am. Chem. Soc.*, 2008, **130**, 5654–5655.
54. X. Wu, Z. Ma, Z. Ye, S. Qian and G. Zhao, *Adv. Synth. Catal.*, 2009, **351**, 158–162.
55. S. S. V. Ramasastry, K. Albertshofer, N. Utsumi and C. F. Barbas, *Org. Lett.*, 2008, **10**, 1621–1624.
56. C.-S. Da, L.-P. Che, Q.-P. Guo, F.-C. Wu, X. Ma and Y.-N. Jia, *J. Org. Chem.*, 2009, **74**, 2541–2546.
57. M.-K. Zhu, X.-Y. Xu and L.-Z. Gong, *Adv. Synth. Catal.*, 2008, **350**, 1390–1396.
58. Y.-C. Teo and G.-L. Chua, *Tetrahedron Lett.*, 2008, **49**, 4235–4238.
59. J. Zhou, V. Wakchaure, P. Kraft and B. List, *Ang. Chem., Int. Ed. Engl.*, 2008, **47**, 7656–7658.
60. T. Kano and K. Maruoka, *Chem. Commun.*, 2008, 5465–5473.
61. Z.-H. Tzeng, H.-Y. Chen, C.-T. Huang and K. Chen, *Tetrahedron Lett.*, 2008, **49**, 4134–4137.
62. C. Isart, J. Burés and J. Vilarrasa, *Tetrahedron Lett.*, 2008, **49**, 5414–5418.
63. M. Laars, K. Kriis, T. Kailas, T. Müürisepp, T. Pehk, T. Kanger and M. Lopp, *Tetrahedron: Asymmetry*, 2008, **19**, 641–645.
64. J. Paradowska, M. Rogozinska and J. Mlynarski, *Tetrahedron Lett.*, 2009, **50**, 1639–1641.
65. Z.-X. Xu, G.-K. Li, C.-F. Chen and Z.-T. Huang, *Tetrahedron*, 2008, **64**, 8668–8675.
66. T. Miyabe, Y. Hase, H. Iida, K. Maeda and E. Yashima, *Chirality*, 2009, **21**, 44–50.

67. X. Chen, J. Wang, Y. Zhu, D. Shang, B. Gao, X. Liu, X. Feng, Z. Su and C. Hu, *Chem. Eur. J.*, 2008, **14**, 10,896–10,899.
68. K. Takada, N. Takemura, K. Cho, Y. Sohtome and K. Nagasawa, *Tetrahedron Lett.*, 2008, **49**, 1623–1626.
69. M. Bandini, R. Sinisi and A. Umani-Ronchi, *Chem. Commun.*, 2008, 4360–4362.
70. M. Hernandez-Rodriguez, C. G. Avila-Ortiz, J. M. del Campo, D. Hernandez-Romero, M. J. Rosales-Hoz and E. Juaristi, *Aust. J. Chem.*, 2008, **61**, 364–375.
71. S. E. Denmark and J. Fu, *Chem. Rev.*, 2003, **103**, 2763–2793.
72. G. Chelucci, S. Baldino, G. A. Pinna, M. Benaglia, L. Buffa and S. Guizzetti, *Tetrahedron*, 2008, **64**, 7574–7582.
73. A. V. Malkov, M.-M. Westwater, A. Gutnov, P. Ramirez-Lopez, Γ. Friscourt, A. Kadlcikova, J. Hodacova, Z. Rankovic, M. Kotora and P. Kocovsky, *Tetrahedron*, 2008, **64**, 11,335–11,348.
74. V. Simonini, M. Benaglia and T. Benincori, *Adv. Synth. Catal.*, 2008, **350**, 561–564.
75. A. Massa, M. R. Acocella, V. De Sio, R. Villano and A. Scettri, *Tetrahedron: Asymmetry*, 2009, **20**, 202–204.
76. M. Rueping, B. J. Nachtsheim, S. A. Moreth and M. Bolte, *Ang. Chem., Int. Ed. Engl.*, 2008, **47**, 593–596.
77. (a) M. Pohl, B. Lingen and M. Müller, *Chem. Eur. J.*, 2002, **8**, 5289–5295; (b) H. Stetter and H. Kuhlmann, *Org. React.*, 1991, **40**, 407–496.
78. G. Bredig, *Biochem. Z.*, 1912, **35**, 324–325.
79. D. Enders and J. Han, *Tetrahedron: Asymmetry*, 2008, **19**, 1367–1371.
80. F. Wang, X. Liu, X. Cui, Y. Xiong, X. Zhou and X. Feng, *Chem. Eur. J.*, 2009, **15**, 589–592.
81. (a) K. Morita, Z. Suzuki and H. Hirose, *Bull. Chem. Soc. Jpn.*, 1968, **41**, 2815–2815; (b) P. Langer, *Ang. Chem., Int. Ed. Engl.*, 2000, **39**, 3049–3052; (c) D. Basavaiah, A. J. Rao and T. Satyanarayana, *Chem. Rev.*, 2003, **103**, 811–892; (d) V. Singh and S. Batra, *Tetrahedron*, 2008, **64**, 4511–4574.
82. M. Gruttadauria, F. Giacalone, P. Lo Meo, A. M. Marculescu, S. Riela and R. Noto, *Eur. J. Org. Chem.*, 2008, 1589–1596.
83. Γ. Giacalone, M. Gruttadauria, Λ. M. Marculescu, F. D'Anna and R. Noto, *Chem. Commun.*, 2008, **9**, 1477–1481.
84. H. Tang, G. Zhao, Z. Zhou, P. Gao, L. He and C. Tang, *Eur. J. Org. Chem.*, 2008, 126–135.
85. Y. Sohtome, N. Takemura, R. Takagi, Y. Hashimoto and K. Nagasawa, *Tetrahedron*, 2008, **64**, 9423–9429.
86. M. Shi and X.-G. Liu, *Org. Lett.*, 2008, **10**, 1043–1046.
87. K. Yuan, L. Zhang, H.-L. Song, Y. Hu and X.-Y. Wu, *Tetrahedron Lett.*, 2008, **49**, 6262–6264.
88. C. Najera and J. M. Sansano, *Chem. Rev.*, 2007, **107**, 4584–4671.
89. S. Lou and S. E. Schaus, *J. Am. Chem. Soc.*, 2008, **130**, 6922–6923.

90. H. Miyabe and Y. Takemoto, *Bull. Chem. Soc. Jpn.*, 2008, **81**, 785–795.
91. Y.-Y. Wu, Z. Chai, X.-Y. Liu, G. Zhao and S.-W. Wang, *Eur. J. Org. Chem.*, 2009, 904–911.
92. R. Gonzalez-Olvera, P. Demare, I. Regla and E. Juaristi, *Arkivoc*, 2008, **vi**, 61–72.
93. J. M. Goss and S. E. Schauss, *J. Org. Chem.*, 2008, **73**, 7651–7656.

Nucleophilic Additions to C=N Double Bonds

Chiral α-branched amines are common substructures within biologically active materials and hence attract broad interest, particularly in the areas of synthetic methodology, bioorganic and medicinal chemistry and natural product synthesis. Additions of carbon fragments to C=N bonds of imines and related compounds build up the carbon framework in the same way as asymmetric induction, so this approach is one of the more attractive entries to chiral amines.[1]

3.1 Mannich Reactions

The Mannich reaction, a widely applied means of producing β-amino carbonyl compounds starting from cheap and readily available substrates, involves an aldehyde, an amine and a ketone reacting in a three-component, one-pot synthesis.[2] As an alternative, the reaction can be performed as a nucleophilic addition of a C-nucleophile to a preformed imine, which is prepared starting from the aldehyde and an amine source. The Mannich reaction tolerates a wide range of acceptors, donors and amine reagents, and can be carried out in a large variety of polar solvents. Organocatalytic Mannich reactions can be carried out either as three-component, one-pot reactions or as reactions of preformed imines with aldol donors. Among a wide variety of chiral organocatalysts that have been used in the asymmetric Mannich reaction,[3] the most widely used is proline. In 2008, several asymmetric L-proline-catalysed indirect Mannich reactions were reported, such as those involving the reaction of acetaldehyde with N-Boc imines, providing the corresponding β-amino aldehydes in moderate yields and exceptional enantioselectivities (96–99% ee).[4] Potential side reactions usually observed in L-proline-catalysed Mannich reactions of acetaldehyde, such as aldol condensations of acetaldehyde itself producing polymers, or subsequent reactions of the thus-formed Mannich products with an

RSC Catalysis Series No. 3
Recent Developments in Asymmetric Organocatalysis
By Hélène Pellissier
Published by the Royal Society of Chemistry, www.rsc.org

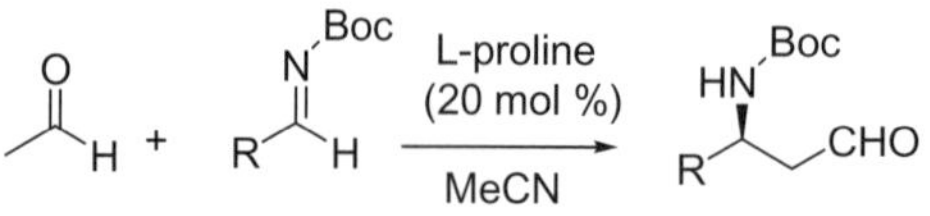

R = Ph: 54% ee > 98%
R = 2-Naph: 40% ee > 98%
R = *p*-Tol: 58% ee = 96%
R = *p*-CF$_3$C$_6$H$_4$: 42% ee = 98%
R = *m*-NO$_2$C$_6$H$_4$: 42% ee > 98%
R = 2-Thio: 30% ee = 98%
R = *n*-Bu: 55% ee > 98%
R = Et: 23% ee > 98%

Scheme 3.1 L-Proline-catalysed Mannich reactions of acetaldehyde with *N*-Boc imines.

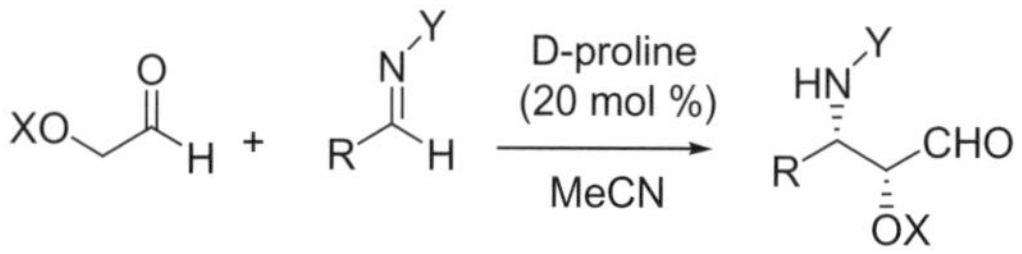

R = Ph, X = TBS, Y = Boc: 56% de > 90% ee = 99%
R = Ph, X = Bn, Y = Boc: 60% de > 90% ee = 99%
R = *p*-Tol, X = TBS, Y = Boc: 56% de = 50% ee = 99%
R = *p*-MeOC$_6$H$_4$, X = Bn, Y = Boc: 52% de = 80% ee = 99%
R = Ph, X = Bn, Y = Bz: 68% de = 90% ee = 99%
R = Ph, X = TBS, Y = Bz: 68% de = 60% ee = 98%
R = *p*-MeOC$_6$H$_4$, X = Bn, Y = Bz: 65% de = 90% ee = 99%
R = *p*-Tol, X = Bn, Y = Bz: 61% de = 80% ee = 98%
R = 2-Naph, X = Bn, Y = Bz: 60% de = 80% ee = 96%

Scheme 3.2 D-Proline-catalysed *syn*-Mannich reactions of protected α-hydroxyaldehydes with *N*-Boc imines or *N*-(phenylmethylene)-benzamides.

additional imine equivalent, could be avoided by using a large excess of acetaldehyde (5–10 equivalents). The results obtained with a range of imines are collected in Scheme 3.1.

In the same area, Cordova *et al.* have developed the highly enantioselective D-proline-catalysed Mannich reaction of *N*-Boc imines with α-oxyaldehydes.[5] The corresponding orthogonally protected α-oxy-β-aminoaldehydes were formed in good yields and diastereoselectivities of up to 90% de combined with excellent enantioselectivities of 99% ee (Scheme 3.2). The importance of this transformation as an entry to the synthesis of α-hydroxy-β-amino acids was further exemplified by the highly stereoselective synthesis of the side chain of taxotere. As an extension of this methodology, these authors have developed the D-proline-catalysed enantioselective addition of protected α-hydroxyaldehydes to *N*-(phenylmethylene)-benzamide, providing the corresponding *syn*-protected α-hydroxy-β-aminoaldehydes in good yields, moderate to good diastereoselectivities (50–90% de) and excellent enantioselectivities (92–99% ee).[6]

5,5-Disubstituted 5,6-dihydro-1,4-oxazin-2-ones were employed as other substrates in L-proline-catalysed asymmetric Mannich reactions with a wide

R^1, R^2 = (CH$_2$)$_4$, R^3 = Me: 80% de = 82% ee = 95%
R^1, R^2 = (CH$_2$)$_4$, R^3 = Ph: 91% de > 90% ee > 99%
R^1, R^2 = (CH$_2$)$_5$, R^3 = Ph: 60% de > 90% ee = 97%
R^1 = Me, R^2 = (CH$_2$)$_3$, R^3 = Ph: 30% de > 90% ee = 99%
R^1 = R^2 = Me, R^3 = Ph: 70% de > 90% ee > 99%
R^1 = Me, R^2 = Et, R^3 = Ph: 90% de > 90% ee = 94%
R^1 = Me, R^2 = CH$_2$CH=CH$_2$, R^3 = Ph: 91% de > 90% ee > 99%
R^1 = Me, R^2 = Bn, R^3 = Ph: 70% de > 90% ee > 99%

Scheme 3.3 L-Proline-catalysed Mannich reactions of cyclic imines with ketones.

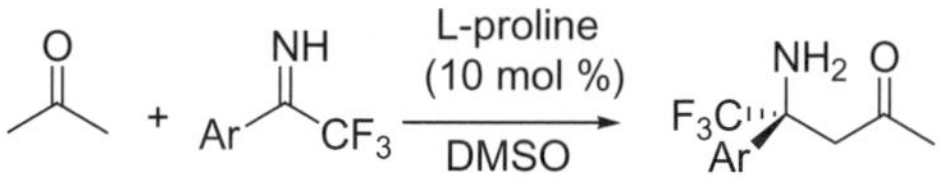

Ar = Ph: 86% ee = 80%
Ar = *m*-Tol: 80% ee = 92%
Ar = *p*-FC$_6$H$_4$: 82% ee = 84%
Ar = *p*-MeOC$_6$H$_4$: 75% ee = 78%
Ar = *p*-Tol: 82% ee = 74%

Scheme 3.4 L-Proline-catalysed Mannich reactions of cyclic imines with ketones.

range of unactivated ketones (Scheme 3.3).[7] These highly enantio- and *anti*-selective reactions enabled the highly stereoselective synthesis of chiral 3-substituted 1,4-morpholin-2-ones, which constituted versatile building blocks for the synthesis of chiral α-amino acids.

Vovk *et al.* have described the first synthesis of chiral β-aryl-β-trifluoromethyl-β-aminoketones based on an L-proline-catalysed Mannich reaction between aryl trifluoromethyl ketimines and acetone.[8] As shown in Scheme 3.4, the desired fluorinated aminoketones were isolated in high yields and enantioselectivities of up to 92% ee.

On the other hand, a direct asymmetric Mannich reaction was reported, in 2009, by Rutjes *et al.* in the course of developing a synthesis of the bioactive quinolizidine alkaloid lasubine II.[9] Therefore, the key step of this synthesis was the L-proline-catalysed Mannich reaction between acetone, *para*-nitro-benzaldehyde and veratryl aldehyde, providing the expected Mannich adduct in moderate yield and almost complete enantioselectivity (Scheme 3.5). This chiral β-amino ketone was further converted into the desired (+)-lasubine II.

In addition, a highly diastereo- and enantioselective synthesis of 2,3-disubstituted tetrahydropyridines was accomplished *via* a proline-mediated cascade Mannich-type/intramolecular cyclisation reaction from preformed

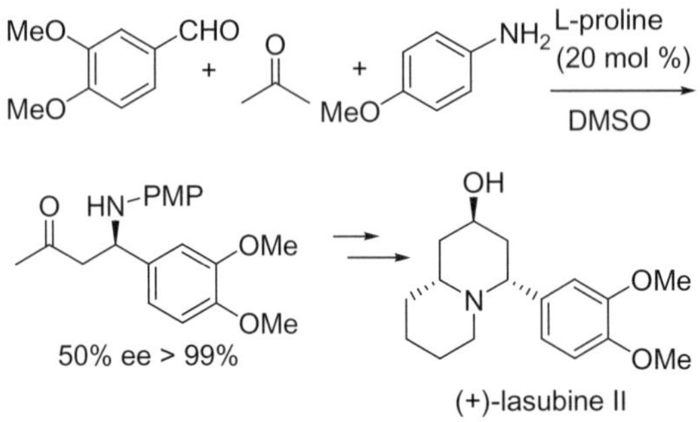

Scheme 3.5 L-Proline-catalysed Mannich reaction of cyclic imine with acetone.

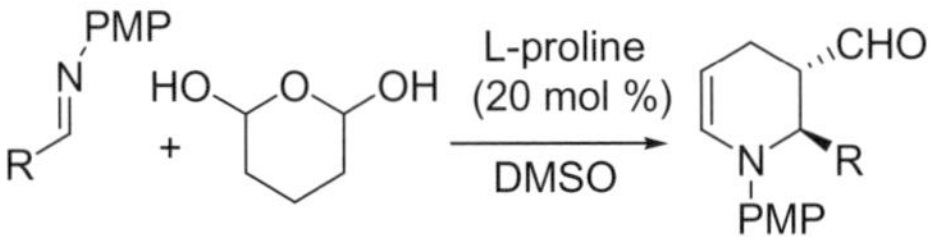

R = *p*-NO$_2$C$_6$H$_4$: 74% de > 92% ee = 98%
R = *m*-NO$_2$C$_6$H$_4$: 65% de > 92% ee = 98%
R = *p*-CNC$_6$H$_4$: 63% de > 92% ee = 98%
R = *m*-ClC$_6$H$_4$: 61% de > 92% ee = 94%
R = *o*-BrC$_6$H$_4$: 39% de > 92% ee = 95%
R = *o*-FC$_6$H$_4$: 61% de > 92% ee > 99%
R = *m*-FC$_6$H$_4$: 60% de > 92% ee = 96%
R = 2-Naph: 54% de > 92% ee = 91%
R = Ph: 56% de > 92% ee = 92%

Scheme 3.6 L-Proline-catalysed domino Mannich-type/intramolecular cyclisation reactions.

N-(*para*-methoxyphenyl) aldimines and inexpensive aqueous tetrahydro-2-*H*-pyran-2,6-diol.[10] These excellent results are collected in Scheme 3.6.

Several variously substituted derivatives of proline have been investigated as chiral organocatalysts for the asymmetric Mannich reaction. As an example, Wu *et al.* have shown that (2*S*,5*S*)-pyrrolidine-2,5-dicarboxylic acid was an efficient organocatalyst for the Mannich reaction between hydroxyacetone, anilines and benzaldehydes, providing the corresponding *syn*-1,2-amino alcohols in good yields and high enantioselectivities of up to 96% ee (Scheme 3.7).[11] Similar results were also obtained by using the corresponding (2*S*,5*S*)-5-(methoxycarbonyl)pyrrolidine-2-carboxylic acid as an organocatalyst for the same reactions.[12]

Finally, Barbas *et al.* have developed enantioselective *anti*-selective Mannich reactions of enolisable aldehydes and ketones with imines catalysed by (*R*)-3-pyrrolidinecarboxylic acid.[13] In the case of the reactions of aldehydes, the Mannich products were isolated with both high *anti*-selectivity and enantioselectivity combined with good to high yields (Scheme 3.8). On the other hand,

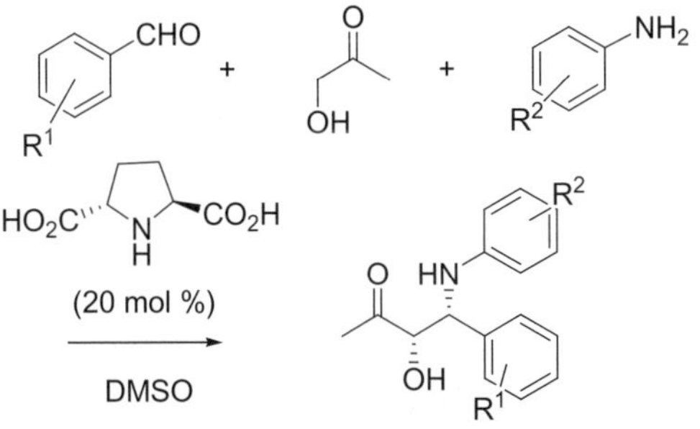

R^1 = 2-NO$_2$, R^2 = 4-MeO: 86% de = 80% ee = 94%
R^1 = 4-CN, R^2 = 4-MeO: 71% de = 80% ee = 88%
R^1 = 3-CN, R^2 = 4-MeO: 64% de = 84% ee = 86%
R^1 = 4-NO$_2$, R^2 = 3-MeO: 95% de = 78% ee = 86%
R^1 = 4-NO$_2$, R^2 = 4-Cl: 90% de = 82% ee = 89%
R^1 = 4-NO$_2$, R^2 = H: 80% de = 84% ee = 96%

Scheme 3.7   (2S,5S)-pyrrolidine-2,5-dicarboxylic acid-catalysed *syn*-Mannich reactions.

the *anti*-Mannich products from ketones were obtained in good yields, variable diastereoselectivities of up to 98% de and good to high enantioselectivities of up to 98% ee (Scheme 3.8). According to an evaluation of a series of pyrrolidine-based catalysts, it was demonstrated that the 3-acid group on pyrrolidine was essential for the efficient acceleration of the C–C bond formation and for the stereocontrol as it engaged proton transfer to the imine nitrogen at the C–C bond-forming step.

These reactions were also performed in DMF by Blanchet *et al.* in the presence of (*R*)-3-trifluoromethanesulfonamidopyrrolidine as the organocatalyst, providing comparative results.[14] In the case of using aldehydes as the substrates, the *anti*-Mannich products were obtained in comparative yields and enantioselectivities, albeit with lower diastereoselectivities (60–88% de), while the products arisen from the reaction of ketones were produced in slightly higher enantioselectivities (>98% ee), albeit with lower diastereoselectivities (60–90% de). A solvent-free direct Mannich reaction was developed by Hayashi *et al.* in the presence of water by using a highly reactive siloxytetrazole hybrid catalyst (Scheme 3.9).[15] This novel process involved dimethoxyacetaldehyde, *para*-nitrobenzaldehyde and a ketone, which led to the corresponding *syn*-Mannich products in high yields, moderate to high diastereoselectivities (64–90% de) combined with high enantioselectivities (83–97% ee). In addition, Kim and Park completed a short and efficient synthesis of biologically active (+)-*epi*-cytoxazone, based on the organocatalytic asymmetric Mannich reaction between *para*-methoxybenzaldehyde *N*-Boc-imine and benzyloxyacetaldehyde using a novel chiral 2-pyrrole-derived imidazolidinone as the catalyst.[16] This reaction, performed in CHCl$_3$ in the presence of TFA as an additive and at 20 mol % of catalyst loading, afforded the corresponding *syn*-Mannich product in 90% yield, with a moderate diastereoselectivity of 60% de

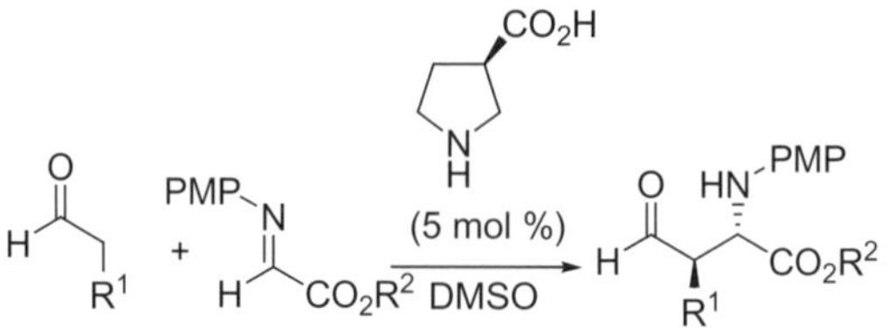

R¹ = Me, R² = Et: 75% de = 86% ee = 96%
R¹ = n-Bu, R² = Et: 60% de = 98% ee = 95%
R¹ = n-Pent, R² = Et: 80% de = 98% ee > 97%
R¹ = CH₂CH=CH₂, R² = Et: 78% de = 98% ee > 97%
R¹ = CH₂CH=CH(n-Pent), R² = Et: 83% de = 96% ee = 99%
R¹ = Bn, R² = Et: 87% de = 92% ee = 99%
R¹ = i-Pr, R² = t-Bu: 82% de = 98% ee = 94%
R¹ = i-Pr, R² = CH₂CH=CH₂: 85% de = 96% ee = 95%

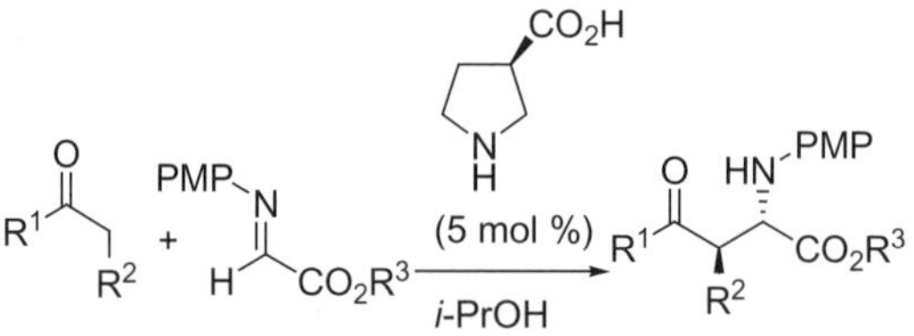

R¹ = R³ = Et, R² = Me: 91% de = 94% ee = 97%
R¹ = Et, R² = Me, R³ = t-Bu: 93% de > 98% ee = 95%
R¹ = n-Pr, R² = R³ = Et: 76% de > 98% ee = 82%
R¹ = Me, R² = CH₂CH=CH₂, R³ = Et: 85% de > 90% ee = 91%
R¹,R² = (CH₂)₄, R³ = Et: 96% de > 98% ee = 96%
R¹,R² = (CH₂)₄, R³ = i-Pr: 94% de > 98% ee = 94%
R¹,R² = (CH₂)₄, R³ = t-Bu: 92% de > 98% ee = 95%
R¹,R² = (CH₂)₅, R³ = Et: 80% de > 90% ee = 84%

Scheme 3.8 (*R*)-3-Pyrrolidinecarboxylic acid-catalysed Mannich reactions.

combined with an enantioselectivity of 82% ee. This product was further converted in three steps into the desired (+)-*epi*-cytoxazone.

On the other hand, diaryl prolinol silyl ethers have been applied as organocatalysts for the asymmetric Mannich reaction of acetaldehyde with *N*-benzoyl-, *N*-Boc- and *N*-Ts-imines in the presence of *para*-nitrobenzoic acid in THF.[17] The generated β-amino aldehyde products were isolated in good to high yields and excellent enantioselectivities (95–98% ee) after conversion into the corresponding alcohols by reduction with LiAlH₄ (Scheme 3.10).

This catalyst was also employed by Melchiorre *et al.* to promote the first aminocatalysed *anti*-selective Mannich reaction of aldehydes with *N*-Cbz- and *N*-Boc-protected imines generated *in situ* from stable α-amido sulfones (Scheme 3.11).[18] Besides the high level of efficiency and stereocontrol achieved, this approach introduced important synthetic advantages, by avoiding the requirement to perform the *N*-carbamoyl imines.

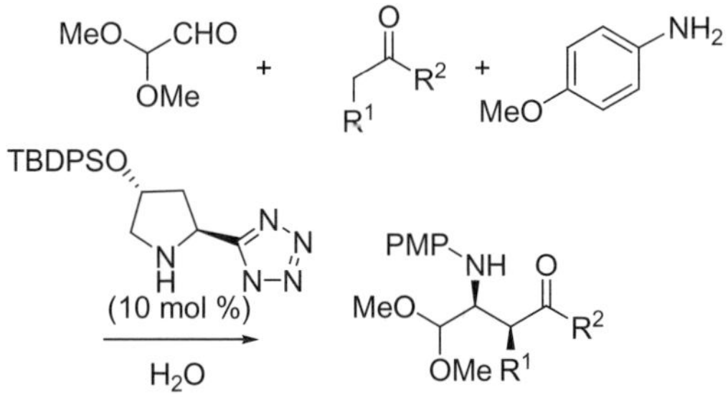

R^1,R^2 = (CH$_2$)$_4$: 93% de = 64% ee = 95%
R^1,R^2 = (CH$_2$)$_5$: 78% de = 82% ee = 95%
R^1 = Et, R^2 = Me: 58% de > 90% ee = 95%
R^1,R^2 = OC(Me)$_2$O: 63% de > 90% ee = 83%

Scheme 3.9 (*R*)-2-Pyrrolidinetetrazole-catalysed *syn*-Mannich reactions in water.

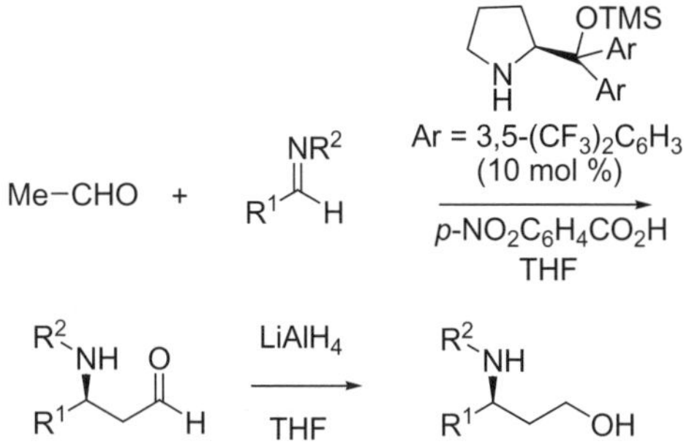

R^1 = Ph, R^2 = Bz: 87% ee = 97%
R^1 = 2-Naph, R^2 = Bz: 77% ee = 98%
R^1 = *p*-Tol, R^2 = Bz: 78% ee = 98%
R^1 = *p*-ClC$_6$H$_5$, R^2 = Bz: 65% ee = 98%
R^1 = *p*-MeOC$_6$H$_5$, R^2 = Bz: 80% ee = 95%
R^1 = Ph, R^2 = Boc: 58% ee = 98%
R^1 = *p*-BrC$_6$H$_5$, R^2 = Ts: 74% ee = 80%

Scheme 3.10 Mannich reactions of acetaldehyde catalysed by diaryl prolinol silyl ether.

Excellent enantioselectivities have also been observed by Maruoka *et al.* by using several other chiral secondary amines as organocatalysts for the *anti*-Mannich reaction between various aldehydes and *N*-protected imines including α-imino esters. Therefore, the involvement of a chiral binaphthyl-based amino sulfonamide as the organocatalyst allowed a series of *anti*-Mannich products to be obtained from both *N*-Boc-protected imines and *N*-PMP-protected α-imino esters by reaction with aldehydes.[19–20] As shown in Scheme 3.12, excellent yields and enantioselectivities (97–99% ee) were observed along with moderate to high *anti* diastereoselectivities of up to 90% de.

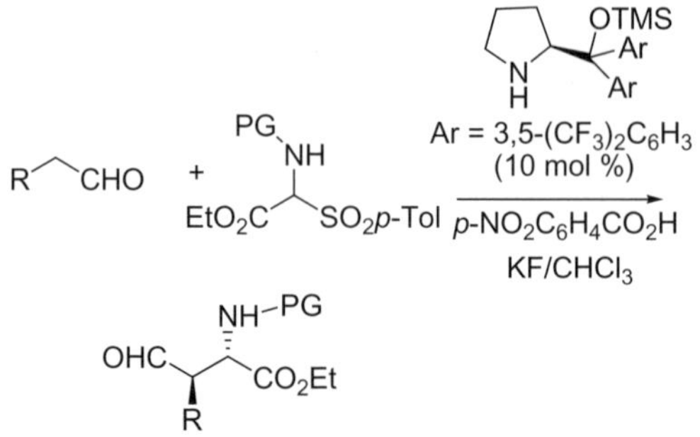

R = Bn, PG = Boc: 87% de = 88% ee = 96%
R = Me, PG = Boc: 92% de = 82% ee = 94%
R = *i*-Pr, PG = Boc: 65% de = 96% ee = 95%
R = Et, PG = Cbz: 95% de = 86% ee = 96%
R = *n*-Bu, PG = Cbz: 96% de = 84% ee = 98%
R = *i*-Pr, PG = Cbz: 85% de = 84% ee = 95%

Scheme 3.11 Mannich reactions of *in-situ*-generated imines catalysed by diaryl prolinol silyl ether.

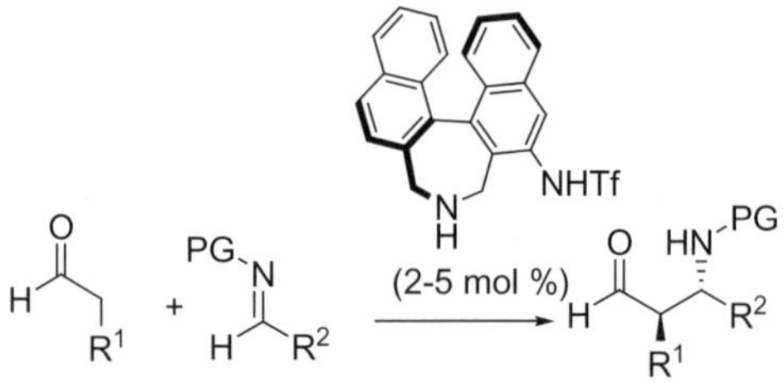

R^1 = Me, R^2 = CO_2Et, PG = PMP: 93% de = 86% ee > 99%
R^1 = *n*-Bu, R^2 = CO_2Et, PG = PMP: 93% de > 90% ee = 99%
R^1 = *n*-Bu, R^2 = Ph, PG = Boc: 93% de = 88% ee = 99%
R^1 = *n*-Bu, R^2 = Ph, PG = Boc: 93% de = 88% ee = 99%
R^1 = Bn, R^2 = Ph, PG = Boc: 80% de = 88% ee = 99%
R^1 = *n*-Bu, R^2 = 3-Py, PG = Boc: 92% de = 88% ee = 99%
R^1 = Me, R^2 = Cy, PG = Boc: 66% de > 90% ee = 99%

Scheme 3.12 Mannich reactions catalysed by binaphthyl-based amino sulfonamide.

The same authors have designed novel chiral pyrrolidine-based amino sulfonamides, which were applied to promote *anti*-Mannich reactions of *N*-PMP-protected α-imino esters with aldehydes, giving comparable results to those obtained with the binaphthyl-based amino sulfonamide catalyst.[21] However, these novel pyrrolidine-based amino sulfonamides were found to be capable of promoting *anti*-Mannich reactions between *N*-PMP-protected α-imino esters and a range of cyclic as well as acyclic ketones with excellent yields, enantioselectivities combined with moderate to high diastereoselectivities of up to 90% de (Scheme 3.13).

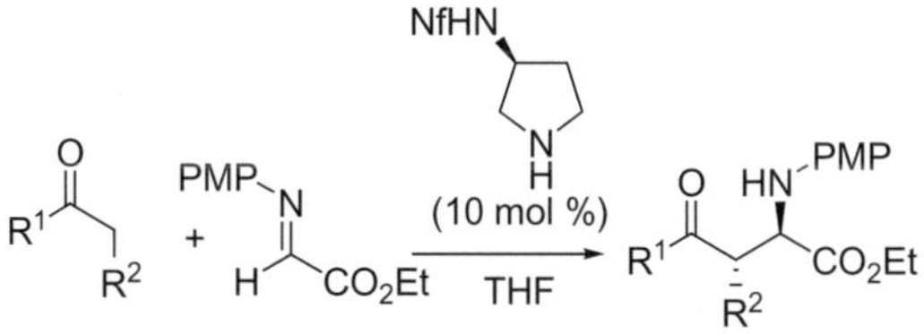

R^1,R^2 = (CH$_2$)$_4$: 98% de > 90% ee = 99%
R^1,R^2 = (CH$_2$)$_2$OCH$_2$: 99% de = 88% ee = 94%
R^1,R^2 = (CH$_2$)$_2$SCH$_2$: 99% de > 90% ee = 97%
R^1 = Me, R^2 = *n*-Pr: 82% de = 70% ee = 98%
R^1 = Me, R^2 = *n*-Pent: 22% de = 24% ee = 90%
R^1 = Me, R^2 = Et: 70% de = 64% ee = 90%

Scheme 3.13   Mannich reactions catalysed by pyrrolidine-based amino sulfonamide.

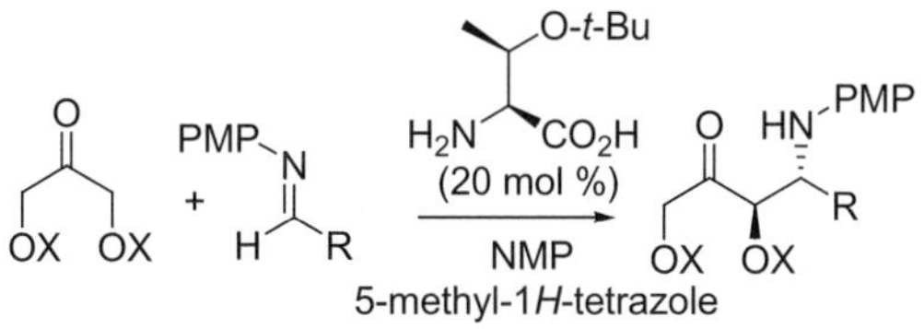

X = H, R = *p*-NO$_2$C$_6$H$_4$: 77% de = 60% ee = 90%
X = H, R = *p*-CNC$_6$H$_4$: 75% de = 50% ee = 88%
X = H, R = BnCH$_2$: 65% de = 34% ee = 72%
X = Bn, R = CO$_2$Et: 78% de = 34% ee = 89%
X = Bn, R = *p*-NO$_2$C$_6$H$_4$: 91% de = 60% ee = 97%
X = Bn, R = *p*-CF$_3$C$_6$H$_4$: 77% de = 60% ee = 94%
X = Bn, R = *p*-BrC$_6$H$_4$: 85% de = 60% ee = 88%
X = MOM, R = *p*-NO$_2$C$_6$H$_4$: 75% de = 42% ee = 90%
X = MOM, R = *p*-CF$_3$C$_6$H$_4$: 71% de = 34% ee = 87%

Scheme 3.14   Mannich reactions catalysed by L-threonine derivative.

In 2008, Barbas *et al.* reported the first primary amine-containing amino acid-catalysed indirect *anti*-Mannich reactions of dihydroxyacetone and acyclic protected dihydroxyacetone derivatives with a variety of imines derived from both aliphatic and aromatic aldehydes.[22] In spite of moderate diastereoselectivities, good to high yields and enantioselectivities were obtained by using an L-threonine derivative as the organocatalyst in *N*-methylpyrrolidinone as the solvent and 5-methyl-1-*H*-tetrazole as an additive, as shown in Scheme 3.14.

In the same context, a direct asymmetric *syn*-Mannich reaction was developed by Teo *et al.* on the basis of using a chiral siloxy-L-serine organocatalyst.[23] The three-component Mannich reaction between a cyclic or an acyclic ketone, an aldehyde and *para*-anisidine performed in the presence of water *via* a two-phase system afforded the corresponding *syn*-Mannich products in good yields, high enantioselectivities of up to 92% ee and moderate diastereoselectivities (Scheme 3.15).

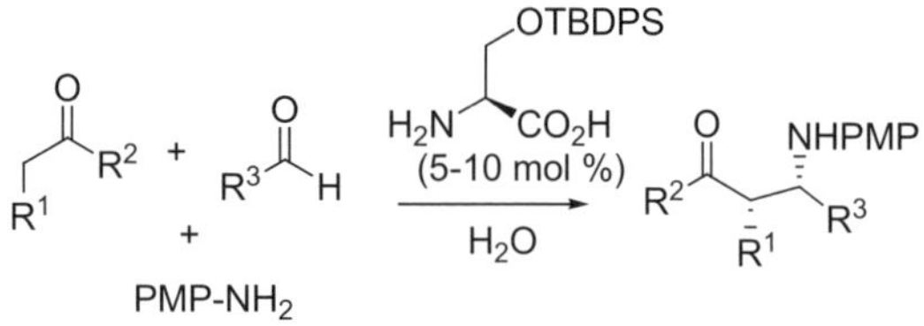

R^1, R^2 = (CH$_2$)$_4$, R^3 = p-NO$_2$C$_6$H$_4$: 86% de = 68% ee = 82%
R^1, R^2 = (CH$_2$)$_4$, R^3 = p-BrC$_6$H$_4$: 78% de = 48% ee = 74%
R^1, R^2 = (CH$_2$)$_4$, R^3 = CO$_2$Et: 71% de = 44% ee = 86%
R^1 = OBn, R^2 = Me, R^3 = p-NO$_2$C$_6$H$_4$: 81% de = 72% ee = 90%
R^1 = OBn, R^2 = Me, R^3 = p-CNC$_6$H$_4$: 79% de = 72% ee = 92%
R^1 = OBn, R^2 = Me, R^3 = 2-Naph: 78% de = 32% ee = 81%

Scheme 3.15   *Syn*-Mannich reactions catalysed by L-serine derivative in water.

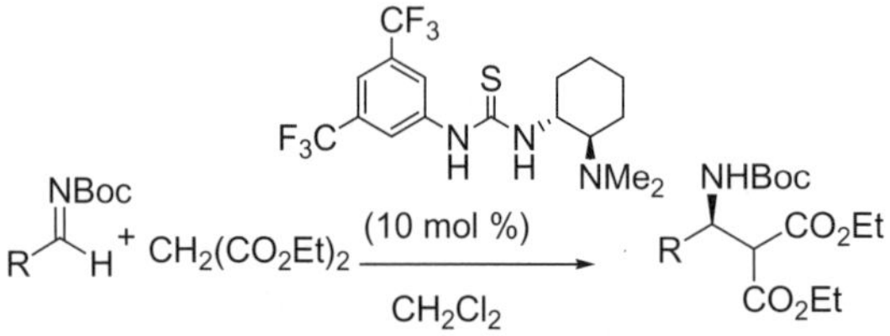

R = Ph: 84% ee = 94%
R = p-MeOC$_6$H$_5$: 82% ee = 93%
R = p-CF$_3$C$_6$H$_5$: 91% ee = 96%
R = p-BrC$_6$H$_5$: 84% ee = 98%
R = 3-Py: 73% ee = 97%

Scheme 3.16   Thiourea-catalysed Mannich reactions of ethyl malonate with *N*-Boc arylimines.

In addition, various chiral amine-thioureas have been successfully applied to promote asymmetric Mannich reactions. As an example, Takemoto and Miyabe have employed a chiral bifunctional organocatalyst possessing a thiourea moiety and a tertiary amino group as catalyst of the Mannich reaction between ethyl malonate and *N*-Boc arylimines, which provided the corresponding products in excellent yields and enantioselectivities (93–98% ee), as shown in Scheme 3.16.[24] The degree of enantioselectivity was shown to be dependent on the reaction temperature, with the best results obtained at low temperature.

A chiral primary amine-thiourea organocatalyst was successfully applied for the first time by Tsogoeva *et al.* to the asymmetric Mannich reaction of ketones with readily available and stable α-hydrazonoesters, which proceeded with good yields and high enantioselectivities of up to 99% ee (Scheme 3.17).[25] Interestingly, whereas acyclic ketones gave *anti*-Mannich products, an excess of *syn* diastereomers was observed with the cyclic ketones. In both cases, only low

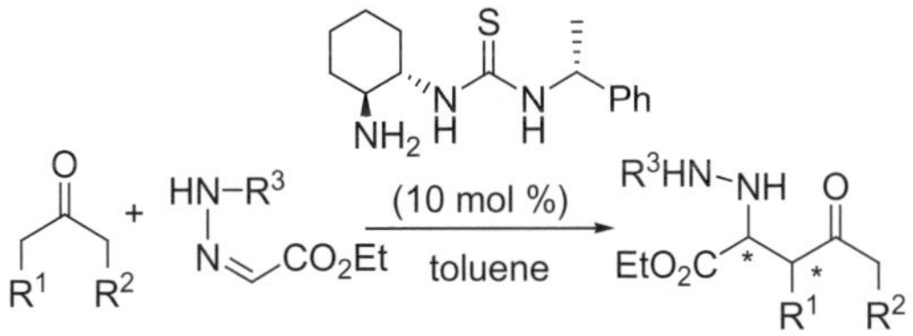

R^1 = R^2 = H, R^3 = Bz: 50% ee > 99%
R^1 = H, R^2 = Me, R^3 = Bz: 86% ee > 99%
R^1 = R^2 = Me, R^3 = Bz: 82% de = 72% (*anti*)
ee (*anti*) = 99% ee (*syn*) = 90%
R^1 = H, R^2 = Et, R^3 = Bz: 80% ee > 99%
R^1,R^2 = (CH$_2$)$_3$, R^3 = Bz: 87% de = 8% (*syn*)
ee (*syn*) > 96% ee (*anti*) > 94%
R^1,R^2 = CH$_2$SCH$_2$, R^3 = Bz: 88% de = 19% (*syn*)
ee (*syn*) > 92% ee (*anti*) > 82%
R^1,R^2 = (CH$_2$)$_3$, R^3 = *p*-MeO-C$_6$H$_4$CO: 83% de = 9% (*syn*)
ee (*syn*) > 93% ee (*anti*) > 92%

Scheme 3.17 Thiourea-catalysed Mannich reactions of ketones with α-hydrazonoesters.

to moderate diastereoselectivities were obtained. Furthermore, the authors provided some of the first computational evidence that the preferred mechanism involved enol, rather than enamine intermediates.

The asymmetric Mannich reaction of 3-substituted oxindoles and *N*-Boc imines was investigated for the first time by Chen *et al.*, employing a thiourea-tertiary amine organocatalyst based on DPEN scaffold.[26] This highly efficient novel procedure exhibited high diastereoselectivities (84–90% de) and the Mannich adducts bearing adjacent quaternary and tertiary chiral centres were generally obtained in good to excellent enantioselectivities of up to 95% ee, as shown in Scheme 3.18.

In addition, a chiral bis-thiourea was selected by Chen *et al.* among a series of various chiral organocatalysts to promote the enantioselective Mannich reaction of stabilised phosphorus ylides with *N*-Boc-protected aldimines.[27] A subsequent reaction with formaldehyde provided a facile access to chiral *N*-Boc-β-amino-α-methylene carboxylic esters in good to excellent enantioselectivities of up to 96% ee (Scheme 3.19).

A second method to achieve the synthesis of the enantioenriched dihydropyrimidone core of SNAP-7941 was based on the cinchona alkaloid-catalysed Mannich reaction of β-keto esters with acylimines.[28] Therefore, the reaction of a β-keto ester with an α-amido sulfone performed in the presence of cinchonine produced the Mannich product as a mixture of two diastereomers in an excellent yield (Scheme 3.20). This diastereomeric mixture was transformed in high yield into a key intermediate of the synthesis of SNAP-7941, showing a good enantioselectivity. This compound was finally converted into desired SNAP-7941, which is an inhibitor of MCH1-R in a G protein-coupled receptor.

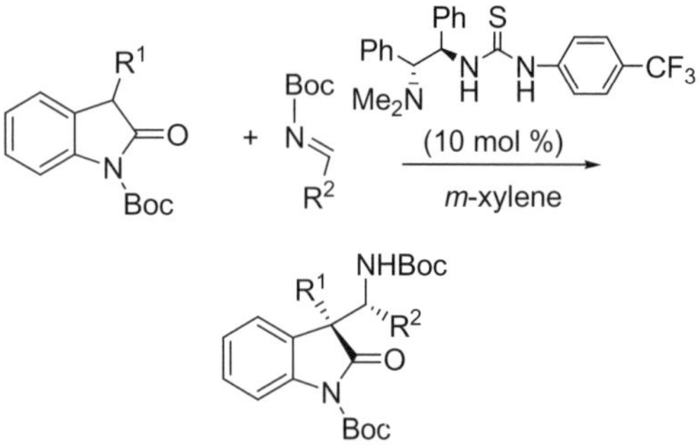

R¹ = Bn, R² = Ph: 94% de > 90% ee = 95%
R¹ = Bn, R² = p-FC₆H₄: 95% de > 90% ee = 91%
R¹ = Bn, R² = m-ClC₆H₄: 90% de = 84% ee = 93%
R¹ = Bn, R² = p-Tol: 89% de = 84% ee = 85%
R¹ = Bn, R² = 2-thienyl: 90% de = 86% ee = 94%
R¹ = Ph, R² = p-FC₆H₄: 95% de > 90% ee = 95%
R¹ = n-Pr, R² = Ph: 76% de > 88% ee = 92%

Scheme 3.18 Thiourea-catalysed Mannich reactions of 3-substituted oxindoles with *N*-Boc imines.

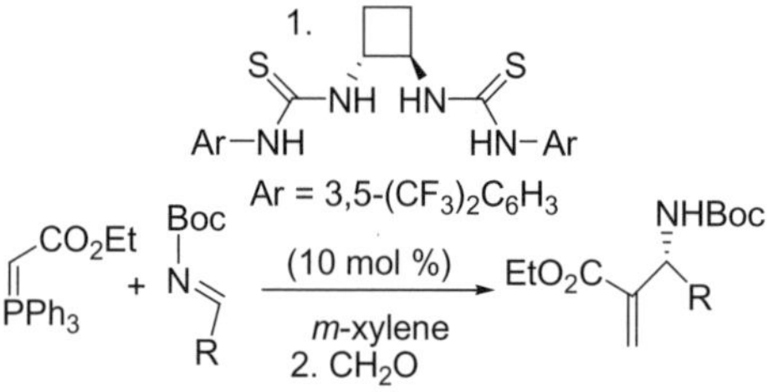

R = Ph: 87% ee = 89%
R = p-FC₆H₄: 67% ee = 87%
R = p-ClC₆H₄: 65% ee = 90%
R = p-BrC₆H₄: 80% ee = 94%
R = p-Tol: 78% ee = 95%
R = m-Tol: 84% ee = 93%
R = Cy: 52% ee = 91%
R = i-Pr: 35% ee = 96%

Scheme 3.19 Bis-thiourea-catalysed Mannich reactions of phosphorus ylides with *N*-Boc imines.

Another cinchona alkaloid-catalysed Mannich reaction was reported by Barbas *et al.*, occurring between a thioester and an α-amido sulfone, leading in the presence of KOH to the *anti*-Mannich product in 79% yield, with both moderate diastereo- and enantioselectivity of 64% de and 45% ee, respectively.[29]

Finally, Akiyama *et al.* have developed a new method for the enantioselective synthesis of γ-butenolide derivatives, which involved the vinylogous Mannich-type reaction catalysed by a novel chiral phosphoric acid bearing iodine groups at the 6,6′-positions.[30] Aliphatic as well as aromatic aldimines

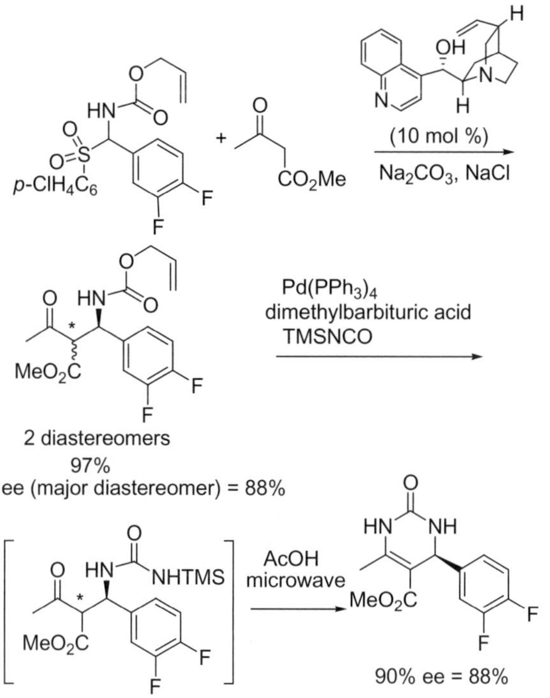

Scheme 3.20 Cinchonine-catalysed Mannich reaction of β-keto ester with α-amido sulfone.

turned out to be good substrates and the corresponding γ-butenolide derivatives were obtained in high yields, with good to high diastereo- and enantio-selectivities, as shown in Scheme 3.21.

3.2 Aza-Henry Reactions

Nitroalkyl anions can serve as the nucleophile component in addition to imino compounds, and these nitro Mannich or aza-Henry reactions afford adducts with an orthogonal nitrogen functionality on neighbouring carbons having some interesting synthetic utility.[31] Although the asymmetric version of this reaction has provoked much interest in the last years, there are only a few organocatalytic asymmetric aza-Henry reactions reported which can afford both an excellent enantioselectivity and a high *anti*-selectivity of greater than 82% de for a broad scope of the reaction partners. In this context, several chiral bifunctional thiourea catalysts have been studied as potential organocatalysts. As an example, Chen *et al.* have designed novel readily accessible thiourea-secondary amines, which were successfully applied as organocatalysts for the highly stereoselective aza-Henry reaction of simple esters of α-substituted nitroacids with *N*-Boc imines.[32] As shown in Scheme 3.22, the aza-Henry

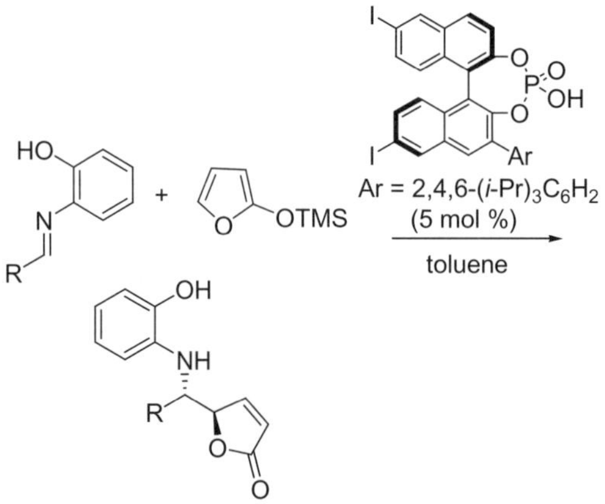

R = *p*-FC$_6$H$_4$: 100% de = 90% ee = 87%
R = Ph: 100% de = 82% ee = 82%
R = *p*-NO$_2$C$_6$H$_4$: 85% de = 94% ee = 96%
R = *m*-NO$_2$C$_6$H$_4$: 86% de = 36% ee = 96%
R = *o*-NO$_2$C$_6$H$_4$: 100% de = 96% ee = 92%
R = *p*-CF$_3$C$_6$H$_4$: 95% de = 38% ee = 99%
R = 4-Py: 30% de = 88% ee = 98%
R = 2-Fu: 77% de = 36% ee = 89%
R = Cy: 77% de = 76% ee = 90%
R = *i*-Pr: 84% de = 76% ee = 92%

Scheme 3.21   Phosphoric acid-catalysed vinylogous Mannich reactions.

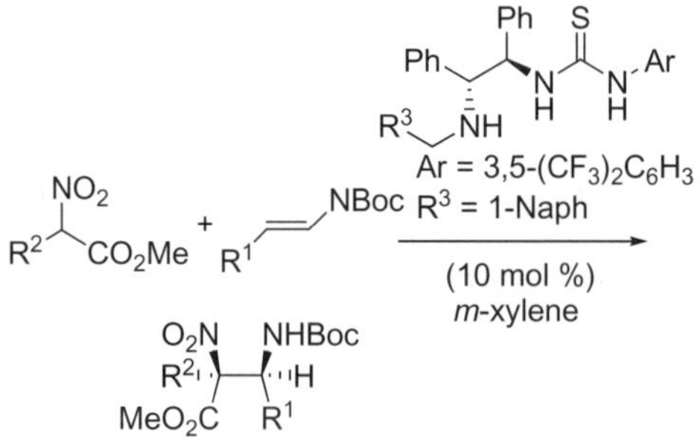

R^1 = Ph, R^2 = Me: 86% de = 90% ee = 96%
R^1 = *p*-FC$_6$H$_4$, R^2 = Me: 78% de = 82% ee = 96%
R^1 = *m*-ClC$_6$H$_4$, R^2 = Me: 83% de = 82% ee = 95%
R^1 = *o*-ClC$_6$H$_4$, R^2 = Me: 79% de = 68% ee = 91%
R^1 = *p*-Tol, R^2 = Me: 86% de = 86% ee = 96%
R^1 = *m*-Tol, R^2 = Me: 85% de = 88% ee = 94%
R^1 = 2-Fu, R^2 = Me: 85% de = 80% ee = 95%
R^1 = Ph, R^2 = *i*-Pr: 38% de = 68% ee = 96%

Scheme 3.22   Aza-Henry reactions catalysed by thiourea-secondary amine.

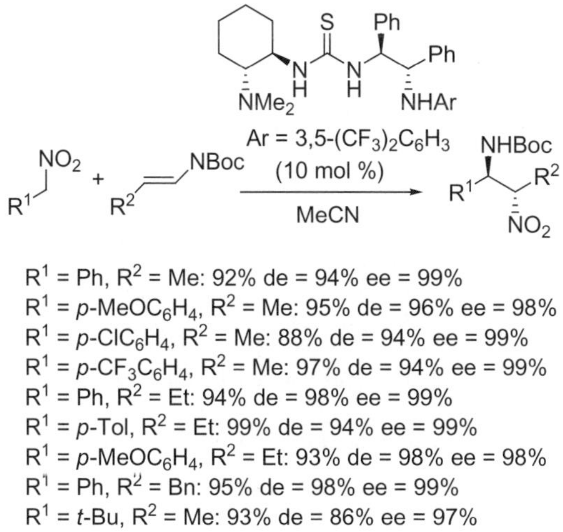
S
Ph
Ph
N
N
H
H
NMe2
NHAr
Ar = 3,5-(CF3)2C6H3
(10 mol %)
MeCN
NO2
R1
R2
NBoc
NHBoc
R2
R1
NO2
R1 = Ph, R2 = Me: 92% de = 94% ee = 99%
R1 = p-MeOC6H4, R2 = Me: 95% de = 96% ee = 98%
R1 = p-ClC6H4, R2 = Me: 88% de = 94% ee = 99%
R1 = p-CF3C6H4, R2 = Me: 97% de = 94% ee = 99%
R1 = Ph, R2 = Et: 94% de = 98% ee = 99%
R1 = p-Tol, R2 = Et: 99% de = 94% ee = 99%
R1 = p-MeOC6H4, R2 = Et: 93% de = 98% ee = 98%
R1 = Ph, R2 = Bn: 95% de = 98% ee = 99%
R1 = t-Bu, R2 = Me: 93% de = 86% ee = 97%

Scheme 3.23 Aza-Henry reactions catalysed by thiourea-secondary amine.

products were obtained in high yields and enantioselectivities of up to 96% ee and good to high diastereoselectivities (58–86% de).

The application of another bifunctional thiourea-secondary amine derived from *trans*-cyclohexane diamine to the asymmetric aza-Henry reaction of *N*-Boc imines with nitroalkanes was reported by Wang *et al.*, providing the corresponding aza-Henry adducts with excellent enantioselectivities (96–99% ee) and high *anti*-selectivities (86–98% de) for a broad scope of substrates (Scheme 3.23).[33]

In 2008, Zhou *et al.* reported the synthesis of a novel bifunctional chiral thiourea organocatalyst bearing a glycosyl scaffold and a tertiary amino group starting from readily available α-D-glucose.[34] This catalyst was proved to be an effective organocatalyst for the asymmetric aza-Henry reaction between *N*-Boc imines and nitroalkanes. Indeed, the corresponding adducts were obtained in good to excellent yields and with excellent enantioselectivities (83–99% ee), as shown in Scheme 3.24.

In addition, Wulff *et al.* have reported the first aza-Henry reaction catalysed by a bis-thiourea catalyst, which was based on the 2,2′-diaminobinaphthalene (BINAM) chiral scaffold.[35] The aza-Henry adducts derived from *N*-Boc imines and nitromethane were isolated in moderate to good yields and good to high enantioselectivities of up to 91% ee (Scheme 3.25).

On the other hand, Palomo *et al.* have developed efficient organocatalytic asymmetric aza-Henry reactions under phase-transfer conditions.[36] This method was based on the reaction of a nitroalkane with an azomethine generated from an α-amido sulfone promoted by $CsOH.H_2O$ as a base in toluene and in the presence of cinchonine-derived ammonium catalysts. The corresponding *syn*-products were obtained in good yields, moderate to good diastereoselectivities (10–86% de) and moderate to excellent enantioselectivities

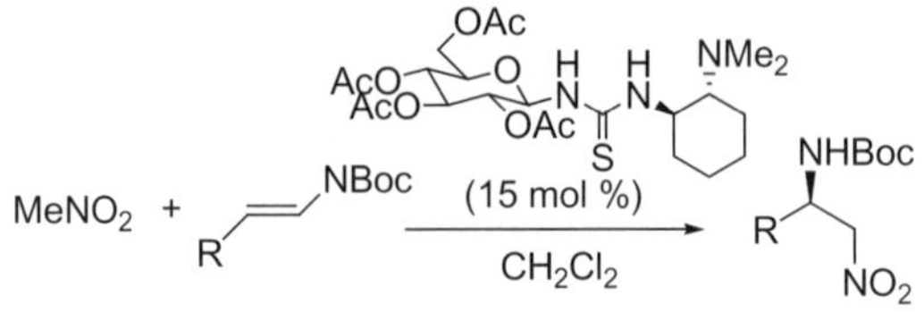

R = Ph: 86% ee > 99%
R = *p*-MeOC₆H₄: 94% ee = 94%
R = *p*-ClC₆H₄: 93% ee > 99%
R = *m*-FC₆H₄: 87% ee > 99%
R = *p*-FC₆H₄: 91% ee > 99%
R = 1-Naph: 95% ee > 99%
R = *o*-CF₃C₆H₄: 84% ee = 96%

Scheme 3.24   Aza-Henry reactions catalysed by D-glucose-derived thiourea-tertiary amine.

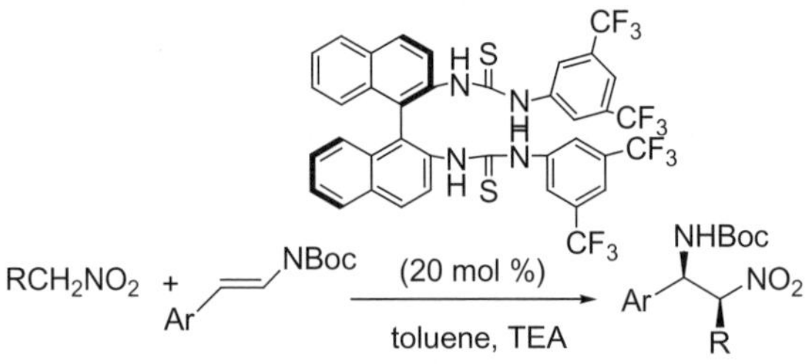

Ar = Ph, R = Me: 55% ee = 86%
Ar = *p*-ClC₆H₄, R = Me: 62% ee = 85%
Ar = *m*-ClC₆H₄, R = Me: 53% ee = 91%
Ar = *p*-MeOC₆H₄, R = Me: 50% ee = 89%
Ar = *p*-Tol, R = Me: 48% ee = 86%
Ar = 1-Naph, R = Me: 65% ee = 85%
Ar = Ph, R = Et: 63% ee = 80%

Scheme 3.25   Aza-Henry reactions catalysed by bis-thiourea.

(30–99% ee), as shown in Scheme 3.26. This methodology presented as interesting features its validity for both non-enolisable and enolisable aldehyde-derived azomethines and the tolerance of nitroalkanes other than nitromethane.

Interestingly, an unsymmetrical bifunctional protonated catalyst was developed by Johnston and Singh to promote the asymmetric aza-Henry reaction of α-alkyl α-nitroesters with *N*-Boc imines, providing the corresponding chiral α-substituted *syn*-α,β-diamino acid derivatives of phenyl alanine in high yields, excellent enantioselectivities of up to 99% ee and moderate to high diastereoselectivities of up to 90% de, as shown in Scheme 3.27.[37]

Finally, a BINOL-derived phosphoric acid was demonstrated by Rueping and Antonchick to be an efficient organocatalyst to promote the asymmetric

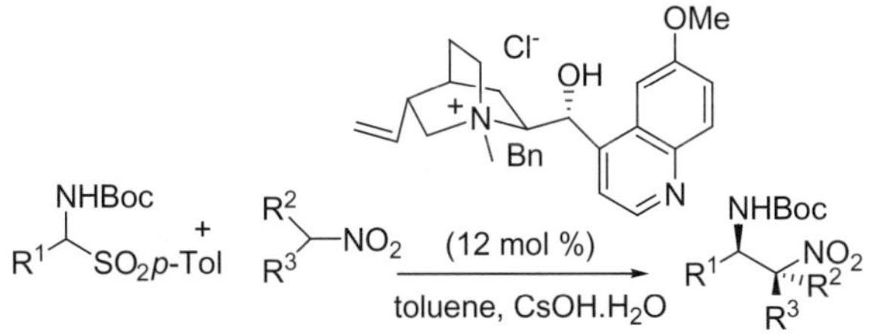

R^1 = Ph, R^2 = H, R^3 = Et: 88% de = 82% ee = 94%
R^1 = *p*-MeOC$_6$H$_4$, R^2 = H, R^3 = CH$_2$=CH(CH$_2$)$_2$: 80%
de = 52% ee = 95%
R^1 = BnCH$_2$, R^2 = H, R^3 = CH(EtO)$_2$: 66% de = 34% ee = 96%
R^1 = *i*-Bu, R^2 = H, R^3 = Et: 91% de = 86% ee = 97%
R^1 = *i*-Bu, R^2 = H, R^3 = CH(EtO)$_2$: 70% de = 80% ee = 99%
R^1 = *i*-Bu, R^2 = H, R^3 = CH$_2$CO$_2$Et: 70% de = 50% ee = 97%

Scheme 3.26 Aza-Henry reactions catalysed by cinchonine-derived ammonium catalyst.

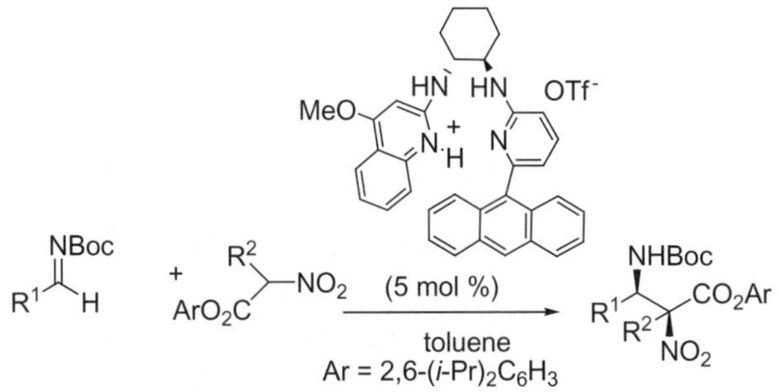

R^1 = *p*-ClC$_6$H$_4$, R^2 = Et: 83% de > 90% ee = 98%
R^1 = *p*-MeSC$_6$H$_4$, R^2 = Et: 81% de = 86% ee = 98%
R^1 = *p*-Tol, R^2 = Et: 80% de > 90% ee = 96%
R^1 = *p*-ClC$_6$H$_4$, R^2 = *t*-Bu: 88% de = 88% ee = 97%

Scheme 3.27 Aza-Henry reactions catalysed by bifunctional protonated catalyst.

aza-Henry reaction of α-imino esters with various nitroalkanes, providing the corresponding chiral *anti*-β-nitro-α-amino acid esters.[38] Under strong base-free conditions, these products were produced in good yields and high enantio-selectivities of up to 92% ee, albeit with moderate to good diastereoselectivities (34–86% de) as shown in Scheme 3.28.

3.3 Aza-Morita–Baylis–Hillman Reactions

The aza-Morita–Baylis–Hillman reaction is a C–C bond-forming reaction of activated alkenes with imines. An asymmetric version of this reaction was

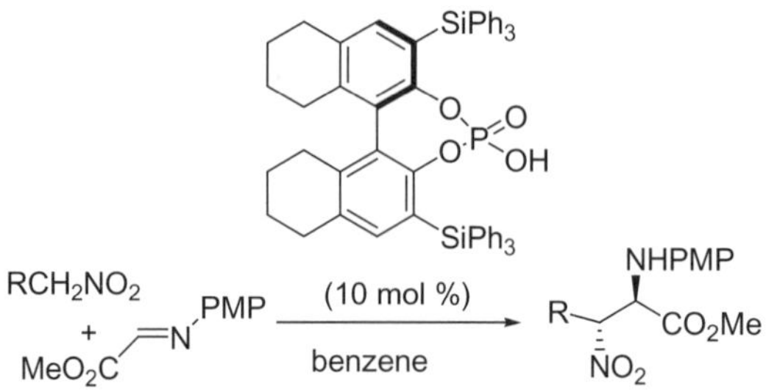

R = Et: 64% de = 80% ee = 92%
R = Me: 61% de = 82% ee = 92%
R = *n*-Pent: 65% de = 80% ee = 92%
R = Bn: 93% de = 86% ee = 88%
R = *p*-BrC$_6$H$_4$: 84% de = 82% ee = 86%
R = *p*-MeOC$_6$H$_4$: 73% de = 86% ee = 88%
R = CH$_2$-2-thienyl: 69% de = 86% ee = 84%
R = (CH$_2$)$_2$Bn: 57% de = 74% ee = 92%

Scheme 3.28 Aza-Henry reactions catalysed by BINOL-derived phosphoric acid.

developed by Masson *et al.* by using a bifunctional β-isocupreidine derivative as a chiral organocatalyst.[39] Therefore, the application of this cinchona alkaloid to the reaction of aromatic imines with β-naphthyl acrylate in the presence of β-naphthol as an additive led to the corresponding aza-Morita–Baylis–Hillman products with uniformly high yields and excellent enantioselectivities of up to 97% ee in the case of aromatic imines. In addition, the aliphatic *N*-sulfinyl imines have been successfully employed in this process for the first time, leading to the corresponding adducts in over 85% ee (Scheme 3.29).

On the other hand, an axially chiral phosphine alcohol organocatalyst was successfully employed by Shi and Liu to promote the aza-Morita–Baylis–Hillman reaction between *N*-(arylmethylidene)arylsulfonamide and methyl vinyl ketone, providing the corresponding adducts in moderate to good yields (26–85%) and high enantioselectivities of up to 94% ee, when used at 10 mol% of catalyst loading in THF.[40] Furthermore, these authors have shown that the corresponding polyether dendritic-supported chiral phosphine was even more efficient than the non-supported one for the aza-Morita–Baylis–Hillman reaction of *N*-sulfonated imines with methyl vinyl ketone, ethyl vinyl ketone, as well as acrolein. Indeed, the products of these reactions catalysed by the dendritic phosphine were achieved with excellent yields and enantioselectivities of up to 97% ee, as shown in Scheme 3.30.

In the same area, these authors have also investigated the efficiency of various chiral sterically congested phosphane-amide bifunctional organocatalysts with a binaphthyl scaffold in asymmetric aza-Morita–Baylis–Hillman reactions of *N*-sulfonated imines with activated olefins such as methyl and ethyl vinyl ketones.[41] The corresponding aza-Morita–Baylis–Hillman adducts could be obtained in moderate to excellent yields (37–98%) and moderate to good

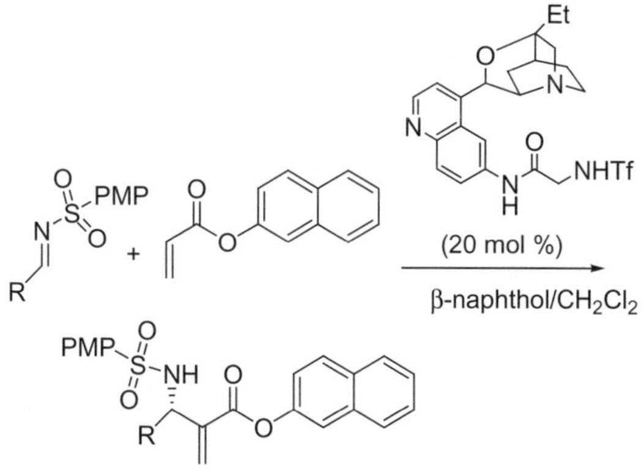

R = *p*-ClC_6H_4: 81% ee = 94%
R = *p*-$NO_2C_6H_4$: 67% ee = 97%
R = *p*-$MeOC_6H_4$: 84% ee = 93%
R = *m*-BrC_6H_4: 84% ee = 92%
R = *m*-Tol: 95% ee = 92%
R = 3-Fu: 80% ee = 98%
R = 2-Naph: 81% ee = 92%
R = *i*-Bu: 57% ee = 87%
R = Cy-CH_2: 45% ee = 84%

Scheme 3.29 Aza-Morita–Baylis–Hillman reactions catalysed by β-isocupreidine derivative.

enantioselectivities (35–93% ee). Comparable results were observed for the same reactions by using a simpler, closely related chiral phosphine amide as the organocatalyst.[42] Indeed, in these conditions, the aza-Morita–Baylis–Hillman adducts were isolated in uniformly good to excellent yields (80–99%) and good to excellent enantioselectivities (51–95% ee), as shown in Scheme 3.31.

In addition, Liu and Garnier have designed novel chiral trifunctional organocatalysts which were successfully applied as promoters to the aza-Morita–Baylis–Hillman reaction of methyl vinyl ketone with various aromatic *N*-tosyl imines in the presence of benzoic acid.[43] In these conditions, the reaction yielded the products in moderate to high yields (31–95%) and general high enantioselectivities (88–94% ee), as shown in Scheme 3.32.

3.4 Strecker Reactions

The Strecker reaction starting from an aldehyde, ammonia and a cyanide source is an efficient method for the preparation of α-amino acids and their derivatives. A popular version for asymmetric purposes is based on the use of preformed imines and a subsequent nucleophilic addition of HCN or TMSCN in the presence of a chiral catalyst.[44] An intense investigation of the asymmetric Strecker-type reaction has continued over many years, due to the importance of α-amino acid building blocks in medicinal chemistry.[45] Interestingly, a number

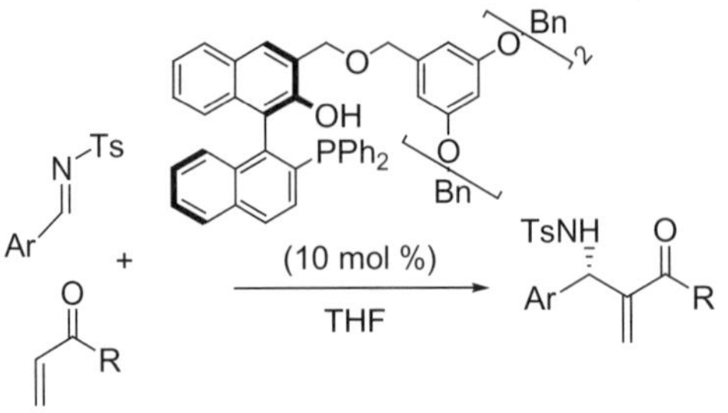

Ar = *p*-ClC$_6$H$_4$, R = Et: 77% ee = 97%
Ar = *p*-BrC$_6$H$_4$, R = Et: 73% ee = 97%
Ar = *p*-NO$_2$C$_6$H$_4$, R = Et: 97% ee = 95%
Ar = *m*-ClC$_6$H$_4$, R = Et: 99% ee = 94%
Ar = *m*-FC$_6$H$_4$, R = Et: 85% ee = 97%
Ar = *p*-FC$_6$H$_4$, R = Et: 90% ee = 95%
Ar = Ph, R = Me: 95% ee = 91%
Ar = *p*-MeOC$_6$H$_4$, R = Me: 90% ee = 95%
Ar = *p*-ClC$_6$H$_4$, R = Me: 99% ee = 97%
Ar = *m*-FC$_6$H$_4$, R = Me: 80% ee = 94%
Ar = *p*-NO$_2$C$_6$H$_4$, R = Me: 99% ee = 97%
Ar = *m*-NO$_2$C$_6$H$_4$, R = Me: 99% ee = 92%
Ar = *p*-ClC$_6$H$_4$, R = H: 86% ee = 89%
Ar = *p*-FC$_6$H$_4$, R = H: 83% ee = 94%
Ar = *p*-BrC$_6$H$_4$, R = H: 90% ee = 90%

Scheme 3.30   Aza-Morita–Baylis–Hillman reactions catalysed by polyether dendritic phosphine.

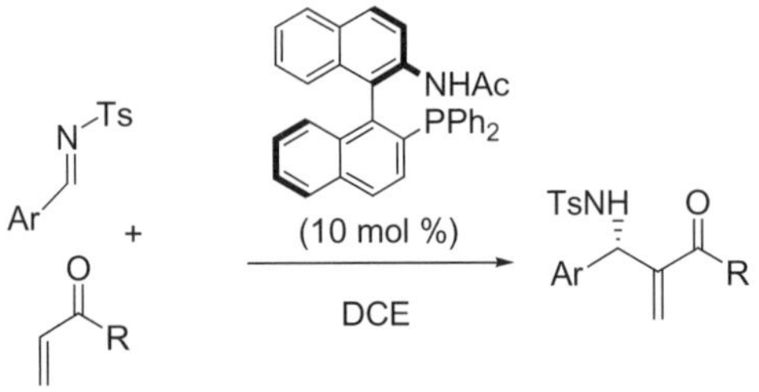

Ar = *p*-BrC$_6$H$_4$, R = Me: 86% ee = 90%
Ar = *p*-NO$_2$C$_6$H$_4$, R = Me: 99% ee = 90%
Ar = *o*-NO$_2$C$_6$H$_4$, R = Me: 85% ee = 61%
Ar = *m*-NO$_2$C$_6$H$_4$, R = Me: 99% ee = 82%
Ar = *p*-ClC$_6$H$_4$, R = Me: 99% ee = 90%
Ar = *m*-ClC$_6$H$_4$, R = Me: 90% ee = 89%
Ar = *p*-FC$_6$H$_4$, R = Me: 90% ee = 91%
Ar = *m*-FC$_6$H$_4$, R = Me: 91% ee = 80%
Ar = *p*-MeOC$_6$H$_4$, R = Me: 98% ee = 82%
Ar = *p*-Tol, R = Me: 99% ee = 90%
Ar = Ph, R = Et: 95% ee = 74%
Ar = *p*-BrC$_6$H$_4$, R = Et: 80% ee = 65%

Scheme 3.31   Aza-Morita–Baylis–Hillman reactions catalysed by phosphine amide.

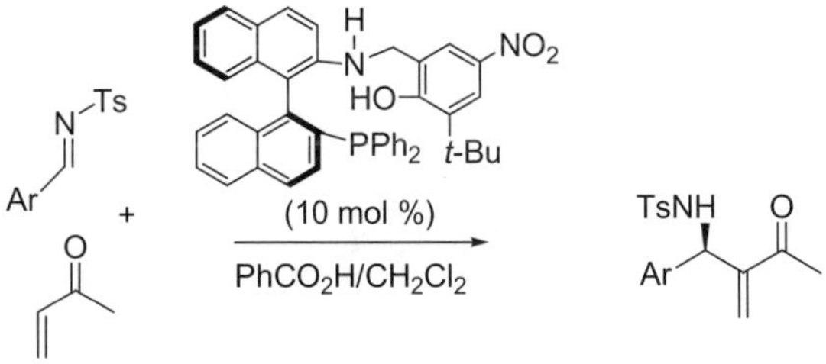

Ar = *m*-$NO_2C_6H_4$: 89% ee = 87%
Ar = *p*-BrC_6H_4: 41% ee = 92%
Ar = *p*-ClC_6H_4: 71% ee = 94%
Ar = *o*-ClC_6H_4: 39% ee = 89%
Ar = *p*-FC_6H_4: 87% ee = 90%
Ar = *p*-$NO_2C_6H_4$: > 95% ee = 87%
Ar = *m*-$MeOC_6H_4$: 31% ee = 94%

Scheme 3.32 Aza-Morita–Baylis–Hillman reactions catalysed by trifunctional catalyst.

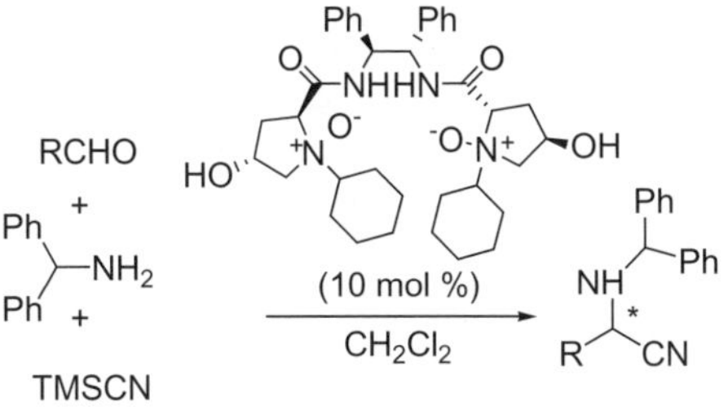

R = Ph: 93% ee = 90%
R = *p*-ClC_6H_4: 95% ee = 87%
R = *p*-BrC_6H_4: 94% ee = 87%
R = 2,3-$Cl_2C_6H_3$: 82% ee = 89%
R = 2-Naph: 96% ee = 91%
R = *n*-Bu: 95% ee = 90%
R = *i*-Pr: 98% ee = 87%
R = *i*-Pent: 86% ee = 95%

Scheme 3.33 Strecker reactions catalysed by L-proline derived *N,N′*-dioxide.

of completely different types of chiral organocatalysts have been found to have catalytic hydrocyanation properties.[46] As an example, an asymmetric, three-component Strecker reaction was developed by Feng *et al.* by using novel *trans*-4-hydroxy-L-proline-derived *N,N′*-dioxides as efficient chiral organocatalysts.[47] Therefore, these organocatalysts were shown to be highly efficient to promote the asymmetric Strecker reaction of both aromatic and aliphatic aldehydes with (1,1-diphenyl)methylamine and TMSCN, providing the corresponding α-amino nitriles in uniformly high yields and enantioselectivities of up to 95% ee (Scheme 3.33).

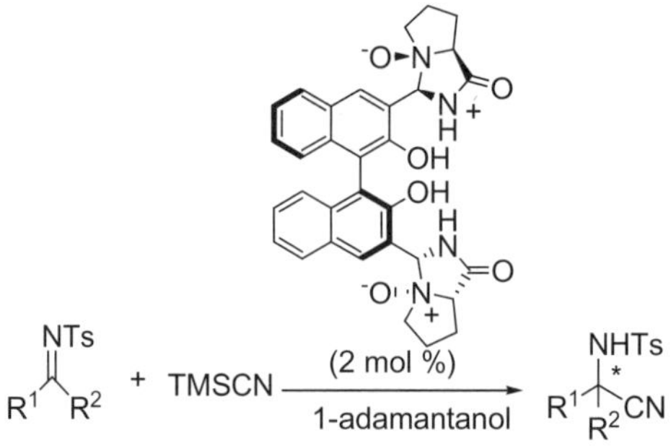

$R^1 = Ph, R^2 = Me$: 93% ee = 97%
$R^1 = o\text{-}FC_6H_4, R^2 = Me$: 95% ee = 94%
$R^1 = m\text{-}ClC_6H_4, R^2 = Me$: 93% ee = 94%
$R^1 = p\text{-}ClC_6H_4, R^2 = Me$: 91% ee = 93%
$R^1 = p\text{-}Tol, R^2 = Me$: 88% ee = 98%
$R^1 = o\text{-}MeOC_6H_4, R^2 = Me$: 82% ee = 97%
$R^1 = m\text{-}MeOC_6H_4, R^2 = Me$: 81% ee = 96%
$R^1 = p\text{-}MeOC_6H_4, R^2 = Me$: 80% ee = 99%
$R^1 = 2\text{-}Naph, R^2 = Me$: 82% ee = 97%
$R^1 = 2\text{-}Fu, R^2 = Me$: 88% ee = 95%
$R^1 = t\text{-}Bu, R^2 = Me$: 96% ee = 93%
$R^1 = Ph, R^2 = Et$: 86% ee = 98%
$R^1 = Ph, R^2 = n\text{-}Pr$: 74% ee = 95%
$R^1 = o\text{-}ClC_6H_4, R^2 = Ph$: 76% ee = 90%

Scheme 3.34   Strecker reactions catalysed by N,N'-dioxide derived from (*S*)-BINOL and L-prolinamide.

However, in contrast to the relatively well-developed cyanation of aldimines, limited reports were related to the cyanation of ketoimines. In this context, Feng *et al.* have designed a novel N,N'-dioxide catalyst derived from BINOL and prolinamide in order to be applied as organocatalyst for the asymmetric Strecker reaction of ketoimines with fairly wide substrate scope and excellent enantioselectivities of up to 99% ee (Scheme 3.34).[48] A low catalyst loading of 2 mol % combined with mild reaction conditions and an operational simplicity made this strategy facile to be used for the synthesis of pharmaceutically important chiral disubstituted α-amino nitriles.

In addition, the thiourea catalysis concept[49] has been applied by List and Pan to the Strecker reaction of imines as well as to the three-component Strecker reaction of aldehydes.[50] Therefore, a chiral thiourea catalyst was used by these authors to develop the first organocatalytic asymmetric three-component Strecker reaction with excellent enantioselectivities of up to 94% ee combined with high yields. The best results concerning this methodology and involving acetyl cyanide as the cyanation agent are collected in Scheme 3.35. Moreover, the same catalyst allowed the asymmetric Strecker of aliphatic as well as aromatic aldimines with acetyl cyanide to be achieved with comparable excellent yields and enantioselectivities of up to 98% ee, as shown in Scheme 3.35.

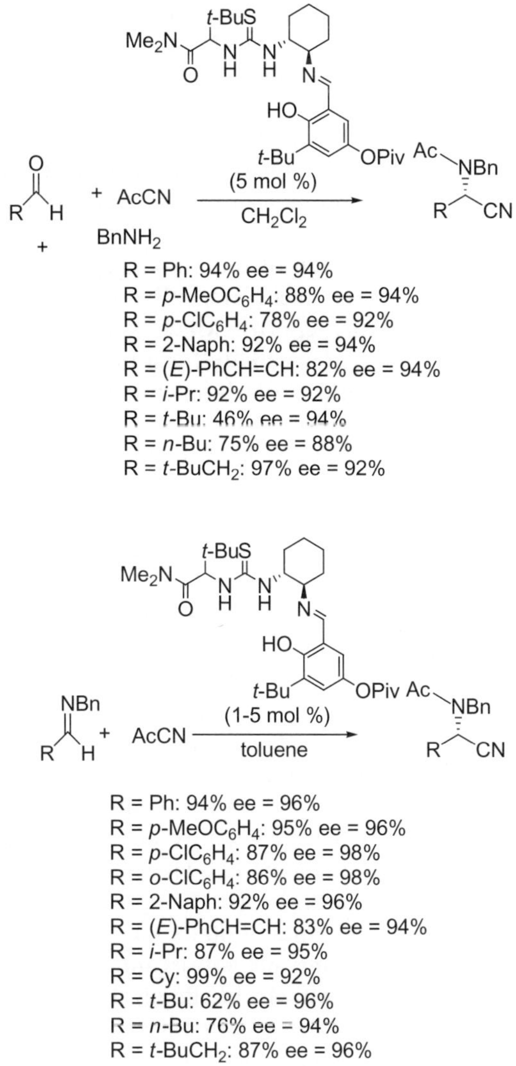
t-BuS
Me2N
N
H
N
H
N
O
HO
t-Bu
OPiv
(5 mol %)
CH2Cl2
O
R H
+ AcCN
+ BnNH2
Ac
NBn
R CN
R = Ph: 94% ee = 94%
R = p-MeOC6H4: 88% ee = 94%
R = p-ClC6H4: 78% ee = 92%
R = 2-Naph: 92% ee = 94%
R = (E)-PhCH=CH: 82% ee = 94%
R = i-Pr: 92% ee = 92%
R = t-Bu: 46% ee = 94%
R = n-Bu: 75% ee = 88%
R = t-BuCH2: 97% ee = 92%
(1-5 mol %)
toluene
NBn
R H
+ AcCN
R = Ph: 94% ee = 96%
R = p-MeOC6H4: 95% ee = 96%
R = p-ClC6H4: 87% ee = 98%
R = o-ClC6H4: 86% ee = 98%
R = 2-Naph: 92% ee = 96%
R = (E)-PhCH=CH: 83% ee = 94%
R = i-Pr: 87% ee = 95%
R = Cy: 99% ee = 92%
R = t-Bu: 62% ee = 96%
R = n-Bu: 76% ee = 94%
R = t-BuCH2: 87% ee = 96%

Scheme 3.35 Thiourea-catalysed Strecker reactions.

3.5 Conclusions

Among a wide variety of chiral organocatalysts that have been used in the asymmetric Mannich reaction, one of the most widely used remains proline itself, which generally provided excellent enantioselectivities for the Mannich products arisen from either three-component, one-pot reactions or reactions of preformed imines with aldol donors. While these reactions were mostly performed at a catalyst loading of 10 mol %, Mannich reactions of enolisable aldehydes and ketones with imines catalysed by (*R*)-3-pyrrolidinecarboxylic

acid gave comparable results at a lower catalyst loading of 5 mol%. This catalyst loading was also applied to vinylogous Mannich-type reactions of aliphatic
as well as aromatic aldimines catalysed by a novel chiral phosphoric acid
bearing iodine groups at the 6,6′-positions, allowing various γ-butenolide
derivatives to be synthesised in both high yields and enantioselectivities. In
addition, various chiral amine-thioureas have been successfully applied as
organocatalysts to efficiently promote various other asymmetric Mannich
reactions.

Although the asymmetric version of the aza-Henry reaction has provoked
much interest in the last years, there are still only a few organocatalytic
asymmetric aza-Henry reactions reported which can afford both an excellent
enantioselectivity and a high *anti*-selectivity for a broad scope of the reaction
partners. In the last year, several chiral bifunctional thiourea catalysts were
studied as potential organocatalysts. Among them, a bifunctional thiourea-
secondary amine derived from *trans*-cyclohexane diamine, used at 10 mol% of
catalyst loading, was shown to efficiently promote the asymmetric aza-Henry
reaction of *N*-Boc imines with nitroalkanes, providing the corresponding aza-
Henry adducts with excellent enantioselectivities (96–99% ee) and high *anti*-
selectivities (86–98% de) for a broad scope of substrates. A lower catalyst
loading (5 mol %) of an unsymmetrical bifunctional protonated catalyst
derived from *trans*-cyclohexane diamine was demonstrated to promote the
asymmetric aza-Henry reaction of α-alkyl α-nitroesters with *N*-Boc imines,
providing the corresponding chiral α-substituted *syn*-α,β-diamino acid derivatives of phenyl alanine in high yields, excellent enantioselectivities of up to 99%
ee and moderate to high diastereoselectivities.

Several interesting results dealing with the asymmetric organocatalytic aza-
Morita–Baylis–Hillman reaction have recently been reported, such as those
concerning the reactions of aromatic imines with β-naphthyl acrylate, leading
to the corresponding aza-Morita–Baylis–Hillman products with uniformly high
yields and excellent enantioselectivities of up to 97% ee in the case of aromatic
imines. An axially chiral phosphine alcohol organocatalyst was also successfully employed to promote the aza-Morita–Baylis–Hillman reaction between
N-(arylmethylidene)arylsulfonamides and methyl vinyl ketone, providing the
corresponding adducts in moderate to good yields and high enantioselectivities
of up to 94% ee, when used at 10 mol % of catalyst loading. More interestingly,
the authors have shown that the corresponding polyether dendritic-supported
chiral phosphine was even more efficient than the non-supported one for the
aza-Morita–Baylis–Hillman reaction of *N*-sulfonated imines with methyl vinyl
ketone and ethyl vinyl ketone, as well as acrolein, since the products of these
reactions catalysed by the dendritic phosphine were achieved with excellent
yields and enantioselectivities of up to 97% ee at the same catalyst loading.

In the area of the asymmetric organocatalytic Strecker reaction, a novel
N,N′-dioxide catalyst derived from **BINOL** and prolinamide was successfully
applied at only 2 mol% of catalyst loading as an organocatalyst to the Strecker
reaction of ketoimines with a fairly wide substrate scope, providing excellent
enantioselectivities of up to 99% ee. In addition, a chiral thiourea catalyst was

used to develop the first organocatalytic asymmetric three-component Strecker reaction with excellent enantioselectivities of up to 94% ee combined with high yields. This catalyst was also proved to achieve the asymmetric Strecker reaction of aliphatic as well as aromatic aldimines with acetyl cyanide with excellent yields and enantioselectivities of up to 98% ee at a low catalyst loading of 1 mol %.

References

1. G. K. Friestad and A. K. Mathies, *Tetrahedron*, 2007, **63**, 2541–2569.
2. (a) A. Cordova, *Acc. Chem. Res.*, 2004, **37**, 102–112; (b) M. M. B. Marques, *Ang. Chem., Int. Ed. Engl.*, 2006, **45**, 348–352.
3. J. M. M. Verkade, L. J. C. van Hemert, P. J. I. M. Quaedflieg and F. P. J. T. Rutjes, *Chem. Soc. Rev.*, 2008, **37**, 29–41.
4. J. W. Yang, C. Chandler, M. Stadler, D. Kampen and B. List, *Nature*, 2008, **452**, 453–455.
5. P. Dziedzic, J. Vesely and A. Cordova, *Tetrahedron Lett.*, 2008, **49**, 6631–6634.
6. P. Dziedzic, P. Schyman, M. Kullberg and A. Cordova, *Chem. Eur. J.*, 2009, **15**, 4044–4048.
7. B. T. Hahn, R. Fröhlich, K. Harms and F. Glorius, *Ang. Chem., Int. Ed. Engl.*, 2008, **47**, 9985–9988.
8. V. A. Sukach, N. M. Golovach, V. V. Pirozhenko, E. B. Rusanov and M. V. Vovk, *Tetrahedron: Asymmetry*, 2008, **19**, 761–764.
9. J. M. M. Verkade, F. van der Pijl, M. M. J. H. P. Willems, P. J. M. L. Quaedflieg, F. L. van Delft and F. P. J. T. Rutjes, *J. Org. Chem.*, 2009, **74**, 3207–3210.
10. R.-G. Han, Y. Wang, Y.-Y. Li and P.-F. Xu, *Adv. Synth. Catal.*, 2008, **350**, 1474–1478.
11. Q. Gu, J.-J. Gong, J. Feng, X.-Y. Wu and Q.-L. Zhou, *Chin. J. Chem.*, 2008, **26**, 1902–1906.
12. Q. Gu, L.-X. Jiang, K. Yuan, L. Zhang and X.-Y. Wu, *Synth. Commun.*, 2008, **38**, 4198–4206.
13. H. Zhang, S. Mitsumori, N. Utsumi, M. Imai, N. Garcia-Delgado, M. Mifsud, K. Albertshofer, P. H.-Y. Cheong, K. N. Houk, F. Tanaka and C. F. Barbas, *J. Am. Chem. Soc.*, 2008, **130**, 875–886.
14. M. Pouliquen, J. Blanchet, M.-C. Lasne and J. Rouden, *Org. Lett.*, 2008, **10**, 1029–1032.
15. Y. Hayashi, T. Urushima, S. Aratake, T. Okano and K. Obi, *Org. Lett.*, 2008, **10**, 21–24.
16. S.-G. Kim and T.-H. Park, *Tetrahedron: Asymmetry*, 2008, **19**, 1626–1629.
17. Y. Hayashi, T. Okano, T. Itoh, T. Urushima, H. Ishikawa and T. Uchimaru, *Ang. Chem., Int. Ed. Engl.*, 2008, **47**, 9053–9058.
18. C. Gianelli, L. Sambri, A. Carlone, G. Bartoli and P. Melchiorre, *Ang. Chem., Int. Ed. Engl.*, 2008, **47**, 8700–8702.

19. T. Kano and K. Maruoka, *Chem. Commun.*, 2008, 5465–5473.
20. T. Kano, Y. Yamaguchi and K. Maruoka, *Ang. Chem., Int. Ed. Engl.*, 2009, **48**, 1838–1840.
21. T. Kano, Y. Hato, A. Yamamoto and K. Maruoka, *Tetrahedron*, 2008, **64**, 1197–1203.
22. H. Zhang, S. S. V. Ramasastry, F. Tanaka and C. F. Barbas, *Adv. Synth. Catal.*, 2008, **350**, 791–796.
23. Y.-C. Teo, J.-J. Lau and M.-C. Wu, *Tetrahedron: Asymmetry*, 2008, **19**, 186–190.
24. H. Miyabe and Y. Takemoto, *Bull. Chem. Soc. Jpn.*, 2008, **81**, 785–795.
25. D. A. Yalalov, S. B. Tsogoeva, T. E. Shubina, I. M. Martynova and T. Clark, *Ang. Chem., Int. Ed. Engl.*, 2008, **47**, 6624–6628.
26. X. Tian, K. Jiang, J. Peng, W. Du and Y.-C. Chen, *Org. Lett.*, 2008, **10**, 3583–3586.
27. Y. Zhang, Y.-K. Liu, T.-R. Kang, Z.-K. Hu and Y.-C. Chen, *J. Am. Chem. Soc.*, 2008, **130**, 2456–2457.
28. J. M. Goss and S. E. Schauss, *J. Org. Chem.*, 2008, **73**, 7651–7656.
29. N. Utsumi, S. Kitagaki and C. F. Barbas, *Org. Lett.*, 2008, **10**, 3405–3408.
30. T. Akiyama, Y. Honma, J. Itoh and K. Fuchibe, *Adv. Synth. Catal.*, 2008, **350**, 399–402.
31. B. Westermann, *Ang. Chem., Int. Ed. Engl.*, 2003, **42**, 151–153.
32. B. Han, Q.-P. Liu, R. Li, X. Tian, X.-F. Xiong, J.-G. Deng and Y.-C. Chen, *Chem. Eur. J.*, 2008, **14**, 8094–8097.
33. C.-J. Wang, X.-Q. Dong, Z.-H. Zhang, Z.-Y. Xue and H.-L. Teng, *J. Am. Chem. Soc.*, 2008, **130**, 8606–8607.
34. C. Wang, Z. Zhou and C. Tang, *Org. Lett.*, 2008, **10**, 1707–1710.
35. C. Rampalakos and W. D. Wulff, *Adv. Synth. Catal.*, 2008, **350**, 1785–1790.
36. E. Gomez-Bengoa, A. Linden, R. Lopez, I. Mugica-Mendiola, M. Oiarbide and C. Palomo, *J. Am. Chem. Soc.*, 2008, **130**, 7955–7966.
37. A. Singh and J. N. Johnston, *J. Am. Chem. Soc.*, 2008, **130**, 5866–5867.
38. M. Rueping and A. P. Antonchick, *Org. Lett.*, 2008, **10**, 1731–1734.
39. N. Abermil, G. Masson and J. Zhu, *J. Am. Chem. Soc.*, 2008, **130**, 12596–12597.
40. Y.-h. Liu and M. Shi, *Adv. Synth. Catal.*, 2008, **350**, 122–128.
41. M.-j. Qi, T. Ai, M. Shi and G. Li, *Tetrahedron*, 2008, **64**, 1181–1186.
42. X.-Y. Guan, Y.-Q. Jiang and M. Shi, *Eur. J. Org. Chem.*, 2008, 2150–2155.
43. J.-M. Garnier and F. Liu, *Org. Biomol. Chem.*, 2009, **7**, 1272–1275.
44. (a) H. Gröger, *Chem. Rev.*, 2003, **103**, 2795–2827; (b) C. Spino, *Ang. Chem., Int. Ed. Engl.*, 2004, **43**, 1764–1766; (c) S. J. Connon, *Ang. Chem., Int. Ed. Engl.*, 2008, **47**, 1176–1178; (d) T. Vilaivan, W. Bhanthumnavin and Y. Sritana-Anant, *Curr. Org. Chem.*, 2005, **9**, 1315–1392.
45. (a) H. Groger, *Chem. Rev.*, 2003, **103**, 2795–2827; (b) T. Vilaivan, W. Bhanthumnavin and Y. Sritana-Anant, *Curr. Org. Chem.*, 2005, **9**, 1315–1392; (c) Y. Ohfune and T. Shinada, *Eur. J. Org. Chem.*, 2005, 5127–5143.

46. P. Merino, E. Marques-Lopez, T. Tejero and R. P. Herrera, *Tetrahedron*, 2009, **65**, 1219–1234.
47. Y. Wen, B. Gao, Y. Fu, S. Dong, X. Liu and X. Feng, *Chem. Eur. J.*, 2008, **14**, 6789–6795.
48. Z. Hou, J. Wang, X. Liu and X. Feng, *Chem. Eur. J.*, 2008, **14**, 4484–4486.
49. M. S. Taylor and E. N. Jacobsen, *Ang. Chem., Int. Ed. Engl.*, 2006, **45**, 1520–1543.
50. S. C. Pan and B. List, *Chem. Asian J.*, 2008, **3**, 430–437.

CHAPTER 4

Nucleophilic Additions to Unsaturated Nitrogen

The direct introduction of either a nitrogen or an oxygen atom adjacent to a carbonyl group in a catalytic enantioselective manner using a chiral organo-catalyst has been described only recently. The chiral products of these reactions represent fundamental building blocks for the construction of complex natural products and other important bioactive molecules.[1,2]

4.1 Nucleophilic Additions to N=N Double Bonds

A large variety of natural products and drugs are nitrogen-containing molecules. The enantioselective construction of molecules bearing a carbon–nitrogen bond *via* direct α-amination using readily available starting materials is one of the many challenges for organic chemists, this being one of the simplest and most straightforward strategies to access important chiral molecules, such as α-amino acids, aldehydes and alcohols. In contrast to the large number of addition reactions to C=O and C=N double bonds, only a few examples of nucleophilic addition to N=N double bonds have been investigated. In 2008, Juaristi *et al.* evaluated a series of novel derivatives of (1*S*,4*S*)-2,5-diazabicyclo [2.2.1]heptane as potential organocatalysts for the asymmetric amination of ethyl α-phenyl-α-cyano acetate with di-*tert*-butyl azodicarboxylate.[3] Among them, a bifunctional derivative provided the aminated product in an excellent yield and with a moderate enantioselectivity (40% ee), as shown in Scheme 4.1.

On the other hand, Kim *et al.* have obtained excellent enantioselectivities (97–99% ee) for the aminated products generated by the amination of α-cyanoketones with azodicarboxylates performed in the presence of a chiral thiourea-tertiary amine catalyst.[4] As shown in Scheme 4.2, good to excellent yields were achieved for all the substrates examined in this study at a low catalyst loading of 1 mol %.

RSC Catalysis Series No. 3
Recent Developments in Asymmetric Organocatalysis
By Hélène Pellissier
© Hélène Pellissier 2010
Published by the Royal Society of Chemistry, www.rsc.org

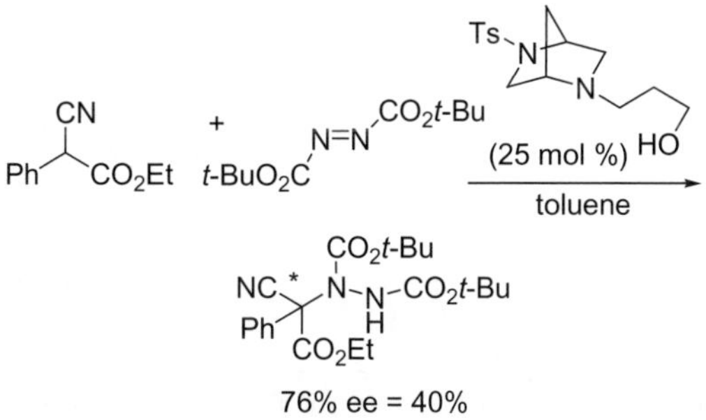

Scheme 4.1 Amination of ethyl α-phenyl-α-cyano acetate.

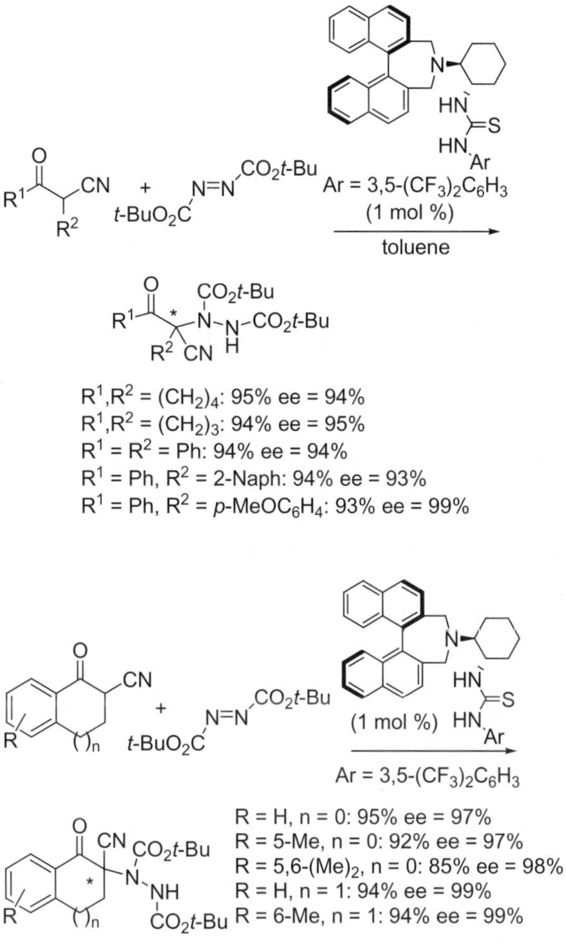

$R^1, R^2 = (CH_2)_4$: 95% ee = 94%
$R^1, R^2 = (CH_2)_3$: 94% ee = 95%
$R^1 = R^2 = Ph$: 94% ee = 94%
$R^1 = Ph, R^2 = 2\text{-Naph}$: 94% ee = 93%
$R^1 = Ph, R^2 = p\text{-MeOC}_6H_4$: 93% ee = 99%

R = H, n = 0: 95% ee = 97%
R = 5-Me, n = 0: 92% ee = 97%
R = 5,6-(Me)$_2$, n = 0: 85% ee = 98%
R = H, n = 1: 94% ee = 99%
R = 6-Me, n = 1: 94% ee = 99%

Scheme 4.2 Aminations of α-cyanoketones.

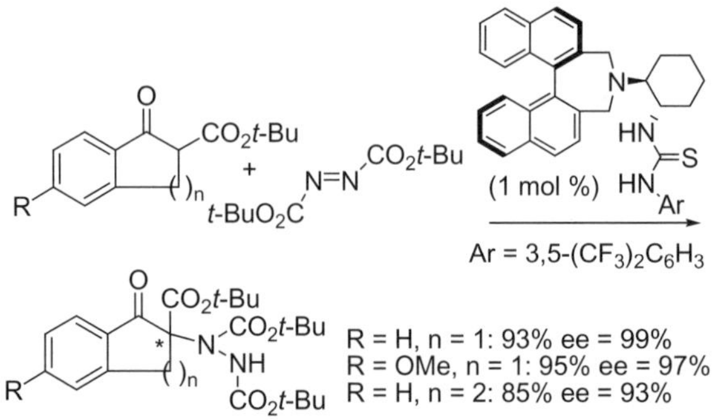

Scheme 4.3   Aminations of β-keto esters.

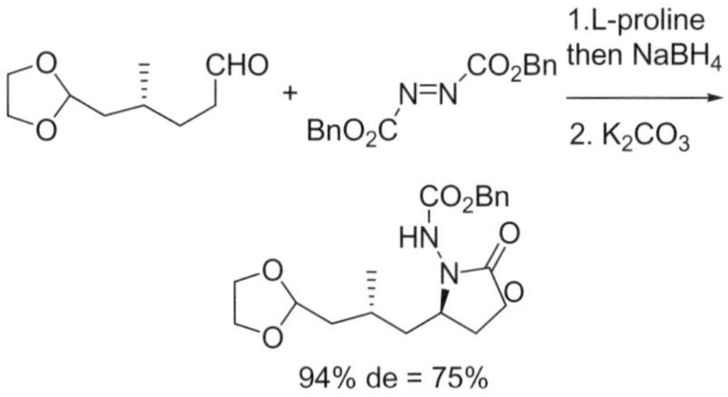

Scheme 4.4   Amination of (+)-citronellal-derived aldehyde.

As an extension of this methodology, these authors could successfully apply the same conditions to the amination of various cyclic β-keto esters.[5] Therefore, the corresponding α-aminated products were obtained in good to high yields and excellent enantioselectivities (93–99% ee), as shown in Scheme 4.3.

Finally, the first total syntheses of two cernuane-type *Lycopodium* alkaloids, (−)-cernuine and (+)-cermizine D, reported by Takayama *et al.*, included as a key step a L-proline-catalysed amination of an aldehyde derived from (+)-citronellal.[6] As shown in Scheme 4.4, this amination performed with dibenzyl azodicarboxylate, followed by an *in-situ* reduction with NaBH₄ and treatment of the resulting mixture with K_2CO_3, furnished the expected oxazolidinone in a high yield and a good diastereoselectivity.

4.2   Nucleophilic Additions to N=O Double Bonds

The α-oxycarbonyl group is a common feature of many natural and biologically active compounds. Furthermore, this functionality is an obvious precursor in the synthesis of other important building blocks such as diols. Among the already existing methods for the asymmetric synthesis of chiral α-hydroxy carbonyl compounds, the direct organocatalysed enantioselective α-aminoxyla-tion of carbonyl compounds is one of the most important strategies for achieving this purpose. Nitroso compounds such as nitrosobenzene are useful electrophiles for performing this type of reaction, although its nitrogen-*versus*-oxygen reactivity should be carefully controlled through the selection of appro-priate catalysts and reaction conditions.[7] The ability of L-proline to control both the O/N-selectivity as well as the enantioselectivity in a variety of solvents and reaction conditions during the catalysis of the direct functionalisation of carbonyl compounds by nitroso compounds has been well documented. In the course of developing a novel route to important chiral tetrahydro-1,2-oxazines, Zhong *et al.* investigated an asymmetric domino α-aminoxylation-aza-Michael reaction catalysed by L-proline.[8] Therefore, the reaction between a nitroalkenal and a nitrosobenzene in the presence of simple L-proline in DMSO led to the expected tetrahydro-1,2-oxazine with moderate to high yields (5–90%) but with a uniformly exceptional level of both diastereo- and enantioselectivity, higher than 98% de and 99% ee, respectively (Scheme 4.5).

A sequence involving an L-proline-catalysed α-aminoxylation of an aldehyde followed by treatment with NaBH$_4$ was used as a key step in a total synthesis of the HRV 3C-protease inhibitor (1*R*,3*S*)-thysanone.[9] This reaction provided the corresponding enantiopure 1,2-diol in a good yield, as shown in Scheme 4.6.

In 2009, Wang *et al.* developed an asymmetric synthesis of 3-hydroxyl-2-alkanones *via* a tandem organocatalytic aminoxylation of aldehydes and a chemoselective diazomethane homologation.[10] An accelerating effect of water was observed for the aminoxylation catalysed by L-proline, while MgCl$_2$ served

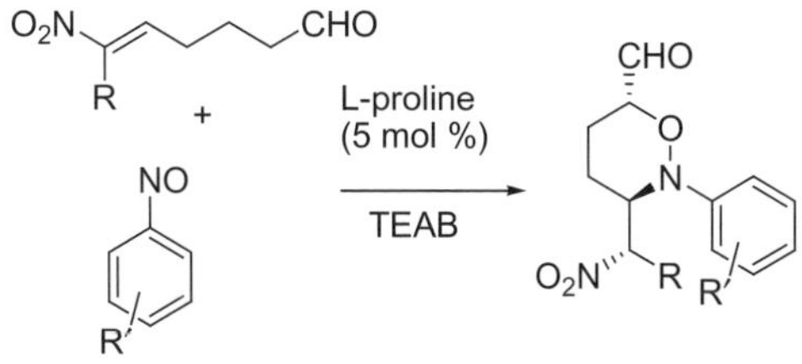

R = R′ = H: 90% de > 98% ee > 99%
R = Me, R′ = H: 75% de > 98% ee > 99%
R = Et, R′ = H: 68% de > 98% ee > 99%
R = Ph, R′ = H: 73% de > 98% ee > 99%
R = Bn, R′ = H: 50% de > 98% ee > 99%
R = *p*-ClC$_6$H$_4$CH$_2$, R′ = H: 78% de > 98% ee > 99%
R = *p*-BrC$_6$H$_4$CH$_2$, R′ = H: 75% de > 98% ee > 99%
R = H, R′ = 2-Me: 79% de > 98% ee = 97%
R = H, R′ = 4-Br: 88% de > 98% ee > 99%
R = H, R′ = 4-Br: 88% de > 98% ee > 99%
R = Me, R′ = 4-Br: 73% de > 98% ee = 99%
R = H, R′ = 4-Me: 90% de > 98% ee > 99%
R = *p*-ClC$_6$H$_4$CH$_2$, R′ = 4-Br: 78% de > 98% ee > 99%

Scheme 4.5   Domino α-aminoxylation-aza-Michael reactions.

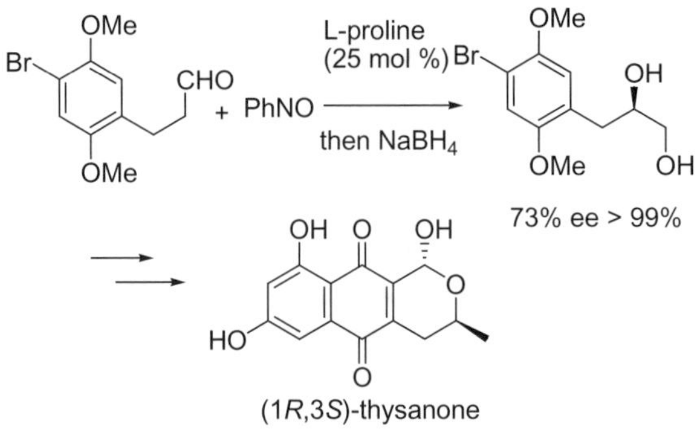
OMe
Br
CHO
+ PhNO
L-proline
(25 mol %)
then NaBH4
OMe
OH
OH
73% ee > 99%
OH O OH
O
HO
O
(1R,3S)-thysanone

Scheme 4.6 Synthesis of (1*R*,3*S*)-thysanone.

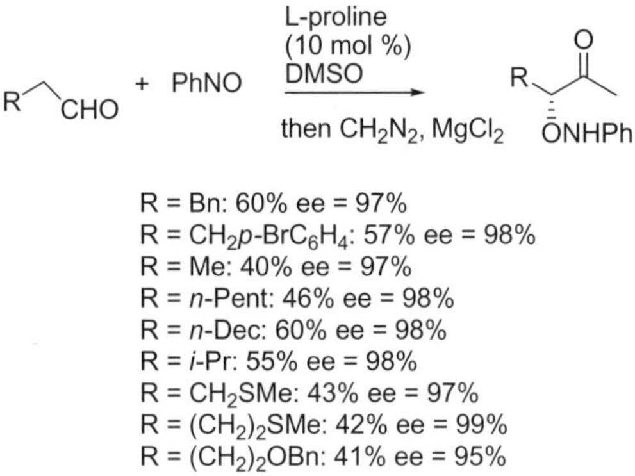
R
CHO
+ PhNO
L-proline
(10 mol %)
DMSO
then CH2N2, MgCl2
O
R
ONHPh
R = Bn: 60% ee = 97%
R = CH2p-BrC6H4: 57% ee = 98%
R = Me: 40% ee = 97%
R = n-Pent: 46% ee = 98%
R = n-Dec: 60% ee = 98%
R = i-Pr: 55% ee = 98%
R = CH2SMe: 43% ee = 97%
R = (CH2)2SMe: 42% ee = 99%
R = (CH2)2OBn: 41% ee = 95%

Scheme 4.7 Synthesis of 3-hydroxyl-2-alkanones.

as an effective Lewis acid additive for the chemoselective homologation of α-aminoxyaldehyde to ketone. The corresponding products were obtained in moderate to good yields and excellent enantioselectivities of up to 99% ee in general (Scheme 4.7). A key intermediate for the synthesis of epothilones was prepared using this approach with a high enantioselectivity.

L-Proline was also used by Zhong *et al.* to promote a highly stereoselective organocatalytic domino aminoxylation-aza-Michael reaction occurring between 2-(5-oxopentylidene) malonate derivatives and aromatic nitroso compounds.[11] This process furnished chiral multifunctionalised tetrahydro-1,2-oxazines in good to high yields, an excellent enantioselectivity of 99% ee in general and exceptional levels of diastereoselectivity of >98% de, as summarised in Scheme 4.8.

In addition, an asymmetric aminoxylation reaction of aldehydes by nitrosobenzene catalysed by a binaphthyl-based chiral amino sulfonamide was reported by Maruoka *et al.*[12,13] As shown in Scheme 4.9, combinations of excellent yields and enantioselectivities were obtained for a wide range of aldehydes. This

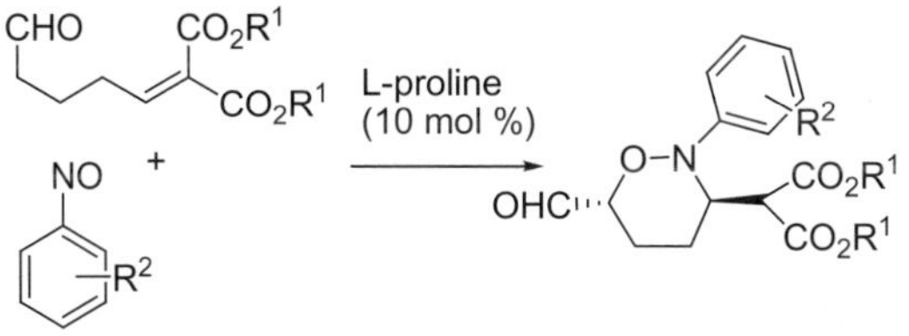

R^1 = Me, R^2 = H: 84% de > 98% ee = 99%
R^1 = Me, R^2 = 3-Me: 63% de > 98% ee = 98%
R^1 = Me, R^2 = 4-Me: 73% de > 98% ee = 99%
R^1 = Me, R^2 = 4-Br: 72% de > 98% ee = 99%
R^1 = n-Pr, R^2 = H: 81% de > 98% ee = 99%
R^1 = n-Bu, R^2 = H: 79% de > 98% ee = 99%
R^1 = n-Hex, R^2 = H: 71% de > 98% ee = 99%
R^1 = n-Pr, R^2 = 2-Me: 72% de > 98% ee = 98%
R^1 = n-Pr, R^2 = 4-Br: 74% de > 98% ee = 99%
R^1 = i-Pr, R^2 = 4-Br: 67% de > 98% ee = 99%
R^1 = n-Bu, R^2 = 4-Br: 71% de > 98% ee = 99%

Scheme 4.8 Synthesis of tetrahydro-1,2-oxazines.

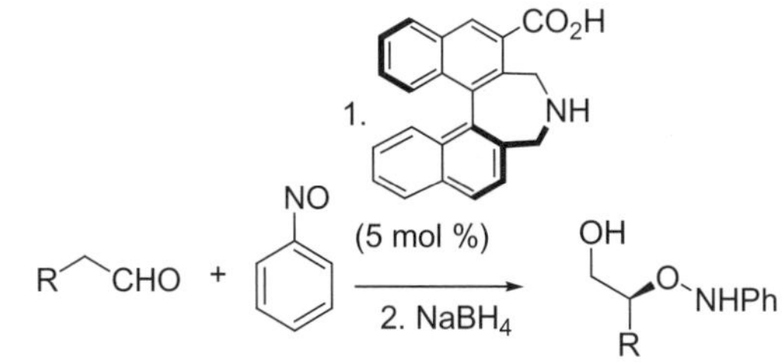

R = Me: 86% ee = 98%
R = Et: 90% ee = 97%
R = n-Bu: 92% ee = 98%
R = CH_2OBn: 92% ee = 97%
R = i-Pr: 96% ee = 98%
R = $CH_2CH=CH_2$: 92% ee = 97%
R = Bn: 88% ee = 97%

Scheme 4.9 α-Aminoxylations of aldehydes.

methodology represented a rare example of the highly enantioselective aminoxylation catalysed by a non-proline type catalyst with a high catalytic performance.

The reactions between nitrosobenzene and enamines as activated carbonyl compounds are known to provide aminoxylation or hydroxyamination products, depending on the catalyst used. In the absence of relatively strong acids such as carboxylic acids, the organocatalytic reaction of aldehydes with nitrosobenzene led to hydroxyamination products exclusively. This phenomenon was checked by Maruoka and Kano by performing the precedent reaction in the presence of a closely related catalyst which did not bear a carboxylic

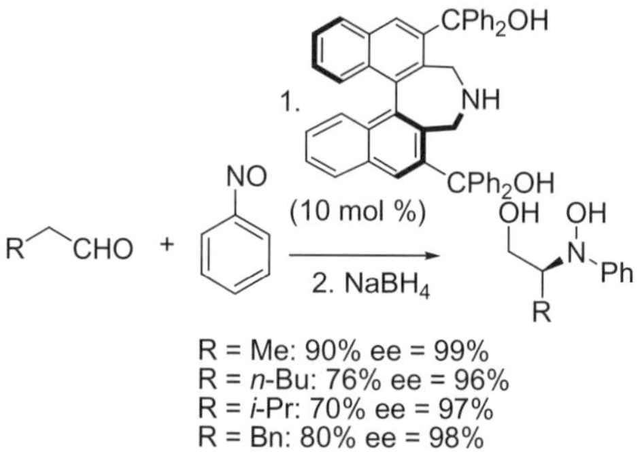

Scheme 4.10   α-Hydroxyaminations of aldehydes.

function.[12] In these conditions, a highly enantioselective hydroxyamination occurred, providing the corresponding products in good yields (>70%) with excellent enantioselectivities (>96% ee) without forming aminoxylation products, as shown in Scheme 4.10.

4.3   Conclusions

In contrast to the large number of addition reactions to C=O and C=N double bonds, only a few examples of nucleophilic addition to N=N double bonds have been investigated. As a recent example, excellent enantioselectivities (97–99% ee) were obtained for the aminated products generated by the asymmetric amination of α-cyanoketones with azodicarboxylates performed in the presence of a chiral thiourea-tertiary amine catalyst used at a low catalyst loading of 1 mol %.

On the other hand, among the already existing methods for the asymmetric synthesis of chiral α-hydroxy carbonyl compounds, the direct organocatalysed enantioselective α-aminoxylation of carbonyl compounds is one of the most important strategies for achieving this purpose. Most of the examples of asymmetric α-aminoxylation of aldehydes reported in the last year involved proline itself as the chiral organocatalyst and allowed exceptional enantioselectivities to be obtained. A rare example of a highly enantioselective aminoxylation catalysed by a non-proline type catalyst with a high catalytic performance is worth highlighting. In this study, the chiral organocatalyst was a binaphthyl-based chiral amino sulfonamide employed at a low catalyst loading of 5 mol %, which provided excellent yields and enantioselectivities for a wide range of aldehydes.

References

1. J. M. Janey, *Ang. Chem., Int. Ed. Engl.*, 2005, **44**, 4292–4300.
2. (a) M. Marigo and K. A. Jorgensen, *Chem. Commun.*, 2006, 2001–2011; (b) G. Guillena and D. J. Ramon, *Tetrahedron: Asymmetry*, 2006, **17**, 1465–1492.

3. R. Melgar-Fernandez, R. Gonzalez-Olvera, J. L. Olivares-Romero, V. Gonzalez-Lopez, L. Romero-Ponce, M. d. R. Ramirez-Zarate, P. Demare, I. Regla and E. Juaristi, *Eur. J. Org. Chem.*, 2008, 655–672.
4. S. M. Kim, J. H. Lee and D. Y. Kim, *Synlett*, 2008, **17**, 2659–2662.
5. S. H. Jung and D. Y. Kim, *Tetrahedron Lett.*, 2008, **49**, 5527–5530.
6. Y. Nishikawa, M. Kitajima and H. Takayama, *Org. Lett.*, 2008, **10**, 1987–1990.
7. H. Yamamoto and N. Momiyama, *Chem. Commun.*, 2005, 3514–3525.
8. M. Lu, D. Zhu, Y. Lu, Y. Hou, B. Tan and G. Zhong, *Ang. Chem., Int. Ed. Engl.*, 2008, **47**, 10187–10191.
9. R. T. Sawant and S. B. Waghmode, *Tetrahedron*, 2009, **65**, 1599–1602.
10. L. Yang, R.-H. Liu, B. Wang, L.-L. Weng and H. Zheng, *Tetrahedron Lett.*, 2009, **50**, 2628–2631.
11 D. Zhu, M. Lu, P. J. Chua, B. Tan, F. Wang, X. Yang and G. Zhong, *Org. Lett.*, 2008, **10**, 4585–4588.
12. T. Kano and K. Maruoka, *Chem. Commun.*, 2008, **1**, 5465–5473.
13. T. Kano, A. Yamamoto and K. Maruoka, *Tetrahedron Lett.*, 2008, **49**, 5369–5371.

CHAPTER 5
Nucleophilic Substitutions at Aliphatic Carbon

5.1 α-Halogenations of Carbonyl Compounds

Halogenated compounds are useful intermediates in organic synthesis, due to the fact that this functionality serves as a lynchpin for further transformations. Indeed, chiral halogen compounds are important in various fields of science, either for the use in further manipulations or because the stereogenic C–halogen centre has a unique property which is of specific importance for a given molecule. The involvement of these functional groups for further stereospecific manipulations and their increasing importance in medicinal chemistry and material sciences have led to an increased search for catalytic asymmetric C–halogen bond-forming reactions.[1,2] Therefore, the enantioselective formation of these compounds is a deserved objective in asymmetric synthesis,[3] with organocatalysis having shown its potential in this type of transformation. All organocatalytic halogenations reported to date are α-halogenations of carbonyl compounds.[4] In particular, fluorine is the most electronegative element in the periodic table and its incorporation in organic compounds alters, sterically and electronically, the properties of molecules, affecting their pK_a, dipole moment and hydrogen-bonding capacity. Furthermore, the carbon–fluorine bond is strong, conferring a special stability and reactivity to fluorinated compounds. All these facts, together with their high metabolic stability, make these compounds ideal candidates for applications in medicinal chemistry.[5] Indeed, a wide range of fluorinated compounds are applied as pharmaceuticals and agrochemicals. Therefore, the enantioselective organocatalysed synthesis of fluorinated products represents an interesting synthetic challenge.[6] In 2009, Lindsley and Fadeyi reported a rapid and general access to chiral β-fluoroamines via organocatalysis,[7] which was based on a powerful extension of the MacMillan enantioselective α-fluorination of aldehydes.[8] This cascade reaction consisted in the treatment of an aldehyde with NFSI in the presence of a chiral imidazolidinone as the organocatalyst, providing the corresponding intermediate

RSC Catalysis Series No. 3
Recent Developments in Asymmetric Organocatalysis
By Hélène Pellissier
© Hélène Pellissier 2010
Published by the Royal Society of Chemistry, www.rsc.org

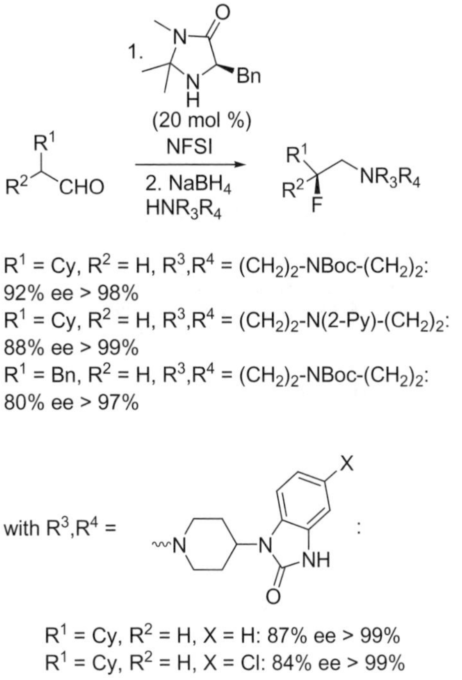
1.
Bn
(20 mol %)
NFSI
2. NaBH4
HNR3R4
R1
R2
CHO
NR3R4
F
R1 = Cy, R2 = H, R3,R4 = (CH2)2-NBoc-(CH2)2:
92% ee > 98%
R1 = Cy, R2 = H, R3,R4 = (CH2)2-N(2-Py)-(CH2)2:
88% ee > 99%
R1 = Bn, R2 = H, R3,R4 = (CH2)2-NBoc-(CH2)2:
80% ee > 97%
with R3,R4 =
X
NH
R1 = Cy, R2 = H, X = H: 87% ee > 99%
R1 = Cy, R2 = H, X = Cl: 84% ee > 99%

Scheme 5.1 Synthesis of β-fluoroamines.

α-fluoroaldehyde, which was subsequently submitted to a reductive amination by treatment with $NaBH_4$ and an amine, yielding the final β-fluoroamine in a good yield and a high enantioselectivity of up to 99% ee, far exceeding that of any other reported method, and without rearranged and dehydrated side products (Scheme 5.1).

A novel route to fluorinated quaternary stereocentres was developed by Shibatomi and Yamamoto based on the asymmetric organocatalytic fluorination of especially reactive α-alkyl α-chlorinated aldehydes performed.[9] Indeed, various α,α-chlorofluoro aldehydes were successfully synthesised with high enantioselectivity from the corresponding α-chloroaldehydes and NFSI in the presence of a chiral silylated diarylprolinol as organocatalyst. These aldehydes were subsequently converted into the corresponding alcohols by treatment with $NaBH_4$ with high yields, as shown in Scheme 5.2.

In 2008, Shibata *et al.* disclosed the first successful catalytic enantioselective fluorination based on the use of cinchona alkaloids.[10] Therefore, it was demonstrated that allyl silanes and silyl enol ethers underwent efficient enantioselective fluorodesilylation with NFSI and a catalytic amount of a bis-cinchona alkaloid in the presence of an excess of base to provide the corresponding fluorinated compounds with an F-substituted quaternary carbon centre with enantioselectivities of up to 95% ee (Scheme 5.3). Furthermore, the authors showed that this catalytic system could be applied to the catalytic

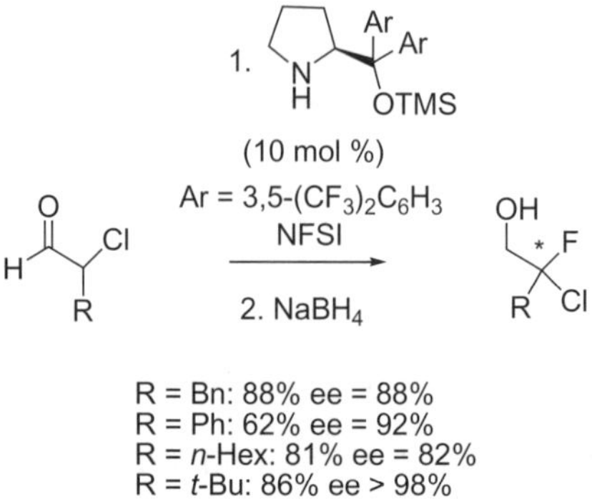
1.
Ar
Ar
N
H
OTMS
(10 mol %)
Ar = 3,5-(CF3)2C6H3
NFSI
2. NaBH4
O
H
Cl
R
OH
F
R
Cl
R = Bn: 88% ee = 88%
R = Ph: 62% ee = 92%
R = n-Hex: 81% ee = 82%
R = t-Bu: 86% ee > 98%

Scheme 5.2 Fluorinations of α-alkyl α-chlorinated aldehydes.

enantioselective fluorination of oxindoles with excellent yields and moderate to high enantioselectivities of up to 86% ee (Scheme 5.3).

Chiral α-iodocarbonyl compounds are especially versatile organic compounds due to their potential for further synthetic transformations. Chiral α-iodoaldehydes are synthetically highly useful since they have characteristic features such as the high leaving group ability and the steric bulk of the iodo group. However, examples of organocatalytic synthesis of these products are especially scarce, probably due to their ease of undesired racemisation. In an effort to address this issue, Maruoka and Kano have designed a novel bifunctional organocatalyst, consisting in a binaphthyl-based amine moiety and hydroxyl groups as activators of an iodination agent.[11,12] This organocatalyst was applied to the iodination of aldehydes with *N*-iodosuccinimide (NIS) in the presence of benzoic acid to yield the corresponding α-iodoaldehydes in good to excellent yields and enantioselectivities (>90% ee), as shown in Scheme 5.4.

5.2 α-Alkylations of Carbonyl Compounds and Derivatives

The direct nucleophilic substitution of alcohols represents a valuable methodology for the preparation of a variety of derivatives, since water is the only by-product of the transformation. In this context, Cozzi *et al.* have reported very recently an organocatalytic alkylation of aldehydes proceeding through an S_N1-type reaction of alcohols.[13] This very simple method to effect the enantioselective direct alkylation of a wide range of aldehydes with unfunctionalised alcohols was catalysed by MacMillan catalyst and provided good to high yields and enantioselectivities of up to 90% ee, as shown in Scheme 5.5.

In 2008, Itsuno *et al.* developed a novel immobilisation method of quaternary ammonium salts through an ionic interaction onto polymer supports.[14] A chiral quaternary ammonium sulfonate could be easily prepared from cinchonidine and a styrene sulfonate monomer was prepared from this

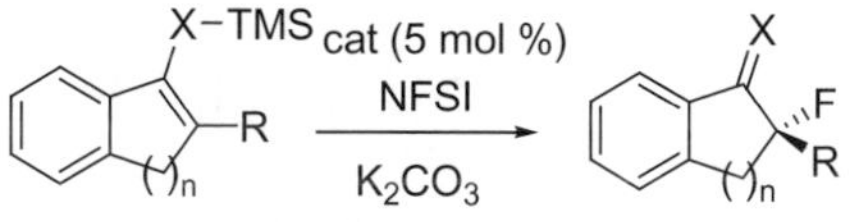

with cat = (DHQ)$_2$PYR:
X = CH$_2$, R = Bn, n = 1: 75% ee = 94%
X = CH$_2$, R = CH$_2$p-Tol, n = 1: 75% ee = 95%
X = CH$_2$, R = p-ClC$_6$H$_4$, n = 1: 81% ee = 94%
X = CH$_2$, R = p-MeOC$_6$H$_4$, n = 1: 65% ee = 90%
X = CH$_2$, R = o-MeOC$_6$H$_4$, n = 1: 58% ee = 93%
X = CH$_2$, R = Bn, n = 2: 74% ee = 81%
with cat = (DHQ)$_2$PHAL:
X = O, R = Bn, n = 2: 79% ee = 86%
X = O, R = CH$_2$p-MeOC$_6$H$_4$, n = 2: 84% ee = 85%
X = O, R = CH$_2$p-ClC$_6$H$_4$, n = 2: 74% ee = 86%

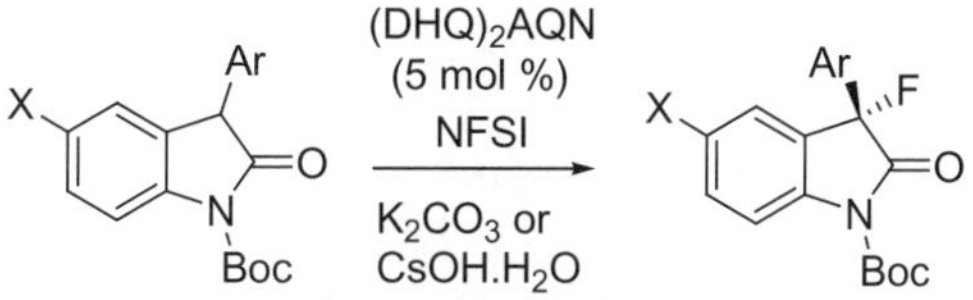

Ar = Ph, X = H: 99% ee = 85%
Ar = p-Tol, X = H: 94% ee = 86%
Ar = Ph, X = OMe: 92% ee = 84%
Ar = Ph, X = OMe: 92% ee = 84%
Ar = p-Tol, X = Me: 86% ee = 84%
Ar = p-Tol, X = OMe: 99% ee = 85%

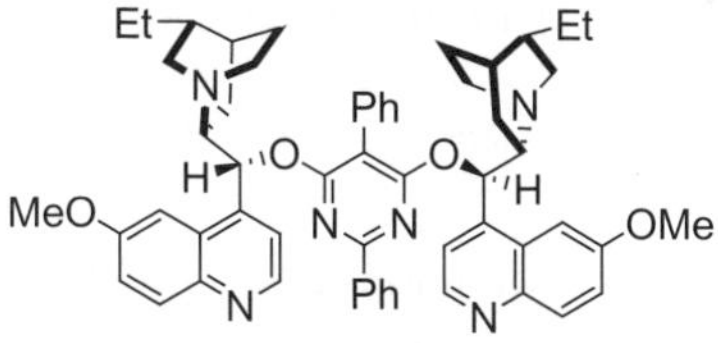

(DHQ)$_2$PYR

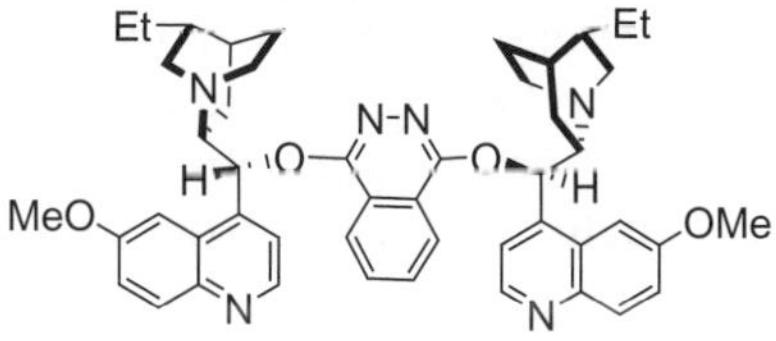

(DHQ)$_2$PHAL

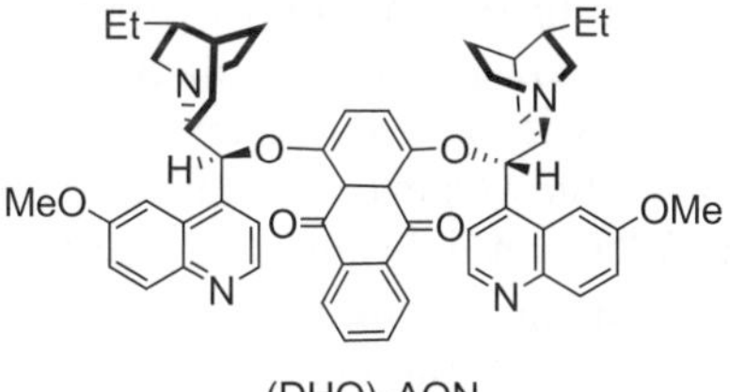

(DHQ)$_2$AQN

Scheme 5.3 Fluorinations of allyl silanes, silyl enols and oxindoles.

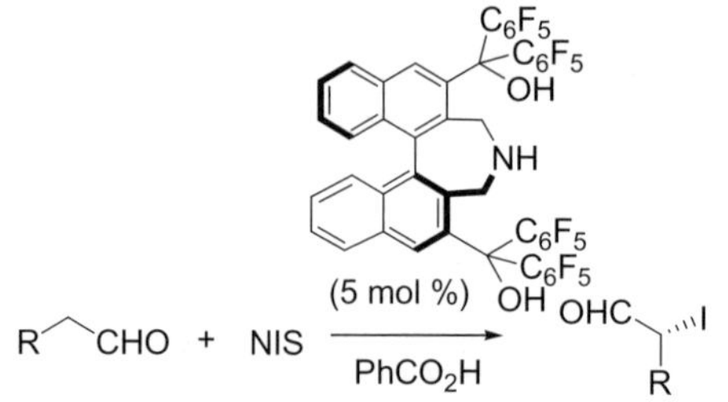

R = Et: 76% ee = 98%
R = Bn: 81% ee = 92%
R = *i*-Pr: 96% ee = 99%
R = CH_2OBn: 97% ee = 93%

Scheme 5.4 Iodinations of aldehydes.

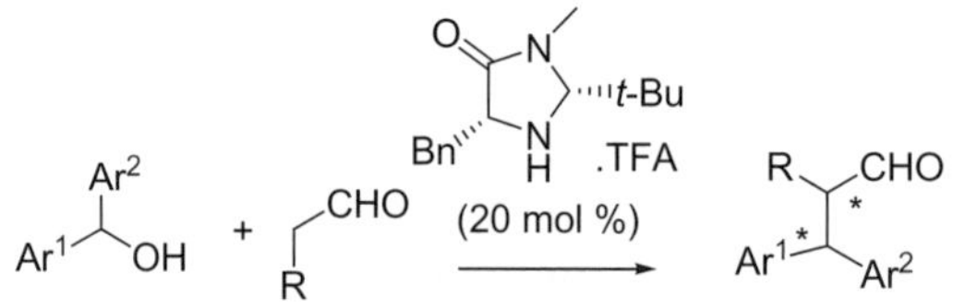

$Ar^1 = Ar^2 = p\text{-}(NMe_2)C_6H_4$, R = Me: 95% ee = 77%
$Ar^1 = Ar^2 = p\text{-}(NMe_2)C_6H_4$, R = Bn: 77% ee = 77%
$Ar^1 = Ar^2 = p\text{-}(NMe_2)C_6H_4$, R = *n*-Hex: 92% ee = 78%
Ar^1 = ferrocenyl, Ar^2 = Ph, R = Me: 48% de = 12% ee = 90%

Scheme 5.5 Alkylations of aldehydes with alcohols.

substrate, which was further copolymerised with styrene and divinylbenzene under radical conditions to give the corresponding polymeric chiral quaternary ammonium salt. This polymer was employed to promote the asymmetric alkylation of *N*-diphenylmethylene glycine *tert*-butyl ester in good yields and high enantioselectivities of up to 98% ee, as shown in Scheme 5.6. Somewhat higher enantioselectivities were obtained in most cases by using this type of polymeric catalyst compared to the unsupported catalysts in homogeneous solution systems.

Another highly performant catalyst for these reactions was developed by Maruoka *et al.* under phase-transfer conditions.[15] Indeed, the application of an (*S*)-binaphthyl phase-transfer catalyst to the alkylation of *N*-diphenylmethylene glycine *tert*-butyl ester with a wide range of alkyl halides yielded the corresponding alkylated products in excellent general yields and enantioselectivities of up to 99% ee, as shown in Scheme 5.7. Remarkably, these excellent results were obtained by using an exceptional low catalyst loading of 0.05 mol %, and it was demonstrated that the enantioselectivities were still excellent when the catalyst loading reached 0.01 mol %.

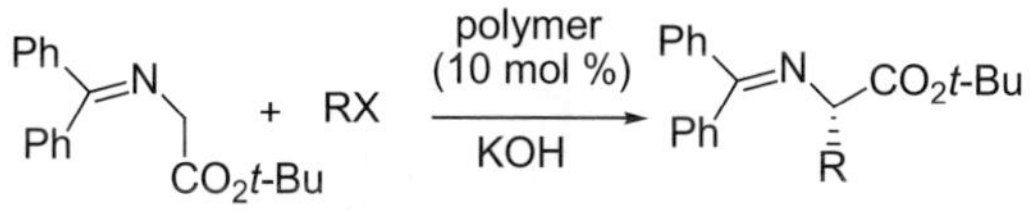

RX = BnBr: 84% ee = 94%
RX = p-Tol-CH$_2$Br: 90% ee = 97%
RX = p-BrC$_6$H$_4$-CH$_2$Br: 95% ee = 91%
RX = CH$_2$=CHCH$_2$Br: 85% ee = 98%
RX = CH$_2$=CHCH$_2$I: 74% ee = 80%
RX = MeI: 45% ee = 70%

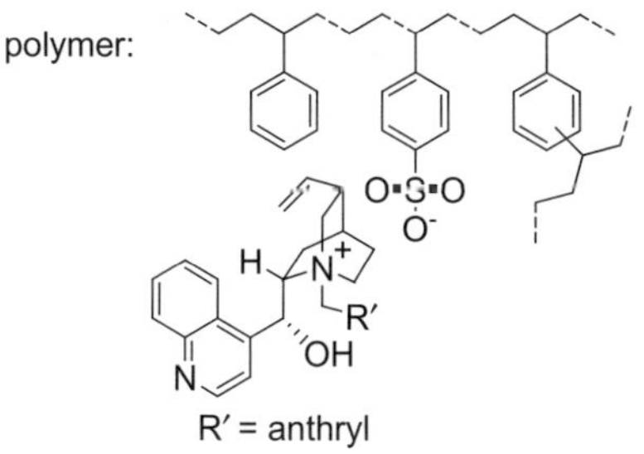

polymer:

R′ = anthryl

Scheme 5.6 Heterogeneous alkylations of *N*-diphenylmethylene glycine *tert*-butyl ester.

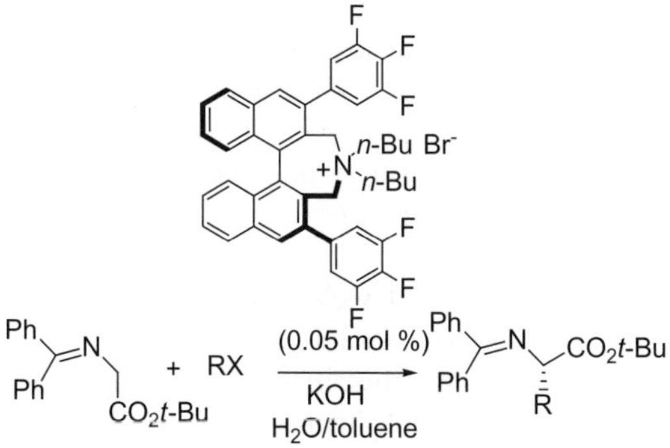

RX = BnBr: 98% ee = 99%
RX = p-Tol-CH$_2$Br: 99% ee = 98%
RX = p-FC$_6$H$_4$-CH$_2$Br: 99% ee = 98%
RX = CH$_2$=CHCH$_2$Br: 87% ee = 98%
RX = 2-NaphCH$_2$Br: 84% ee = 97%
RX = EtI: 67% ee = 99%

Scheme 5.7 Phase-transfer alkylations of *N*-diphenylmethylene glycine *tert*-butyl ester.

In the same context, the catalytic activity of novel calixarene-based chiral phase-transfer catalysts derived from cinchona alkaloids was evaluated by Sirit *et al.* by carrying out the phase-transfer benzylation of *N*-diphenylmethylene glycine ethyl ester with benzyl bromide.[16] Although a good yield (89%) was

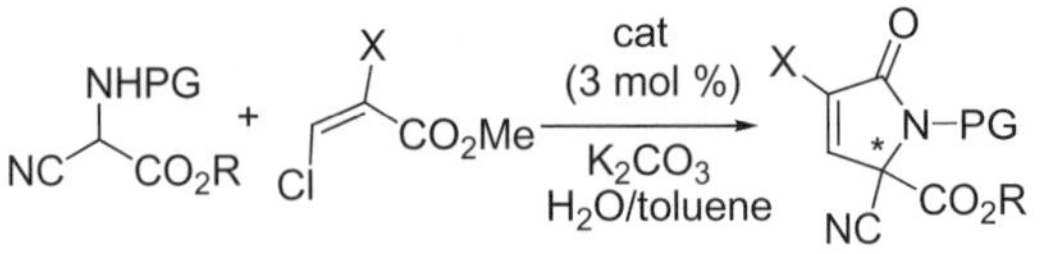

PG = Boc, R = Et, X = I: 73% ee > 99%
PG = Boc, R = *t*-Bu, X = I: 71% ee > 99%
PG = Troc, R = *t*-Bu, X = I: 63% ee > 99%
PG = Cbz, R = *t*-Bu, X = I: 64% ee > 99%
PG = Boc, R = *t*-Bu, R^2 = Me, X = Cl: 90% ee = 85%

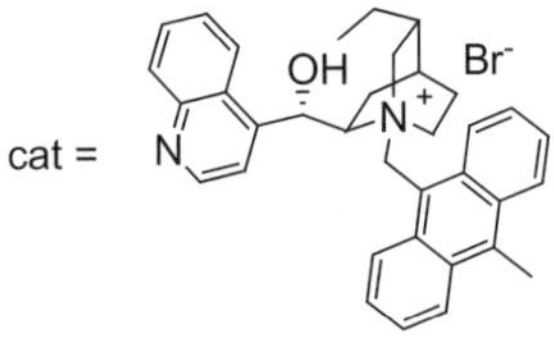

cat =

Scheme 5.8   Synthesis of 3-halo-3-pyrrolidin-2-ones *via* vinylic substitutions.

obtained in the presence of NaOH as a base, the enantioselectivity of the reaction remained moderate ($\leq 46\%$ ee).

A dihydrocinchonine-derived phase-transfer catalyst was employed as an organocatalyst by Jorgensen *et al.* to develop a novel use of the organocatalytic enantioselective vinylic substitution reaction for the single-step construction of C5 quaternary 3-halo-3-pyrrolidin-2-ones, which represents a flexible starting point for the preparation of structurally diverse optically active γ-lactams.[17] A key to the reaction was the use of a stereochemically well-defined α,β-dihalogenated acrylate ester as the electrophile, which reacted with a 1,2-dinucleophile, providing after immediate ring-closure of the intermediate the expected 3-halo-3-pyrrolidin-2-ones in good yields and almost complete enantioselectivities in most cases of substrates (Scheme 5.8).

In 2009, Chen *et al.* reported the first highly enantioselective allylic-allylic alkylation of α,α-dicyanoalkenes with Morita–Baylis–Hillman carbonates by dual organocatalysis of commercially available modified cinchona alkaloids and (*S*)-BINOL.[18] Excellent stereoselectivities were achieved for a broad range of substrates by using hydroquinidine (anthraquinone-1,4-diyl) diether ((DHQD)₂AQN) as the cinchona alkaloid. Indeed, in all the cases studied, only one diastereomer was isolated with both excellent enantioselectivity and yield, as shown in Scheme 5.9.

In the course of developing an easy access to chiral γ-butenolides, Shi *et al.* have established an efficient multifunctional chiral binaphthyl phosphine-catalysed allylic substitution of Morita–Baylis–Hillman acetate with 2-trimethylsiloxy furan.[19] The regiospecific allylic substitution occurred to provide the *syn*-γ-butenolide in good to excellent yield, high regioselectivity and excellent enantioselectivity by using water as an additive. The scope of this reaction could be successfully extended to a variety of Morita–Baylis–Hillman acetates, as shown in Scheme 5.10.

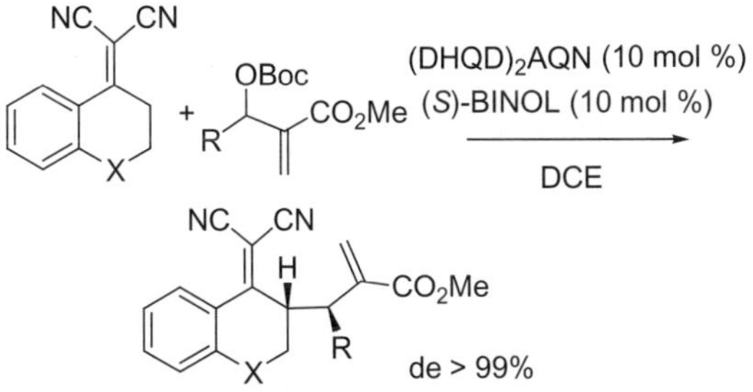

R = Ph, X = S: 95% ee = 97%
R = *p*-ClC$_6$H$_4$, X = S: 96% ee = 96%
R = *m*-ClC$_6$H$_4$, X = S: 95% ee = 96%
R = *o*-ClC$_6$H$_4$, X = S: 96% ee = 92%
R = *p*-FC$_6$H$_4$, X = S: 99% ee = 95%
R = *p*-MeOC$_6$H$_4$, X = S: 93% ee = 93%
R = Ph, X = O: 92% ee = 96%
R = Ph, X = CH$_2$: 92% ee = 96%

Scheme 5.9 Allylic-allylic alkylations of α,α-dicyanoalkenes with Morita–Baylis–Hillman carbonates.

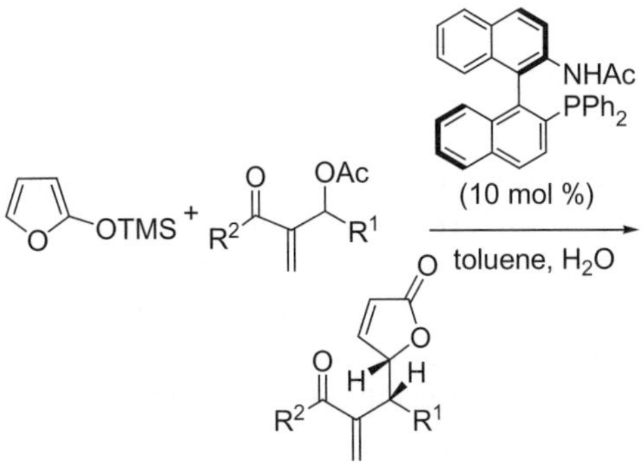

R^1 = Ph, R^2 = Et: 98% ee = 91%
R^1 = *p*-Tol, R^2 = Me: 81% ee = 95%
R^1 = *p*-BrC$_6$H$_4$, R^2 = Me: 85% ee = 95%
R^1 = *p*-ClC$_6$H$_4$, R^2 = Me: 89% ee = 95%
R^1 = *m*-ClC$_6$H$_4$, R^2 = Me: 95% ee = 94%
R^1 = *o*-ClC$_6$H$_4$, R^2 = Me: 89% ee = 94%
R^1 = *p*-NO$_2$C$_6$H$_4$, R^2 = Me: 98% ee = 91%

Scheme 5.10 Allylic substitutions between 2-trimethylsiloxy furan and Morita–Baylis–Hillman acetates.

In another context, a chiral phase-transfer *N*-9-anthracenylmethyl *O*-adamantoyl derivatised cinchonine catalyst was shown to promote the asymmetric ring-opening of *N*-tosyl protected aziridines with cyclic β-keto esters in the presence of an aqueous base, leading to the formation of chiral aminoethyl

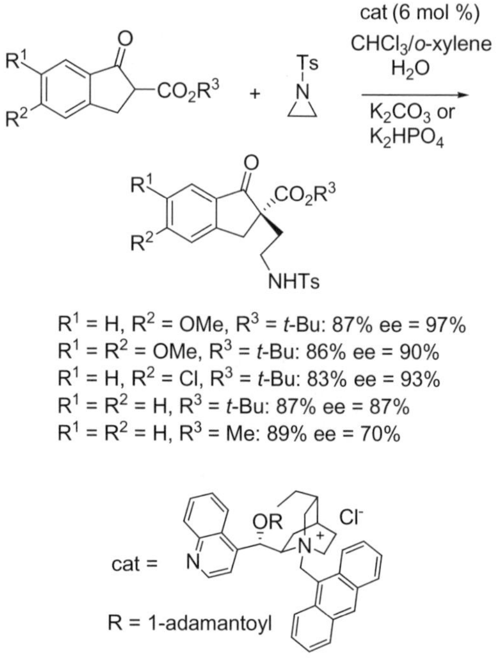

Scheme 5.11 Ring-openings of aziridine.

functionalised compounds with moderate to excellent enantioselectivities of up to 99% ee (Scheme 5.11).[20] This methodology was previously reported by Dixon *et al.* by using *O*-(trifluoromethane)benzenesulfonyl aziridines, which allowed the formation of the corresponding indanone adducts in comparable yields and enantioselectivities.[21] However, the authors have found a significant decrease in the reactivity when using *N*-mesitylene sulfonyl aziridine in similar conditions.

With the aim of studying the reactivity of *N*-methylindole with a β,γ-unsaturated α-keto ester, Rueping *et al.* have found that according to the nature of the organocatalyst employed, the reaction produced either the 1,4-product or the corresponding bisindole derived from a nucleophilic substitution.[22] Therefore, performing the reaction in the presence of a chiral *N*-triflylphosphoramide bearing two 9-phenanthryl groups on the binaphthyl scaffold led to the unprecedented formation of a chiral bisindole through nucleophilic substitution with a moderate enantioselectivity (Scheme 5.12).

Finally, a chiral 2-phenylpyrrolidine-derived thiourea was demonstrated to be an efficient organocatalyst to promote the clean substitution of 1-chloro-isochromans by silyl ketene acetals to provide the corresponding chiral substituted isochromans with high yields and enantioselectivities of up to 90% ee.[23] Actually, the 1-chloroisochromans were prepared from the corresponding more stable methyl acetals, which were directly used without purification in the protocol evolving through a cationic oxocarbenium ion (Scheme 5.13).

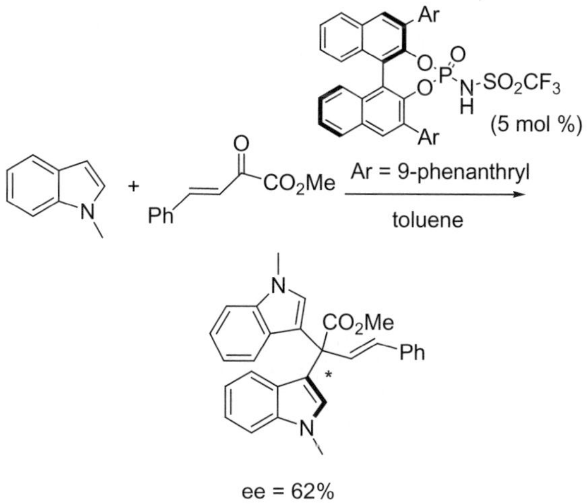

Scheme 5.12 Synthesis of bisindole through nucleophilic substitution.

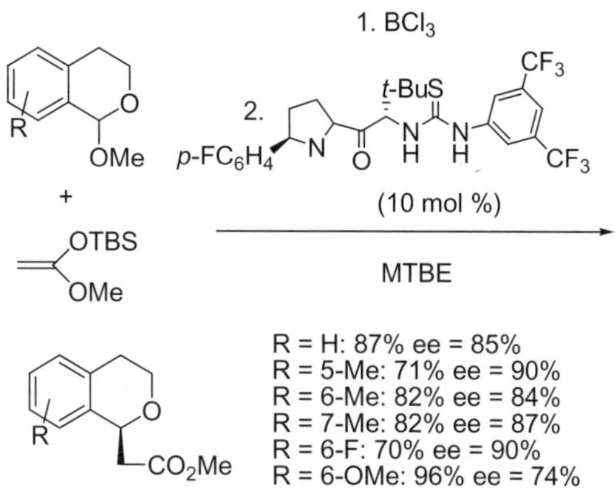

Scheme 5.13 Synthesis of substituted isochromans.

5.3 α-Aminations of Carbonyl Compounds

In 2009, Greck *et al.* reported the five-step syntheses of cyclic amino acids, such as (*R*)-pipecolic acid and D-proline, based on an enantioselective organocatalytic α-amination of functionalised aldehydes.[24] This reaction was catalysed by L-proline and led to the formation of the corresponding hydrazino aldehydes, which were *in situ* reduced after the amination step in order to avoid the epimerisation of the enolisable products, giving the corresponding hydrazine alcohols in good yields and enantioselectivities of up to 94% ee, as shown in Scheme 5.14. It is noteworthy that L-proline was involved in the synthesis of its enantiomer.

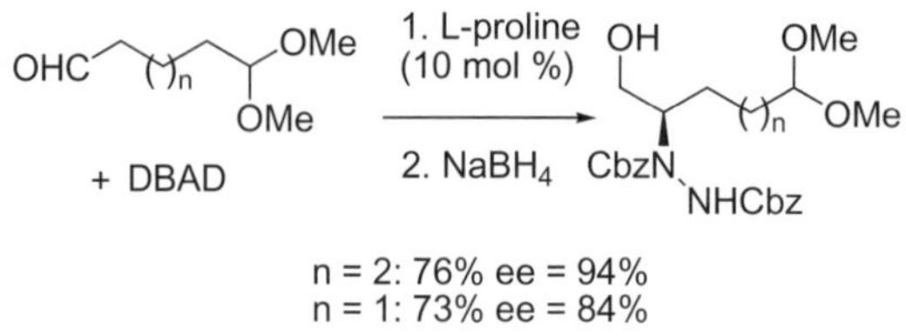

OHC
OMe
OMe
+ DBAD
1. L-proline
(10 mol %)
2. NaBH4
OH
OMe
OMe
CbzN
NHCbz
n = 2: 76% ee = 94%
n = 1: 73% ee = 84%

Scheme 5.14 α-Aminations of aldehydes.

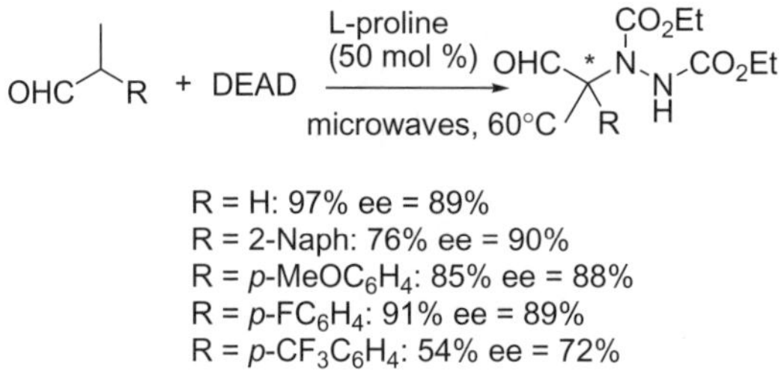

OHC
R
+ DEAD
L-proline
(50 mol %)
microwaves, 60°C
CO2Et
OHC
N
N
CO2Et
H
R
R = H: 97% ee = 89%
R = 2-Naph: 76% ee = 90%
R = p-MeOC6H4: 85% ee = 88%
R = p-FC6H4: 91% ee = 89%
R = p-CF3C6H4: 54% ee = 72%

Scheme 5.15 α-Aminations of disubstituted aldehydes.

In addition, Bräse *et al.* have studied the thermal effects in the L-proline-catalysed asymmetric α-amination of disubstituted aldehydes with azodicarboxylates, demonstrating that this reaction was accelerated under microwave conditions at 60 °C.[25] Compared to the results previously obtained at room temperature, both the yield and the enantioselectivity could be significantly increased and the reaction time considerably reduced. As shown in Scheme 5.15, this improved protocol allowed the fast and efficient synthesis of chiral α,α-disubstituted amino aldehydes with enantioselectivities of up to 90% ee, providing the best results for the α-amination of α-branched aldehydes to date. Significantly, the amination of branched aldehydes is one of the few organocatalytic reactions where good to moderate levels of enantioselection can be achieved at high temperatures (60–70 °C).

5.4 Miscellaneous Reactions

In 2009, Armstrong and Emmerson reported the synthesis of chiral allenamides with high levels of enantioselectivity through a [2,3]-sigmatropic rearrangement of the corresponding propargylic sulfides.[26] The required branched propargylic sulfides were prepared by an enantioselective organocatalytic aldehyde α-sulfenylation followed by a Corey–Fuchs alkynylation. This sequence induced by a diarylprolinol silyl ether occurred with a high level of chirality transfer since the final allenamides were isolated with an enantioselectivity of up to 89% ee, as shown in Scheme 5.16.

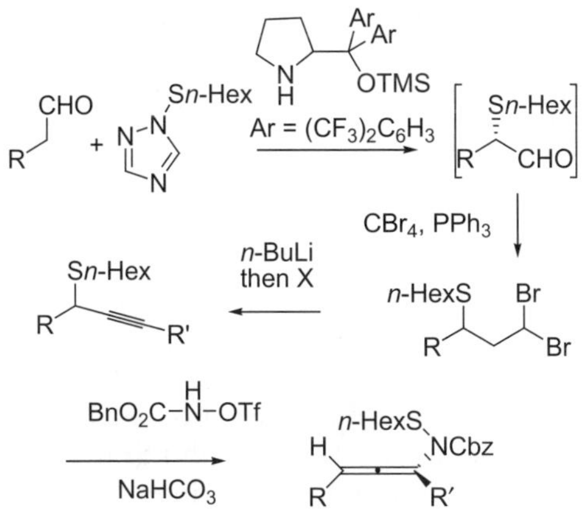

R = Et, R′ = H: 76% ee = 87% (from X = H_2O)
R = Et, R′ = CH_2OH: 83% ee = 88% (from X = CH_2O)
R = i-Pr, R′ = H: 71% ee = 89% (from X = H_2O)
R = Bn, R′ = CH_2OH: 74% ee = 81% (from X = CH_2O)

Scheme 5.16 Synthesis of allenamides *via* α-sulfenylations of aldehydes.

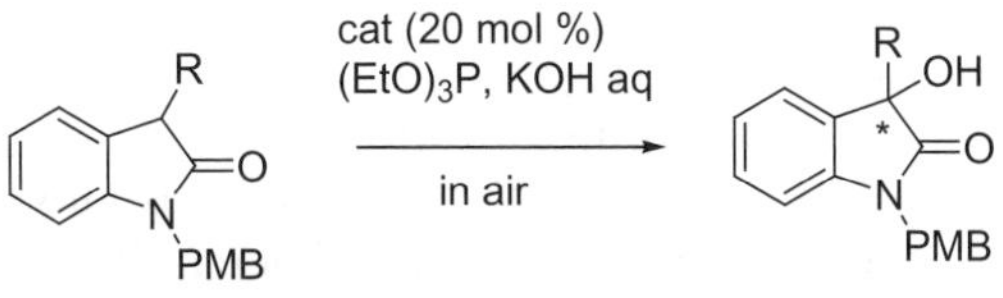

R = $CH_2CH=CH_2$: 98% ee = 85%
R = Bn: 95% ee = 86%
R = $CH_2C≡CH$: 92% ee = 93%
R = n-Hex: 100% ee = 92%
R = CH_2-p-$MeOC_6H_4$: 97% ee = 91%
R = $(CH_2)_3CO_2Et$: 100% ee = 82%

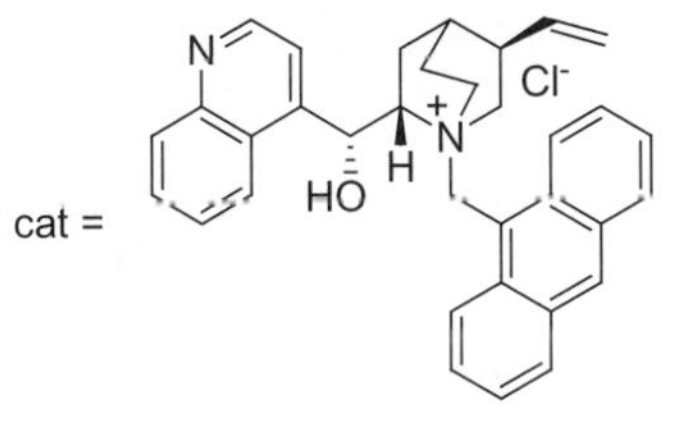

cat =

Scheme 5.17 α-Hydroxylations of oxindoles.

An organocatalytic asymmetric hydroxylation of oxindoles by molecular oxygen as an oxidant using a phase-transfer catalyst was reported by Itoh *et al.*, in 2008.[27] The use of O_2 as the oxidant was a paramount process, because it is inexpensive and environmentally benign. In these conditions, the reaction of a series of 3-substituted oxindoles in the presence of a cinchonidine-derived phase-transfer catalyst provided the corresponding 3-hydroxy derivatives in excellent yields and moderate to high enantioselectivities, as shown in Scheme 5.17. One of these products was converted into a synthetic precursor of the alkaloid CPC-1.

5.5 Conclusions

The enantioselective organocatalysed synthesis of halogenated products represents an interesting synthetic challenge since their functionality serves as a lynchpin for further transformations. In this context, a powerful extension of the MacMillan enantioselective α-fluorination of aldehydes was developed on the basis of the treatment of an aldehyde with NFSI in the presence of a chiral imidazolidinone as the organocatalyst, providing the corresponding inter-mediate α-fluoroaldehyde, which was subsequently submitted to a reductive amination, yielding the final β-fluoroamine in good yield and high enantios-electivity of up to 99% ee.

On the other hand, examples of organocatalytic synthesis of chiral α-iodoaldehydes are especially scarce, probably due to their ease of undesired racemisation. In an effort to address this issue, a novel bifunctional organo-catalyst, consisting in a binaphthyl-based amine moiety and hydroxyl groups, was applied to the iodination of aldehydes with *N*-iodosuccinimide, yielding the corresponding α-iodoaldehydes in good to excellent yields and enantioselec-tivities and at a low catalyst loading of 5 mol %.

In addition, highly enantioselective alkylations of carbonyl compounds were developed under phase-transfer conditions. Therefore, the use of an (*S*)-binaphthyl phase-transfer catalyst for the alkylation of *N*-diphenylmethylene glycine *tert*-butyl ester with a wide range of alkyl halides yielded the corre-sponding alkylated products in excellent general yields and enantioselectivities of up to 99% ee. Remarkably, these excellent results were obtained by using an exceptional low catalyst loading of 0.01–0.05 mol %.

References

1. J. M. Janey, *Ang. Chem., Int. Ed. Engl.*, 2005, **44**, 4292–4300.
2. (a) P. M. Pihko, *Ang. Chem., Int. Ed. Engl.*, 2006, **45**, 544–547; (b) S. France, A. Weatherwax and T. Lectka, *Eur. J. Org. Chem.*, 2005, 475–479; (c) M. Oestreich, *Ang. Chem., Int. Ed. Engl.*, 2005, **44**, 2324–2327.
3. V. A. Brunet and D. O'Hagan, *Ang. Chem., Int. Ed. Engl.*, 2008, **47**, 1179–1182.
4. M. Ueda, T. Kano and K. Maruoka, *Org. Biomol. Chem.*, 2009, **7**, 2005–2012.
5. J.-A. Ma and D. Cahard, *Chem. Rev.*, 2004, **104**, 6119–6146.
6. P. Dinér, A. Kjaersgaard, M. A. Lie and K. A. Jorgensen, *Chem. Eur. J.*, 2008, **14**, 122–127.
7. O. Fadeyi and C. W. Linsley, *Org. Lett.*, 2009, **11**, 943–946.

8. T. D. Beeson and D. W. C. MacMillan, *J. Am. Chem. Soc.*, 2005, **127**, 8826–8828.
9. K. Shibatomi and H. Yamamoto, *Ang. Chem., Int. Ed. Engl.*, 2008, **47**, 5796–5798.
10. T. Ishimaru, N. Shibata, T. Horikawa, N. Yasuda, S. Nakamura, T. Toru and M. Shiro, *Ang. Chem., Int. Ed. Engl.*, 2008, **47**, 4157–4161.
11. T. Kano and K. Maruoka, *Chem. Commun.*, 2008, 5465–5473.
12. T. Kano, M. Ueda and K. Maruoka, *J. Am. Chem. Soc.*, 2008, **130**, 3728–3729.
13. P. G. Cozzi, F. Benfatti and L. Zoli, *Ang. Chem., Int. Ed. Engl.*, 2009, **48**, 1313–1316.
14. Y. Arakawa, N. Haraguchi and S. Itsuno, *Ang. Chem., Int. Ed. Engl.*, 2008, **47**, 8232–8235.
15. M. Kitamura, S. Shirakawa, Y. Arimura, X. Wang and K. Maruoka, *Chem. Asian J.*, 2008, **3**, 1702–1714.
16. S. Bozkurt, M. Durmaz, M. Yilmaz and A. Sirit, *Tetrahedron: Asymmetry*, 2008, **19**, 618–623.
17. T. B. Poulsen, G. Dickmeiss, J. Overgaard and K. A. Jorgensen, *Ang. Chem., Int. Ed. Engl.*, 2008, **47**, 4687–4690.
18. H.-L. Cui, J. Peng, X. Feng, W. Du, K. Jiang and Y.-C. Chen, *Chem. Eur. J.*, 2009, **15**, 1574–1577.
19. Y.-Q. Jiang, Y.-L. Shi and M. Shi, *J. Am. Chem. Soc.*, 2008, **130**, 7202–7203.
20. M. W. Paixao, M. Nielsen, C. B. Jacobsen and K. A. Jorgensen, *Org. Biomol. Chem.*, 2008, **6**, 3467–3470.
21. T. A. Moss, D. R. Fenwick and D. J. Dixon, *J. Am. Chem. Soc.*, 2008, **130**, 10076–10077.
22. M. Rueping, B. J. Nachtsheim, S. A. Moreth and M. Bolte, *Ang. Chem., Int. Ed. Engl.*, 2008, **47**, 593–596.
23. S. E. Reisman, A. G. Doyle and E. N. Jacobsen, *J. Am. Chem. Soc.*, 2008, **130**, 7198–7199.
24. D. Kalch, N. De Rycke, X. Moreau and C. Greck, *Tetrahedron Lett.*, 2009, **50**, 492–494.
25. T. Baumann, M. Bächle, C. Hartmann and S. Bräse, *Eur. J. Org. Chem.*, 2008, 2207–2212.
26. A. Armstrong and D. P. G. Emmerson, *Org. Lett.*, 2009, **11**, 1547–1550.
27. D. Sano, K. Nagata and T. Itoh, *Org. Lett.*, 2008, **10**, 1593–1595.

CHAPTER 6

Cycloaddition Reactions

6.1 Diels–Alder Reactions

The asymmetric Diels–Alder reaction is one of the most important organic transformations and has proved to be a versatile means of synthesis of a large number of important chiral building blocks and intermediates in the total synthesis of natural products.[1] For a long time, it was not known that organocatalysts could be used to promote the Diels–Alder reactions and base-catalysed Diels–Alder reactions, in particular, were regarded as uncommon. In recent years, several different organocatalysts have been developed. As an example, camphor-derived cyclic sulfonylhydrazines have been independently employed as organocatalysts by Lee[2] and Langlois[3] to promote the enantioselective Diels–Alder reaction of cyclopentadiene with dienophiles such as α,β-unsaturated aldehydes. In the presence of trichloroacetic acid as a co-catalyst, they have proved to be efficient to catalyse these reactions with good yields and enantioselectivities of up to 96% ee, albeit with low *endo*-diastereoselectivities (≤42% de), as shown in Scheme 6.1.

On the other hand, several cinchona alkaloids have recently been investigated as organocatalysts for asymmetric Diels–Alder reactions. Therefore, excellent results were obtained by Ricci *et al.* for the Diels–Alder reaction of 3-vinylindoles catalysed by a chiral bifunctional thiourea derived from hydroquinine.[4] In these conditions, the reaction of 3-vinylindoles with maleimides led to the almost exclusive formation of the *endo*-carbazoles in high yields and enantioselectivities of up to 99% ee, as shown in Scheme 6.2. Derivatisation with trifluoroacetic anhydride (TFAA) after the reaction gave additional stability to the cycloadducts, thus facilitating their isolation by chromatography on silica gel. Various 1-substituted-3-vinylindoles were also tested in the reaction with phenylmaleimide in similar conditions, providing the corresponding cycloadducts, albeit in racemic form. These results were tentatively taken to suggest the requirement of an interaction between the basic moiety of the catalyst and the N–H group of the diene. This novel methodology offered a direct approach to chiral tetrahydrocarbazole derivatives.

RSC Catalysis Series No. 3
Recent Developments in Asymmetric Organocatalysis
By Hélène Pellissier
© Hélène Pellissier 2010
Published by the Royal Society of Chemistry, www.rsc.org

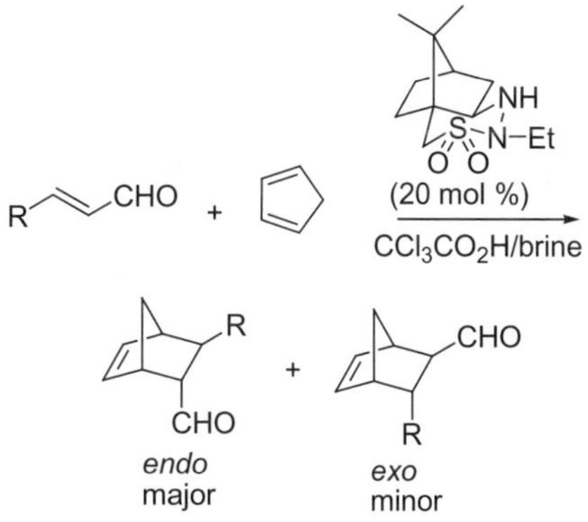

R = Ph: 92% de = 4%
ee (*endo*) = 93% ee (*exo*) = 78%
R = *o*-NO$_2$C$_6$H$_4$: 94% de = 42%
ee (*endo*) = 90% ee (*exo*) = 72%
R = *p*-ClC$_6$H$_4$: 84% de = 4%
ee (*endo*) = 96% ee (*exo*) = 84%
R = *p*-BrC$_6$H$_4$: 86% de = 4%
ee (*endo*) = 96% ee (*exo*) = 86%

Scheme 6.1 Camphor-derived sulfonylhydrazine-catalysed Diels–Alder reactions.

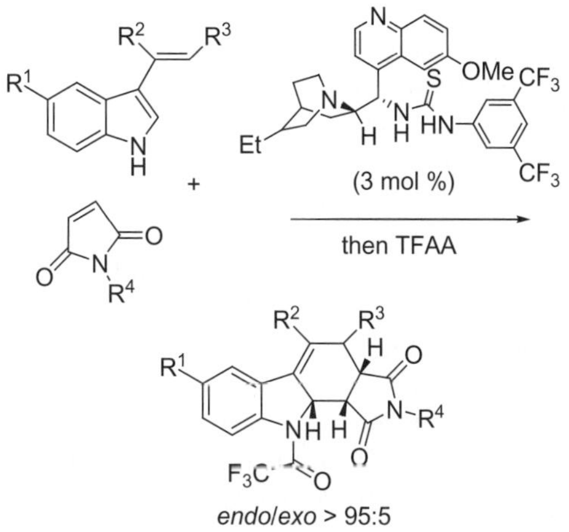

R^1 = R^2 = R^3 = H, R^4 = Ph: 91% ee = 98%
R^1 = Br, R^2 = R^3 = H, R^4 = Ph: 86% ee = 90%
R^1 = OMe, R^2 = R^3 = H, R^4 = Ph: 77% ee = 96%
R^1 = R^3 = H, R^2 = Me, R^4 = Ph: 79% ee = 96%
R^1 = R^2 = H, R^3 = Me, R^4 = Ph: 58% ee = 92%
R^1 = R^2 = R^3 = H, R^4 = Me: 89% ee = 98%
R^1 = R^2 = R^3 = H, R^4 = Bn: 89% ee = 96%

Scheme 6.2 Diels–Alder reactions of 3-vinylindoles catalysed by hydroquinine-derived thiourea.

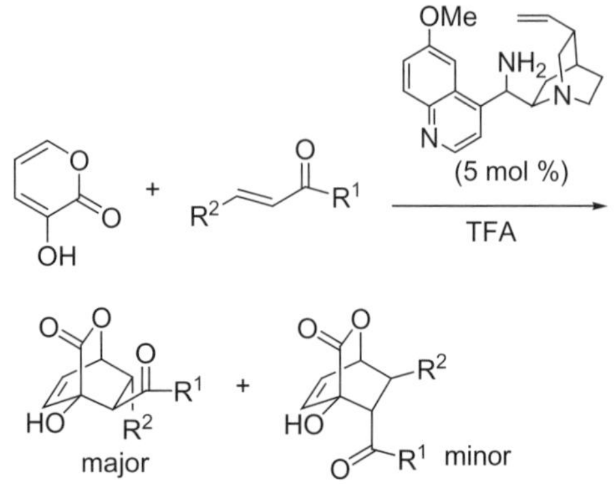

R^1 = Me, R^2 = Ph: 87% de = 60% ee = 98%
R^1 = Me, R^2 = CH_2Bn: 99% de = 88% ee = 99%
R^1 = Et, R^2 = H: 99% de = 94% ee = 99%
R^1 = *n*-Pr, R^2 = H: 90% de = 84% ee = 99%
R^1 = *n*-Pent, R^2 = H: 99% de = 86% ee = 96%

Scheme 6.3 Cinchona alkaloid-catalysed Diels–Alder reactions of 2-pyrones.

A readily available 9-NH_2 cinchona alkaloid has proved to efficiently catalyse the asymmetric Diels–Alder reaction of simple α,β-unsaturated ketones with 2-pyrones.[5] In the presence of TFA as an additive, the reaction afforded the *exo*-cycloadduct as the major product in good enantiomeric excess of up to 99% ee in almost all cases of substrates, as shown in Scheme 6.3. Moderate to excellent diastereoselectivities of 50–94% de were obtained, but it is important to note that both the diastereoselectivity and the enantioselectivity of the reaction did not fluctuate significantly when the aromatic substituent of the α,β-unsaturated ketone was changed to an aliphatic substituent.

The asymmetric Diels–Alder reaction of 2-cyclohexen-1-one with simple α-substituted aldehydes was studied by Bella *et al.*, demonstrating a synergic beneficial effect of two organocatalysts, such as a cinchona alkaloid and a chiral cyclic secondary amine.[6] In this strategy, aldehydes were activated towards electrophilic attack by the secondary amine in combination not with an acid, but instead with a base. The corresponding bicyclo[2.2.2]octan-2-ones were achieved in variable stereoselectivities and moderate to good yields. The best result was obtained by using the combination of quinine with chiral 2,2-dimethyl thiazoline carboxylate for the Diels–Alder reaction of phenylacetaldehyde, which gave a high diastereoselectivity of 90% de combined with a high enantioselectivity of 87% ee, albeit with a moderate yield of 41%.

A chiral bicyclic guanidine was found to be an efficient organocatalyst for the asymmetric Diels–Alder reaction between substituted anthrones and protected *N*-hydroxymaleimides, providing the corresponding anthrone-derived *N*-hydroxyphthalimide analogues in good yields and moderate to high enantioselectivities (12–92% ee), as shown in Scheme 6.4.[7]

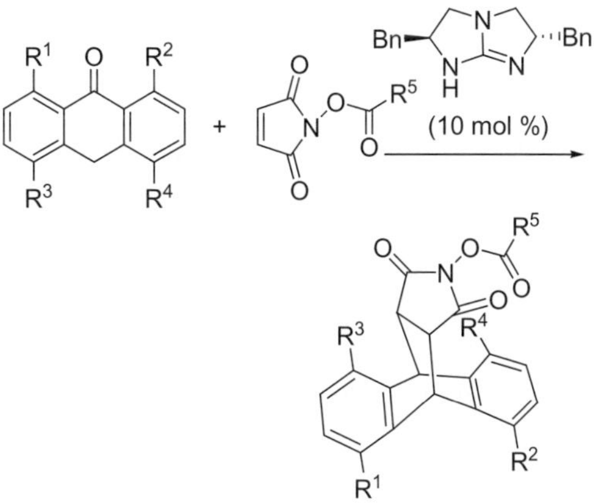

R¹ = R² = Cl, R³ = R⁴ = H, R⁵ = Me: 86% ee = 92%
R¹ = R² = Cl, R³ = R⁴ = H, R⁵ = Ph: 85% ee = 12%
R¹ = R² = H, R³ = R⁴ = Cl, R⁵ = Me: 83% ee = 64%
R¹ = R² = H, R³ = R⁴ = Cl, R⁵ = Ph: 84% ee = 87%

Scheme 6.4 Guanidine-catalysed Diels–Alder reactions of anthrones.

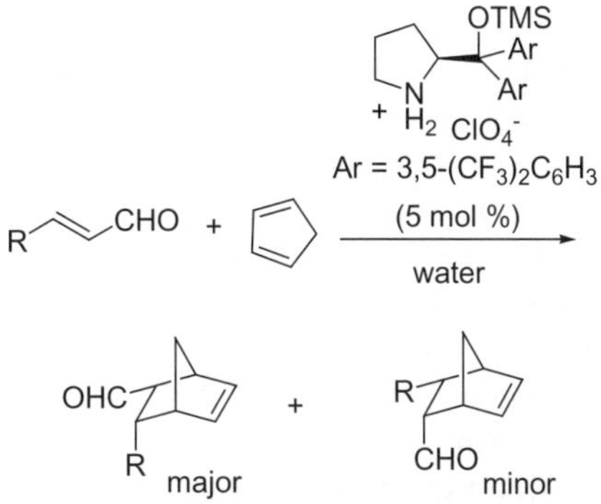

R = Ph: 93% de = 60% ee (*exo*) = 97% ee (*endo*) = 92%
R = Me: 73% de = 44% ee (*exo*) = 99% ee (*endo*) = 99%
R = *n*-Bu: 95% de = 60% ee (*exo*) = 98% ee (*endo*) = 92%
R = Cy: 91% de = 70% ee (*exo*) = 90% ee (*endo*) = 98%
R = H: 85% de = 24% ee (*exo*) = 98% ee (*endo*) = 97%

Scheme 6.5 Diels–Alder reactions in water.

On the other hand, a diarylprolinol silyl ether salt was demonstrated by Hayashi *et al.* to be an efficient organocatalyst for the enantioselective Diels–Alder reaction of cyclopentadiene with α,β-unsaturated aldehydes performed in water.[8] Under these solvent-free conditions, the cycloadducts were obtained in high yields with good *exo* selectivities and excellent enantioselectivities of up to 99% ee (Scheme 6.5). The authors have examined the role of water and demonstrated that it accelerated the reaction and increased the enantioselectivity.

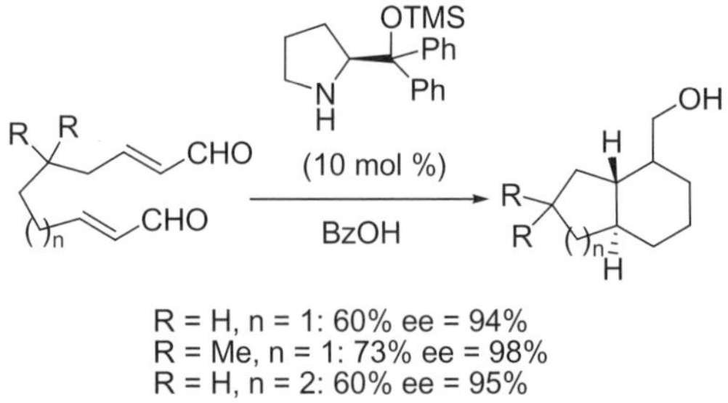

R = H, n = 1: 60% ee = 94%
R = Me, n = 1: 73% ee = 98%
R = H, n = 2: 60% ee = 95%

Scheme 6.6   Intramolecular Diels–Alder reactions.

The scope of this methodology could be extended to other dienes, such as iso-prene or 2,3-dimethylbutadiene, giving similar results with enantioselectivities of up to 94% ee.

This type of catalyst was also employed by Christmann *et al.* to promote the asymmetric intramolecular Diels–Alder reaction of tethered α,β-unsaturated dialdehydes in the presence of benzoic acid as a co-catalyst through vinylogous enamine activation.[9] The corresponding cycloadducts were obtained in good yields and excellent enantioselectivities of up to 98% ee, as shown in Scheme 6.6. When one of the aldehyde functions was replaced by an α,β-unsaturated ketone as the acceptor, no formal [4 + 2] cycloaddition was observed; instead, a direct enantioselective vinylogous Michael addition occurred.

An asymmetric organocatalytic intramolecular Diels–Alder reaction was the key step of a total synthesis of the cytotoxic natural product amaminol B.[10] The cycloaddition of a tetraenal catalysed by a chiral imidazolidinone afforded in the presence of TFA as an additive the indane core of amaminol in good yield and excellent enantioselectivity, as shown in Scheme 6.7.

In addition, the chiral organocatalyst (5S)-2,2,3-trimethyl-5-phenyl-methyl-4-imidazolidinone monohydrochloride was successfully trapped *via* a cation-exchange using montmorillonite clay.[11] This montmorillonite-entrapped organocatalyst acted as a highly efficient heterogeneous catalyst for the asymmetric Diels–Alder reaction of α,β-unsaturated aldehydes with cyclic dienes. The best result was obtained for the cycloaddition of acrolein with cyclohexadiene which provided the cycloadduct in a high yield (82%), a high *endo* diastereoselectivity of 92% de and a high enantioselectivity of 92% ee for the *endo* adduct. Furthermore, the solid montmorillonite-entrapped catalyst could be easily separated from the reaction mixture by simple filtration and was reusable up to four times without any decrease in either the activity or the enantioselectivity.

In 2008, Ishihara *et al.* demonstrated how H-L-Phe-L-Leu-N(CH$_2$CH$_2$)$_2$-reduced triamine and pentafluorobenzenesulfonic acid activated α-phthalimido-acrolein as an aldiminium cation intermediate (Scheme 6.8).[12] The overall process, involving 2,3-dimethylbuta-1,3-diene as the diene partner of the Diels–Alder reaction, provided the corresponding cyclic, *N*-protected α-quaternary

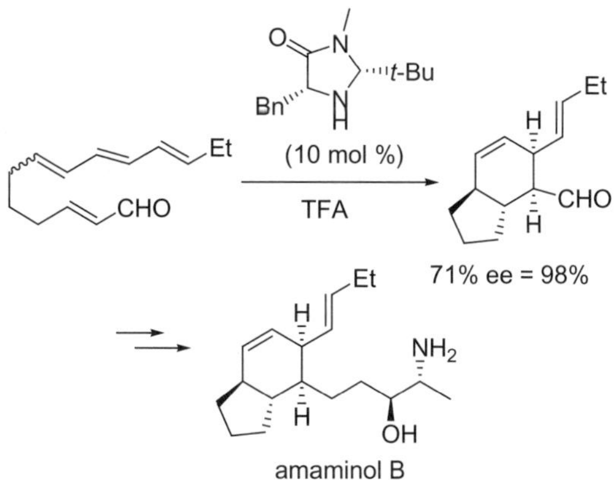

Scheme 6.7 Synthesis of amaminol B.

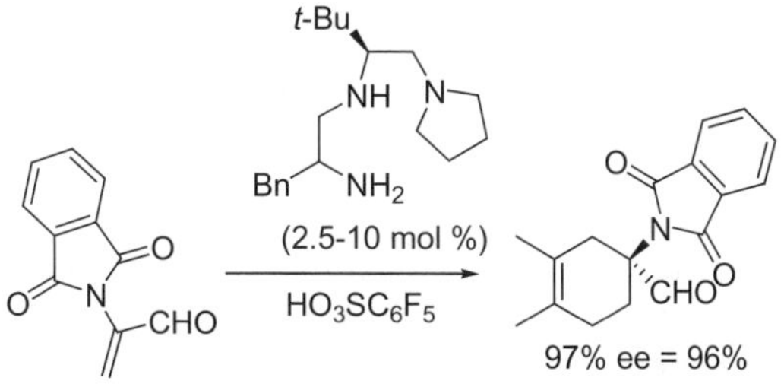

Scheme 6.8 Diels–Alder reaction of *N*-sulfonyl-1-aza-1,3-butadienes with aldehydes.

α-amino acid precursor in excellent yield and enantioselectivity, as shown in Scheme 6.8.

In 2008, Göbel *et al.* reported the Diels–Alder reaction of *N*-substituted maleimides with anthrone derivatives which constituted the first asymmetric catalysis by a bisoxazoline in the absence of any metal ions.[13] These cycloadditions led to the corresponding cycloadducts in excellent yields and moderate enantioselectivities (≤70% ee), as shown in Scheme 6.9.

Compared to the hetero-Diels–Alder reaction of carbonyl compounds and derivatives with dienes, where only a limited number of catalytic and enantioselective reactions have been reported, the number of asymmetric hetero-Diels–Alder reactions in which the ketone or imine functionality is part of a heterodiene is much higher.[14] In contrast, there are only a few examples of using α,β-unsaturated aldehydes in inverse hetero-Diels–Alder reactions. In the case of the inverse electron demand hetero-Diels–Alder reaction, the ketone or imine functionality is part of an α,β-unsaturated system, which reacts in a cycloaddition reaction with an electron-rich alkene. The inverse electron demand hetero-Diels–Alder reaction is primarily controlled by a $LUMO_{diene}$–$HOMO_{dienophile}$ interaction, which can be found, for example, in the reactions

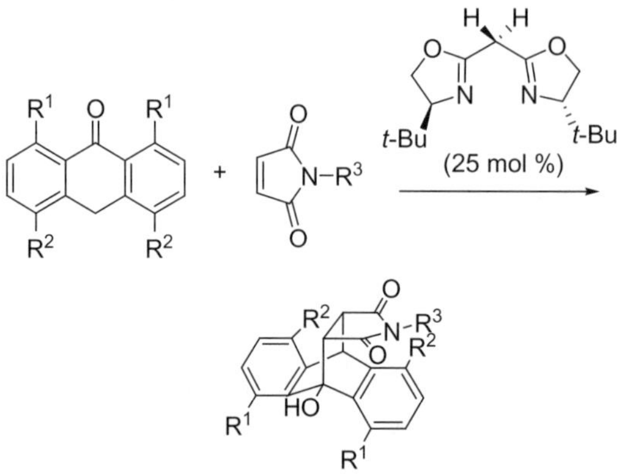

R^1 = Cl, R^2 = H, R^3 = Cy: 94% ee = 70%
R^1 = Cl, R^2 = H, R^3 = Bn: 94% ee = 67%
R^1 = R^2 = H, R^3 = Ph: 85% ee = 47%
R^1 = H, R^2 = Cl, R^3 = Ph: 68% ee = 39%
R^1 = R^2 = H, R^3 = p-MeOC$_6$H$_4$: 76% ee = 46%

Scheme 6.9 Diels–Alder reaction of *N*-substituted maleimides with anthrone derivatives.

of enones and hetero-analogues with alkenes having electron-donating groups, whereas the normal electron demand reaction is a LUMO$_{\text{dienophile}}$–HOMO$_{\text{diene}}$-controlled HDA reaction, which predominantly occurs between electron-rich dienes and electron-deficient dienophiles. Indeed, this reaction is controlled by the LUMO of the α,β-unsaturated ketone or imine interacting with the HOMO of the alkene. This reaction can be catalysed by Lewis acids or organocatalysts, which coordinate to the ketone functionality of the α,β-unsaturated acyl system, thereby lowering the LUMO energy of the dienophile. For these inverse electron demand hetero-Diels–Alder reactions, the ketone or imine carbon atom is converted into a prochiral sp^2-hybridised carbon atom where the chiral centre(s) in the molecule is (are) introduced in the reaction. In this area, several groups have developed organocatalysed asymmetric hetero-Diels–Alder reactions, such as aza-Diels–Alder reactions, which constitute one of the most powerful C–C bond-forming reactions for the preparation of chiral nitrogen-containing compounds, such as piperidine and quinolidine derivatives. In this context, Chen *et al.* have developed asymmetric organocatalytic inverse-electron-demand aza-Diels–Alder reactions of *N*-sulfonyl-1-aza-1,3-butadienes with aldehydes, furnishing the corresponding hemiaminals in good yields and excellent enantioselectivities (93–99% ee), as shown in Scheme 6.10.[15] These reactions, employing chiral α,α-diphenylprolinol trimethylsilyl ether as the organocatalyst combined with acetic acid in a mixture of acetonitrile and water, gave remarkable levels of diastereoselectivity ($>98\%$ de) in all cases of substrates. Moreover, a variety of chiral piperidine derivatives and other useful compounds could be readily prepared from these hemiaminals.

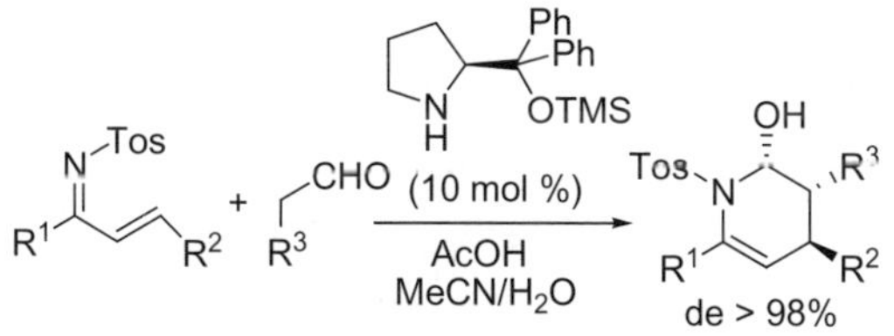

R^1 = R^2 = Ph, R^3 = Et: 88% ee = 97%
R^1 = Ph, R^2 = *p*-ClC$_6$H$_4$, R^3 = Et: 85% ee = 98%
R^1 = Ph, R^2 = *m*-ClC$_6$H$_4$, R^3 = Et: 92% ee = 99%
R^1 = Ph, R^2 = 2-Fu, R^3 = Et: 83% ee = 98%
R^1 = Ph, R^2 = 2-thienyl, R^3 = Et: 87% ee = 98%
R^1 = Ph, R^2 = Me, R^3 = Et: 83% ee = 93%
R^1 = Ph, R^2 = CO$_2$Et, R^3 = Et: 95% ee = 99%
R^1 = *p*-Tol, R^2 = Ph, R^3 = Et: 85% ee = 99%
R^1 = *p*-ClC$_6$H$_4$, R^2 = Ph, R^3 = Et: 82% ee = 98%
R^1 = *m*-BrC$_6$H$_4$, R^2 = Ph, R^3 = Et: 86% ee = 99%
R^1 = 1-Naph, R^2 = Ph, R^3 = Et: 83% ee = 94%
R^1 = R^2 = Ph, R^3 = Me: 92% ee = 98%

Scheme 6.10 Aza-Diels–Alder reactions of *N*-sulfonyl-1-aza-1,3-butadienes with aldehydes.

The organocatalysed asymmetric cycloaddition reaction of α,β-unsaturated ketones and aldehydes is scarcely studied, especially for fluorine-containing substrates. In this area, Liu *et al.* have reported unusual inverse-electron-demand oxa-Diels–Alder reactions of α,β-unsaturated trifluoromethyl ketones with aldehydes catalysed by chiral α,α-diphenylprolinol trimethylsilyl ether.[16] It was shown that the addition of *para*-fluorophenol and silica gel along with this catalyst was necessary, otherwise the reaction was very slow and a poor yield was obtained. Under these optimal conditions, the expected cyclic adducts were obtained in good yields and high *anti*-diastereo- and enantioselectivities, as shown in Scheme 6.11.

There are only few examples of asymmetric inverse hetero-Diels–Alder reactions using α,β-unsaturated aldehydes. As a recent example, a cinchona alkaloid was successfully used to catalyse the oxa-Diels–Alder reaction of heterodienes such as 3-formylchromones with acetylene dicarboxylates, providing the corresponding tricyclic benzopyrones in moderate to high yields (54–87%) and enantioselectivities (46–84% ee), as shown in Scheme 6.12.[17]

6.2 Miscellaneous Cycloadditions

The 1,3-dipolar cycloaddition, also known as the Huisgen cycloaddition,[18] is a classic reaction in organic chemistry consisting of the reaction of a dipolarophile with a 1,3-dipolar compound that allows the production of various 5-membered heterocycles.[19] In particular, the 1,3-dipolar cycloaddition reaction of nitrones with dipolarophiles such as alkenes has received considerable attention in

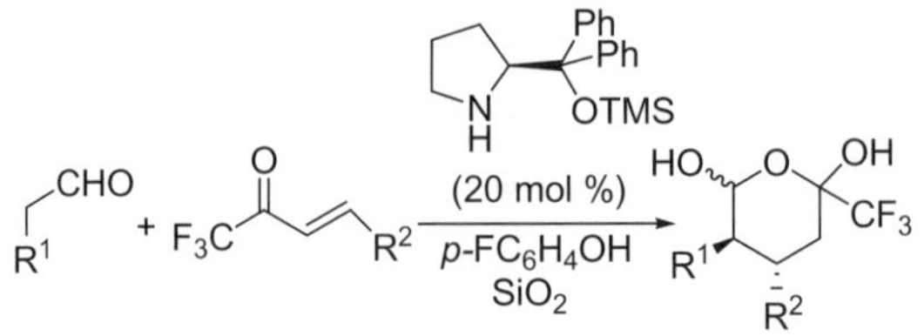

R¹ = Me, R² = Ph: 63% de > 90% ee = 97%
R¹ = Me, R² = *p*-ClC₆H₄: 62% de > 90% ee = 97%
R¹ = Me, R² = *p*-BrC₆H₄: 63% de > 90% ee = 95%
R¹ = Me, R² = *m*-BrC₆H₄: 63% de = 92% ee = 92%
R¹ = Me, R² = *p*-Tol: 76% de = 92% ee = 94%
R¹ = Me, R² = *p*-MeOC₆H₄: 57% de > 90% ee = 93%
R¹ = *i*-Pr, R² = Ph: 71% de = 92% ee = 93%

Scheme 6.11 Oxa-Diels–Alder reactions of α,β-unsaturated trifluoromethyl ketones with aldehydes.

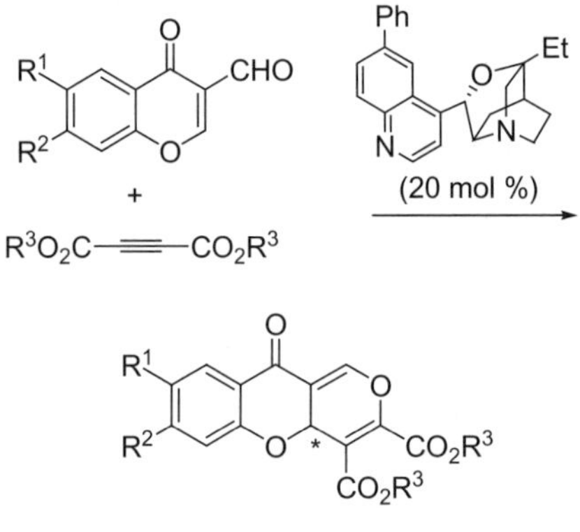

R¹ = R² = H, R³ = Me: 83% ee = 91%
R¹ = Cl, R² = H, R³ = Me: 82% ee = 78%
R¹ = Br, R² = H, R³ = Me: 83% ee = 70%
R¹ = Cl, R² = R³ = Me: 81% ee = 55%
R¹ = *i*-Pr, R² = H, R³ = Et: 85% ee = 67%

Scheme 6.12 Oxa-Diels–Alder reactions of 3-formylchromones with acetylene dicarboxylates.

asymmetric synthesis over the past 20 years.[20] One of the reasons for the success of the synthetic applications of nitrones is that, contrary to the majority of the other 1,3-dipoles, most nitrones are stable compounds that do not require an *in situ* formation. Another synthetic utility of this reaction is the variety of attractive nitrogenated compounds which are available from the thus-formed isoxazolidines. An organocatalytic asymmetric 1,3-dipolar cycloaddition of

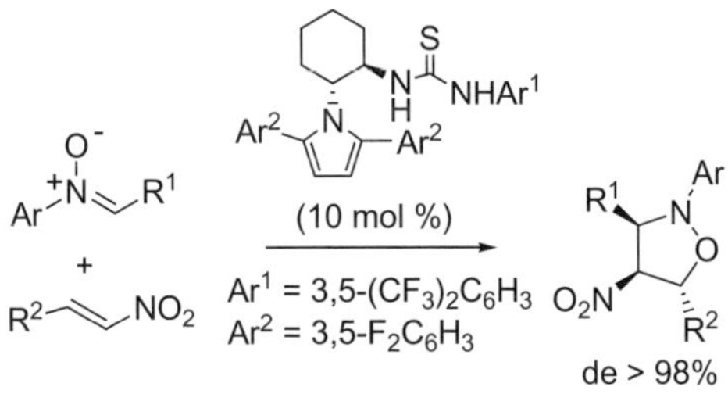

R^1 = Ar = Ph, R^2 = *i*-Pr: 79% ee = 82%
R^1 = Ar = Ph, R^2 = Cy: 85% ee = 83%
R^1 = Ar = Ph, R^2 = Et: 84% ee = 87%
R^1 = Ar = Ph, R^2 = *n*-Pr: 93% ee = 84%
R^1 = *p*-$MeOC_6H_4$, R^2 = Et, Ar = Ph: 78% ee = 88%
R^1 = *p*-Tol, R^2 = Et, Ar = Ph: 71% ee = 86%
R^1 = *p*-ClC_6H_4, R^2 = Et, Ar = Ph: 54% ee = 87%
R^1 = *o*-$MeOC_6H_4$, R^2 = Et, Ar = Ph: 91% ee = 81%
R^1 = *p*-FC_6H_4, R^2 = Et, Ar = Ph: 65% ee = 84%

Scheme 6.13 1,3-Dipolar cycloadditions of nitrones to nitroolefins.

nitrones to nitroolefins was developed by Chen *et al.* by using a novel chiral thiourea-pyrrole organocatalyst derived from (*R*,*R*)-1,2-diaminocyclohexane.[21] In these conditions, the corresponding chiral isoxazolidines were achieved in good yields, moderate to high enantioselectivities (40–88% ee) and excellent diastereoselectivities (generally >98% de), as shown in Scheme 6.13.

The first example of 1,3-dipolar cycloaddition of nitrones catalysed by a chiral organocatalyst was that developed by Yamamoto *et al.*, concerning the cycloaddition with ethyl vinyl ether promoted by a chiral BINOL-derived phosphoramide.[22] These reactions yielded the *endo* products as the major diastereomers with high diastereoselectivities of up to 94% de, excellent yields and high enantioselectivities of up to 93% ee, as shown in Scheme 6.14.

In addition, chiral α,α-diphenylprolinol trimethylsilyl ether was found by Cordova *et al.* to be capable of inducing a one-pot catalytic cascade synthesis of cycloheptane derivatives in which six new bonds and five new stereocentres were formed with an excellent stereocontrol of >92% de and >98% ee.[23] The assembly of the functional cycloheptanes was feasible by employing a three-component intermolecular [3+2] cycloaddition starting from two simple aldehydes and hydroxylamine, followed directly by a two-component intramolecular [3+2] cycloaddition of the nitrone intermediate and another hydroxylamine, as shown in Scheme 6.15. This process constituted an unprecedented example of regiospecific and highly chemoselective one-pot organocatalytic cascade synthesis of bis-oxazolidines bearing a functionalised seven-membered carbocycle core, offering a direct entry to nearly diastereo- and enantiomerically pure cycloheptane derivatives.

In recent years, azomethine ylides have become one of the most investigated classes of 1,3-dipoles and, based on their cycloaddition chemistry, various

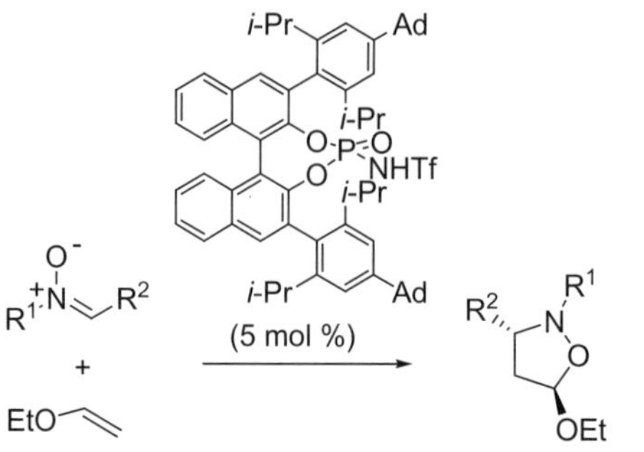

R¹ = R² = Ph: 85% de = 92% ee = 70%
R¹ = Ph, R² = p-ClC$_6$H$_4$: 95% de = 94% ee = 90%
R¹ = Ph, R² = p-CF$_3$C$_6$H$_4$: 69% de = 92% ee = 92%
R¹ = p-ClC$_6$H$_4$, R² = p-CF$_3$C$_6$H$_4$: 66% de = 86% ee = 93%
R¹ = p-ClC$_6$H$_4$, R² = p-NO$_2$C$_6$H$_4$: 98% de = 78% ee = 92%
R¹ = p-ClC$_6$H$_4$, R² = 2-Fu: 95% de = 86% ee = 89%

Scheme 6.14   1,3-Dipolar cycloadditions of nitrones to ethyl vinyl ether.

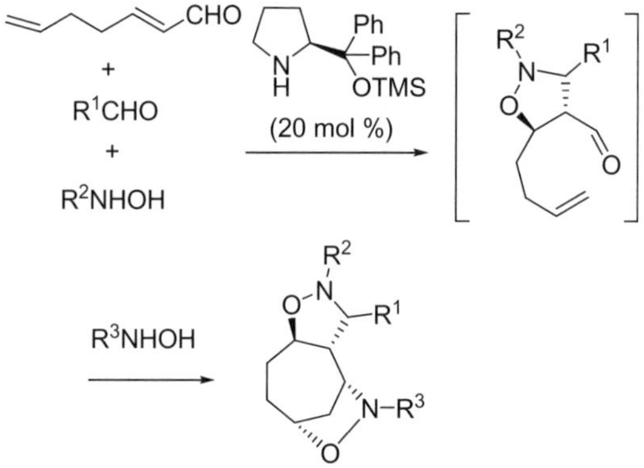

R¹ = R² = R³ = Ph: 42% de > 92% ee = 99%
R¹ = p-BrC$_6$H$_4$, R² = R³ = Ph: 42%
de > 92% ee = 98%
R¹ = p-MeOC$_6$H$_4$, R² = R³ = p-ClC$_6$H$_4$: 49%
de > 92% ee > 99%
R¹ = (E)-PhCH=CH, R² = R³ = Ph: 51%
de > 92% ee = 99%
R¹ = p-BrC$_6$H$_4$, R² = p-ClC$_6$H$_4$, R³ = Ph: 41%
de > 92% ee = 99%
R¹ = (E)-PhCH=CH, R² = R³ = p-ClC$_6$H$_4$: 43%
de > 92% ee = 99%
R¹ = R³ = Ph, R² = Bn: 68% de > 92% ee = 98%

Scheme 6.15   Domino four-component double [3 + 2] cycloaddition reactions.

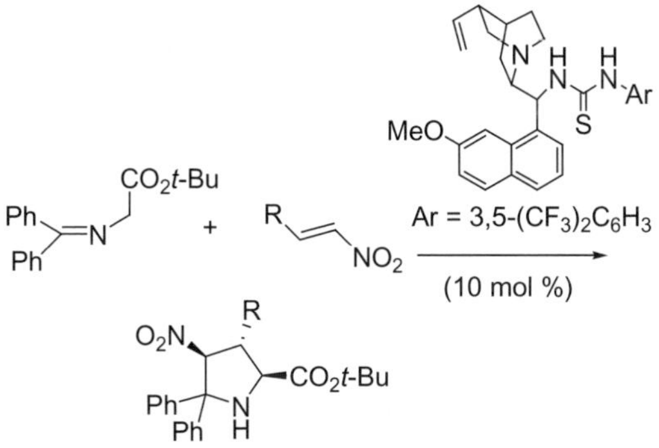

R = Ph: 77% de = 86% ee = 63%
R = *p*-MeOC$_6$H$_4$: 63% de = 78% ee = 65%
R = *p*-Tol: 56% de = 90% ee = 62%
R = *o*-Tol: 68% de = 76% ee = 60%
R = *p*-BrC$_6$H$_4$: 65% de = 88% ee = 49%
R = thienyl-2-yl: 50% de > 98% ee = 59%

Scheme 6.16   1,3-Dipolar cycloadditions of azomethine ylide to nitroalkenes.

methods for the synthesis of pyrrolidine derivatives have been developed. However, organocatalytic approaches to access chiral pyrrolidines are still scarce. In this context, Gong *et al.* have developed the first organocatalytic stereoselective 1,3-dipolar cycloaddition of an azomethine ylide with nitroalkenes.[24] This process catalysed by a cinchona alkaloid thiourea provided a straightforward access to chiral highly substituted pyrrolidines with good yields, moderate enantioselectivities and very high diastereoselectivities of up to 98% de, as shown in Scheme 6.16.

These authors also reported the first catalytic asymmetric 1,3-dipolar cycloaddition between α-substituted isocyanoesters and nitroolefins by using a cinchona alkaloid derivative as the organocatalyst to afford chiral 2,3-dihydropyrroles with high diastereo- and enantioselectivities of up to >90% de and >99% ee, respectively.[25] The best results are collected in Scheme 6.17.

Moreover, these authors described a three-component asymmetric 1,3-dipolar cycloaddition between aldehydes, amino esters and dipolarophiles catalysed by a new biphosphoric acid derived from (*R,R*)-linked BINOL, furnishing multiply substituted pyrrolidines in high yields and with excellent enantioselectivities of up to 99% ee (Scheme 6.18).[26]

A new class of chiral bifunctional thiourea catalysts derived from *trans*-2-amino-1-(diphenylphosphino)cyclohexane was developed by Jacobsen and Fang in order to be applied to a highly enantioselective synthesis of a wide range of 2-aryl-2,5-dihydropyrrole derivatives.[27] This strategy was based on a [3 + 2] cycloaddition between an *N*-phosphinoyl imine and an allene in the presence of TEA and water as additives. High yields combined with excellent enantioselectivities of up to 98% ee were observed in all cases of substrates, as shown in Scheme 6.19.

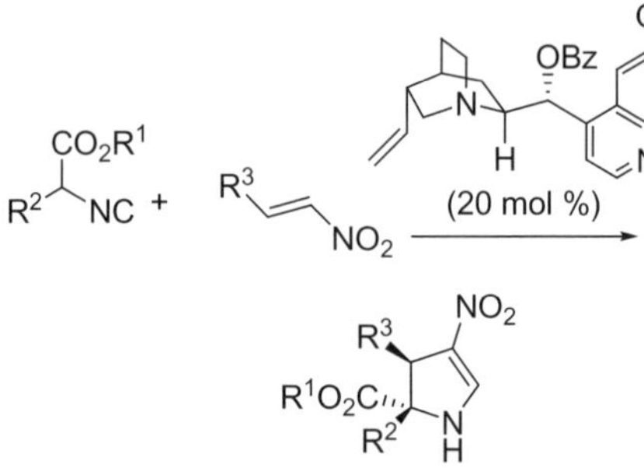

R$_1$ = Bn, R$_2$ = Ph, R$_3$ = 1-Naph: 73%
de = 90% ee = 97%
R$_1$ = Et, R$_2$ = Ph, R$_3$ = 1-Naph: 78%
de > 90% ee = 91%
R$_1$ = Me, R$_2$ = p-MeOC$_6$H$_4$, R$_3$ = p-BrC$_6$H$_4$:
60% de > 90% ee = 97%
R$_1$ = Me, R$_2$ = p-AcOC$_6$H$_4$, R$_3$ = p-BrC$_6$H$_4$:
99% de = 72% ee = 90%

Scheme 6.17   [3 + 2] Cycloadditions of α-substituted isocyanoesters to nitroolefins.

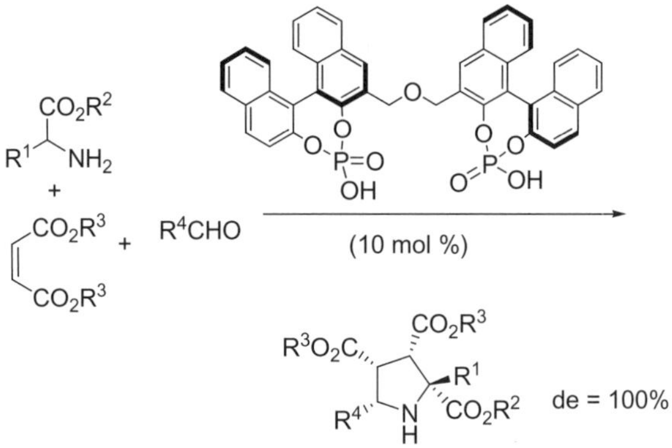

R^1 = CO$_2$Et, R^2 = Et, R^3 = Me, R^4 = m-NO$_2$C$_6$H$_4$: 95% ee = 99%
R^1 = CO$_2$Et, R^2 = Et, R^3 = Me, R^4 = o-NO$_2$C$_6$H$_4$: 97% ee = 96%
R^1 = CO$_2$Et, R^2 = Et, R^3 = Me, R^4 = p-MeOC$_6$H$_4$: 88% ee = 98%
R^1 = CO$_2$Et, R^2 = Et, R^3 = Me, R^4 = p-BrC$_6$H$_4$: 89% ee = 99%
R^1 = CO$_2$Me, R^2 = R^3 = Me, R^4 = p-NO$_2$C$_6$H$_4$: 95% ee = 96%
R^1 = CO$_2$Me, R^2 = Me, R^3 = Et, R^4 = p-NO$_2$C$_6$H$_4$: 92% ee = 91%
R^1 = Ph, R^2 = R^3 = Me, R^4 = p-NO$_2$C$_6$H$_4$: 93% ee = 97%

Scheme 6.18   Three-component 1,3-dipolar cycloadditions.

Even though aza-β-lactams have attracted interest because of their biological activities, only limited progress has been reported with regard to the enantioselective synthesis of this family of heterocycles. The Staudinger [2 + 2] cycloaddition of ketenes with imines is a versatile and efficient route to

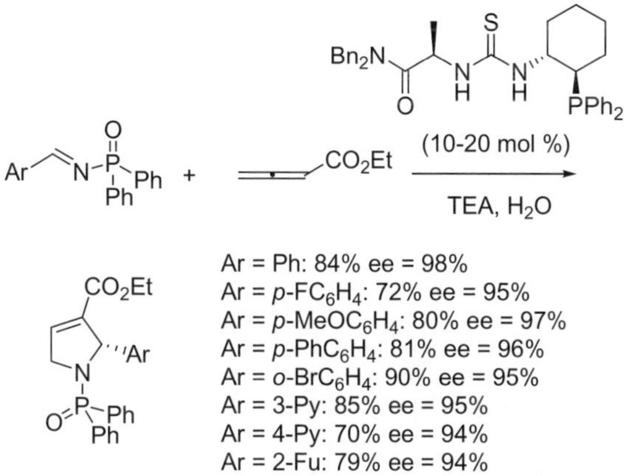

Scheme 6.19 [3 + 2] Cycloadditions of allenes with *N*-phosphinoyl imines.

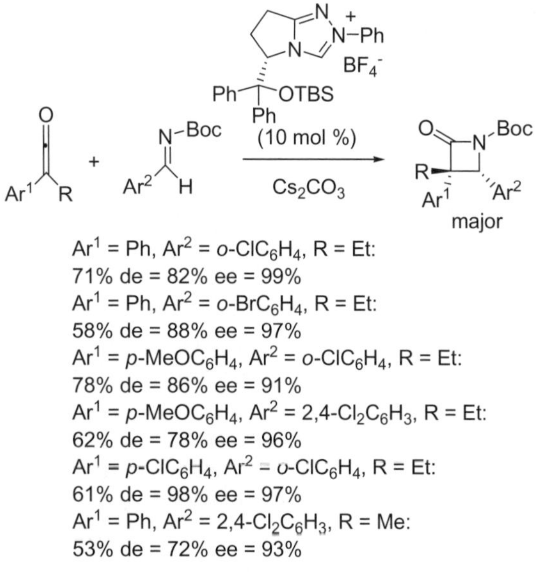

Scheme 6.20 [2 + 2] Cycloadditions of *N*-Boc imines with ketenes.

construct β-lactams. In 2008, Ye *et al.* demonstrated that chiral *N*-heterocyclic carbenes could be efficient catalysts for the asymmetric version of this reaction for the first time.[28] Therefore, in the presence of a chiral triazolium salt derived from L-pyroglutamic acid, the reaction of *N*-Boc imines with ketenes provided the expected *N*-Boc β-lactams in good yields, moderate to high *cis*-diastereoselectivities of up to 98% de and excellent enantioselectivities (91–99% ee), as shown in Scheme 6.20.

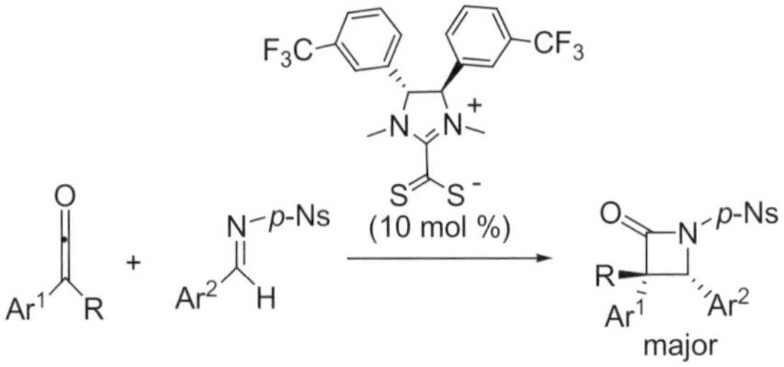

Ar^1 = Ph, Ar^2 = *p*-$CF_3C_6H_4$, R = Et:
96% de = 74% ee = 95%
Ar^1 = Ph, Ar^2 = *p*-CNC_6H_4, R = Et:
99% de = 52% ee = 93%
Ar^1 = Ph, Ar^2 = *p*-FC_6H_4, R = Et:
99% de = 64% ee = 96%
Ar^1 = Ph, Ar^2 = 1-Naph, R = Et:
96% de = 78% ee = 83%
Ar^1 = R = Ph, Ar^2 = *p*-FC_6H_4: 76% ee = 96%

Scheme 6.21 [2 + 2] Cycloadditions of *N-para*-nosyl imines with ketenes.

In the same area, enantiopure imidazolinium-dithiocarboxylates have been found to be efficient organocatalysts for the [2 + 2] cycloaddition of ketenes with *N-para*-nosyl imines.[29] The corresponding *cis-N-para*-nosyl β-lactams were obtained as major diastereomers in excellent yields (96–99%) and moderate to good diastereoselectivities ($\leq$78% de) combined with good enantioselectivities of up to 96% ee, as shown in Scheme 6.21.

On the other hand, Hayashi *et al.* have reported the highly enantioselective formal [3 + 3] cycloaddition of α,β-unsaturated aldehydes with enecarbamates catalysed by diphenylprolinol silyl ether as an organocatalyst.[30] This reaction consisted of four consecutive reactions including an asymmetric ene reaction, an isomerisation from an imine into an enecarbamate, a hydrolysis and a hemiacetal formation in one pot to afford synthetically important chiral piperidine derivatives with excellent enantioselectivities of up to 99% ee, good yields and moderate to good diastereoselectivities, as shown in Scheme 6.22.

Similar conditions were applied by Gong *et al.* to the asymmetric formal [3 + 3] cycloaddition of α,β-unsaturated aldehydes with Nazarov reagents bearing an aryl group at C5, which, after oxidation, afforded the corresponding 3,4-dihydropyranones in good yields combined with excellent enantioselectivities of up to 97% ee.[31] The best results are collected in Scheme 6.23.

Cyclopropanes are useful building blocks for the synthesis of natural and synthetic products due to their unique structures and reactivities. They are also found as a basic structural unit in a wide range of biologically active compounds. As a result, the development of efficient methods for the asymmetric synthesis of cyclopropanes has attracted intensive research interest in recent years.[32] In 2008, Cao *et al.* reported a novel domino cyclopropanation-Wittig

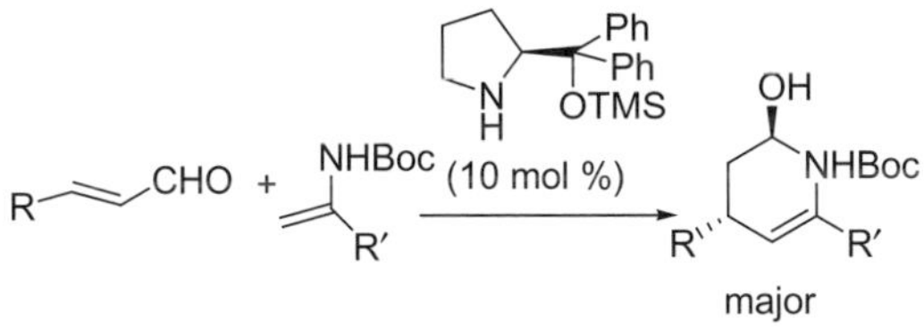

R = 2-Naph, R' = Ph: 83% de = 42% ee = 90%
R = p-NO₂C₆H₄, R' = Ph: 83% de = 42% ee = 97%
R = p-MeOC₆H₄, R' = Ph: 73% de = 62% ee = 97%
R = 2-Fu, R' = Ph: 80% de = 46% ee = 99%
R = Ph, R' = p-BrC₆H₄: 89% de = 72% ee = 90%
R = Ph, R' = 2-Fu: 85% de = 62% ee = 90%

Scheme 6.22 Formal [3 + 3] cycloadditions of α,β-unsaturated aldehydes with enecarbamates.

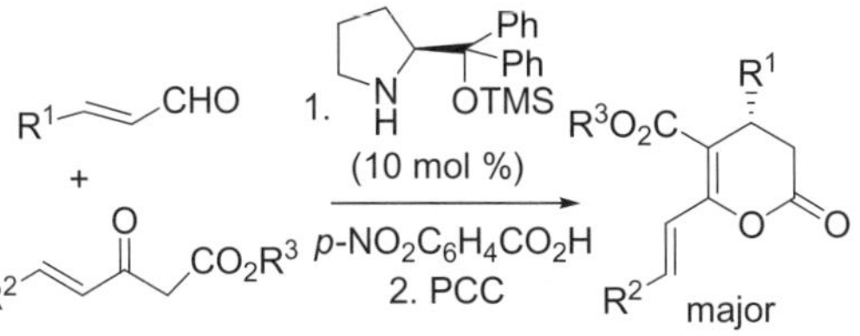

R¹ = p-NO₂C₆H₄, R² = Ph, R³ = Et: 77% ee = 95%
R¹ = m-NO₂C₆H₄, R² = Ph, R³ = Et: 71% ee = 94%
R¹ = p-CF₃C₆H₄, R² = Ph, R³ = Et: 41% ee = 96%
R¹ = CO₂Et, R² = Ph, R³ = Et: 63% ee = 97%
R¹ = p-NO₂C₆H₄, R² = Ph, R³ = t-Bu: 43% ee = 96%
R¹ = CO₂Et, R² = Ph, R³ = Bn: 45% ee = 97%
R¹ = CO₂Et, R² = p-BrC₆H₄, R³ = Et: 78% ee = 96%

Scheme 6.23 Formal [3 + 3] cycloadditions of α,β-unsaturated aldehydes with Nazarov reagents.

reaction of α,β-unsaturated aldehydes with arsonium ylides using a chiral 2-trimethylsilanyloxy-methyl-pyrrolidine-based dendritic catalyst.[33] This process provided the corresponding chiral cyclopropanes with excellent enantioselectivities of up to 99% ee, good to high diastereoselectivities and good yields, as shown in Scheme 6.24. Furthermore, this dendritic catalyst was recoverable and reusable without any loss in the activity.

A total synthesis of eicosanoid was developed by Kumaraswamy and Padmaja in which the genesis of chirality was obtained through an organocatalytic cyclopropanation using a quinidine-based alkaloid as the organocatalyst.[34] Therefore, an enone reacted with *tert*-butyl bromoacetate in the presence of (DHQD)₂Py as the organocatalyst and Cs₂CO₃ as a base upon heating to furnish the enantiopure cyclopropane in good yield (Scheme 6.25).

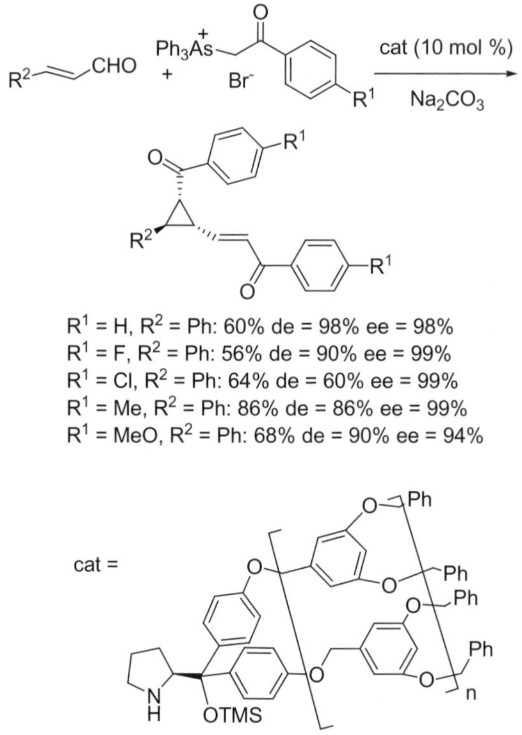
cat (10 mol %)
Na_2CO_3
R^1 = H, R^2 = Ph: 60% de = 98% ee = 98%
R^1 = F, R^2 = Ph: 56% de = 90% ee = 99%
R^1 = Cl, R^2 = Ph: 64% de = 60% ee = 99%
R^1 = Me, R^2 = Ph: 86% de = 86% ee = 99%
R^1 = MeO, R^2 = Ph: 68% de = 90% ee = 94%
cat =
OTMS

Scheme 6.24 Domino cyclopropanation-Wittig reactions of α,β-unsaturated aldehydes with arsonium ylides.

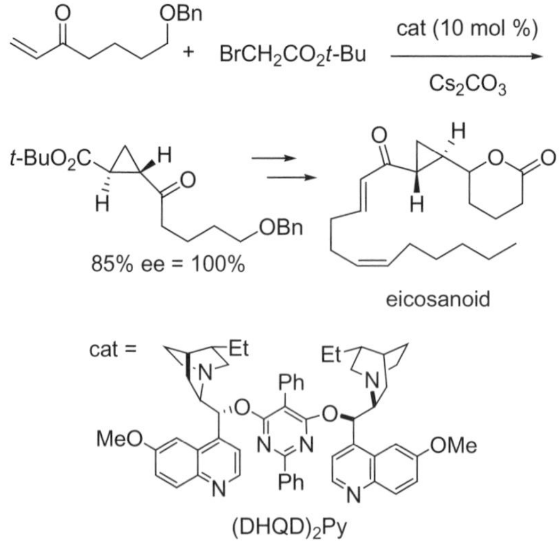
$BrCH_2CO_2t$-Bu
cat (10 mol %)
Cs_2CO_3
85% ee = 100%
eicosanoid
cat =
$(DHQD)_2Py$

Scheme 6.25 Synthesis of eicosanoid.

6.3 Conclusions

Several cinchona alkaloids have recently been demonstrated as efficient chiral organocatalysts for asymmetric Diels–Alder reactions. Therefore, excellent results were obtained for the Diels–Alder reaction of 3-vinylindoles with maleimides catalysed by a chiral bifunctional thiourea derived from hydroquinine, leading almost exclusively to the corresponding *endo*-carbazoles in high yields and enantioselectivities of up to 99% ee and, moreover, by using a low catalyst loading of 3 mol%. In addition, a diarylprolinol silyl ether salt used at 5 mol% of catalyst loading has proved to be an efficient organocatalyst for the enantioselective Diels–Alder reaction of cyclopentadiene with α,β-unsaturated aldehydes performed in water. Under solvent-free conditions, the cycloadducts were obtained in high yields with good *exo* selectivities combined with excellent enantioselectivities of up to 99% ee.

Some chiral organocatalysts were also successfully applied to promote asymmetric inverse-electron-demand hetero-Diels–Alder reactions. As an example, the aza-Diels–Alder reaction of *N*-sulfonyl-1-aza-1,3-butadienes with aldehydes yielded the corresponding hemiaminals in good yields and excellent enantioselectivities in the presence of chiral α,α-diphenylprolinol trimethylsilyl ether as the organocatalyst. In the same area, the oxa-Diels–Alder reaction of α,β-unsaturated trifluoromethyl ketones with aldehydes performed in the presence of the same organocatalyst provided the corresponding cyclic adducts in good yields and high *anti*-diastereo- and enantioselectivities.

Other successful asymmetric organocatalytic cycloadditions were developed in the last year, such as a three-component asymmetric 1,3-dipolar cycloaddition between aldehydes, amino esters and dipolarophiles, which was catalysed by a new biphosphoric acid derived from (*R,R*)-linked BINOL, furnishing multiply substituted pyrrolidines in high yields and with excellent enantioselectivities of up to 99% ee. In addition, a new class of chiral bifunctional thiourea catalysts derived from *trans*-2-amino-1-(diphenylphosphino)cyclohexane was applied to a highly enantioselective synthesis of a wide range of 2-aryl-2,5-dihydropyrrole derivatives on the basis of a [3 + 2] cycloaddition between an *N*-phosphinoyl imine and an allene, providing high yields combined with excellent enantioselectivities of up to 98%.

References

1. E. J. Corey, *Ang. Chem., Int. Ed. Engl.*, 2002, **41**, 1650–1667.
2. H. He, B.-J. Pei, H.-H. Chou, T. Tian, W.-H. Chan and A. W. M. Lee, *Org. Lett.*, 2008, **10**, 2421–2424.
3. Y. Langlois, A. Petit, P. Rémy, M.-C. Scherrmann and C. Kouklovsky, *Tetrahedron Lett.*, 2008, **49**, 5576–5579.
4. C. Gioia, A. Hauville, L. Bernardi, F. Fini and A. Ricci, *Ang. Chem., Int. Ed. Engl.*, 2008, **47**, 9236–9239.

5. R. P. Singh, K. Bartelson, Y. Wang, H. Su, X. Lu and L. Deng, *J. Am. Chem. Soc.*, 2008, **130**, 2422–2423.
6. M. Bella, D. M. S. Schietroma, P. P. Cusella and T. Gasperi, *Chem. Commun.*, 2009, 597–599.
7. J. Shen and C.-H. Tan, *Org. Biomol. Chem.*, 2008, **6**, 4096–4098.
8. Y. Hayashi, S. Samanta, H. Gotoh and H. Ishikawa, *Ang. Chem., Int. Ed. Engl.*, 2008, **47**, 6634–6637.
9. R. M. De Figueiredo, R. Fröhlich and M. Christmann, *Ang. Chem., Int. Ed. Engl.*, 2008, **47**, 1450–1453.
10. W. C. Jacobs and M. Christmann, *Synlett*, 2008, **2**, 247–251.
11. T. Mitsudome, K. Nose, T. Mizugaki, K. Jitsukawa and K. Kaneda, *Tetrahedron Lett.*, 2008, **49**, 5464–5466.
12. K. Ishihara, K. Nakano and M. Akakura, *Org. Lett.*, 2008, **10**, 2893–2896.
13. D. Akalay, G. Dürner and M. W. Göbel, *Eur. J. Org. Chem.*, 2008, 2365–2368.
14. H. Pellissier, *Tetrahedron*, 2009, **65**, 2839–2877.
15. B. Han, J.-L. Li, C. Ma, S.-J. Zhang and Y.-C. Chen, *Ang. Chem., Int. Ed. Engl.*, 2008, **47**, 9971–9974.
16. Y. Zhao, X.-J. Wang and J.-T. Liu, *Synlett*, 2008, **7**, 1017–1020.
17. H. Waldmann, V. Khedkar, H. Dückert, M. Schürmann, I. M. Oppel and K. Kumar, *Ang. Chem., Int. Ed. Engl.*, 2008, **47**, 6869–6872.
18. R. Huisgen, *Ang. Chem., Int. Ed. Engl.*, 1963, **10**, 565–598.
19. H. Pellissier, *Tetrahedron*, 2007, **63**, 3235–3285.
20. (a) M. Frederickson, *Tetrahedron*, 1997, **53**, 403–425; (b) S. Karlsson and H.-E. Högberg, *Org. Prep. Proc. Int.*, 2001, **33**, 103–172; (c) K. V. Gothelf and K. A. Jorgensen, *Chem. Rev.*, 1998, **98**, 863–909.
21. W. Du, Y.-K. Liu, L. Yue and Y.-C. Chen, *Synlett*, 2008, **19**, 2997–3000.
22. P. Jiao, D. Nakashima and H. Yamamoto, *Ang. Chem., Int. Ed. Engl.*, 2008, **47**, 2411–2413.
23. J. Vesely, R. Rios, I. Ibrahem, G.-L. Zhao, L. Eriksson and A. Cordova, *Chem. Eur. J.*, 2008, **14**, 2693–2698.
24. M.-X. Xue, X.-M. Zhang and L.-Z. Gong, *Synlett*, 2008, **5**, 691–694.
25. C. Guo, M.-X. Xue, M.-K. Zhu and L.-Z. Gong, *Ang. Chem., Int. Ed. Engl.*, 2008, **47**, 3414–3417.
26. X.-H. Chen, W.-Q. Zhang and L.-Z. Gong, *J. Am. Chem. Soc.*, 2008, **130**, 5652–5653.
27. Y.-Q. Fang and E. N. Jacobsen, *J. Am. Chem. Soc.*, 2008, **130**, 5660–5661.
28. Y.-R. Zhang, L. He, X. Wu, P.-L. Shao and S. Ye, *Org. Lett.*, 2008, **10**, 277–280.
29. O. Sereda, A. Blanrue and R. Wilhelm, *Chem. Commun.*, 2009, 1040–1042.
30. Y. Hayashi, H. Gotoh, R. Masui and H. Ishikawa, *Ang. Chem., Int. Ed. Engl.*, 2008, **47**, 4012–4015.

31. M.-K. Zhu, Q. Wei and L.-Z. Gong, *Adv. Synth. Catal.*, 2008, **350**, 1281–1285.
32. H. Pellissier, *Tetrahedron*, 2008, **64**, 7041–7095.
33. Y.-H. Zhao, C.-W. Zheng, G. Zhao and W.-G. Cao, *Tetrahedron: Asymmetry*, 2008, **19**, 701–708.
34. G. Kumaraswamy and M. Padmaja, *J. Org. Chem.*, 2008, **73**, 5198–5201.

CHAPTER 7
Oxidations

7.1 Epoxidations of Alkenes

Asymmetric epoxidation is an extremely useful methodology for generating chiral compounds, because chiral epoxides are versatile building blocks in organic synthesis.[1] In addition, many biologically active compounds and natural products contain epoxide functionalities.[2] Various elegant and useful organocatalytic asymmetric epoxidations of alkenes have been developed.[3] In 2008, Page *et al.* studied an organocatalytic asymmetric epoxidation of 1-phenylcyclohexene based on the use of chiral iminium salt catalysts, in which the stoichiometric persulfate oxidant was generated electrochemically.[4] This system offered a comparable level of enantioselectivity of up to 64% ee to the use of commercially available persulfate, as shown in Scheme 7.1.

With the aim of developing a cheaper and greener alternative to oxone, the same authors have also successfully demonstrated that hydrogen peroxide was an efficient stoichiometric oxidant in iminium salt-catalysed asymmetric epoxidation of 1-phenylcyclohexene.[5] This reaction was promoted by a catalytic amount of an inorganic promoter such as carbonate, hydrogen carbonate and hydroxide. The enantioselectivity was, however, generally lower than when using oxone as the oxidant, but was largely independent of the amount of base and catalyst. In contrast, water content and temperature appeared to have the greatest impact on the enantioselectivity and rate. The best enantioselectivity of 56% ee was obtained by using a chiral biphenylazepinium salt, ethereal hydrogen peroxide in acetonitrile in the absence of water at −5 °C, as shown in Scheme 7.2.

7.2 Epoxidations of α,β-Unsaturated Carbonyl Compounds

Although methods for the enantioselective epoxidation of alkenes have been admirably performed over the last 30 years, achievements in the epoxidation of

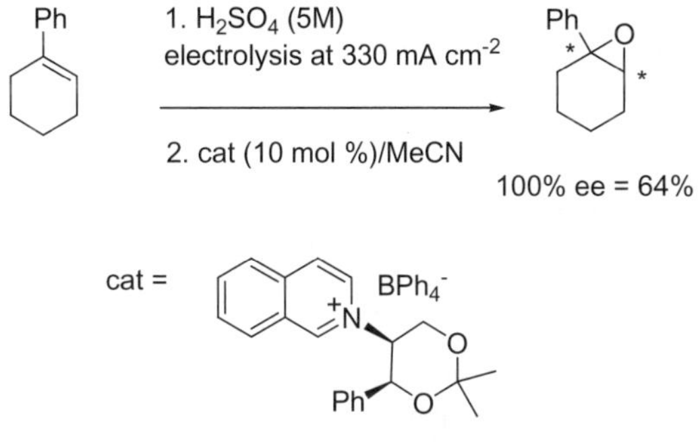

Scheme 7.1 Electrochemical epoxidation of 1-phenylcyclohexene.

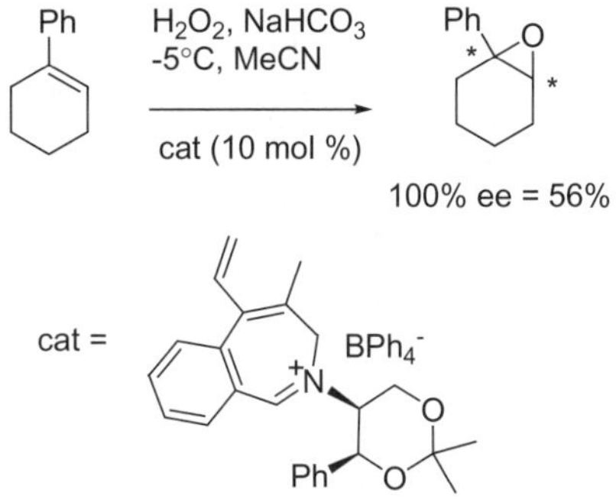

Scheme 7.2 Epoxidation of 1-phenylcyclohexene.

electron-poor alkenes, such as enones, with good results have been less developed. In this case, a nucleophilic oxygen-donor molecule is necessary for carrying out this transformation. Recently, a number of useful combinations of different types of organocatalysts and oxidative reagents have been elaborated. As an example, List and Wang have developed asymmetric epoxidations of α,β-unsaturated aldehydes by using a dibenzyl ammonium salt of a chiral phosphoric acid derived from BINOL.[6] The use of this chiral organocatalyst together with *tert*-butyl hydroperoxide as the oxidant converted a wide range of α,β-unsaturated aldehydes into the corresponding *trans*-epoxides in good to excellent yields (60–95%), high *trans*-diastereoselectivities (88–98% de) and high enantioselectivities of up to 96% ee, as shown in Scheme 7.3. Furthermore, the scope of the process could be extended to 1,2,2-trisubstituted α,β-unsaturated enals with enantioselectivities of up to 94% ee.

In 2009, another organocatalytic asymmetric epoxidation of α,β-unsaturated aldehydes was reported by Gilmour *et al.* based on the use of hydrogen peroxide as the oxidant.[7] Excellent enantioselectivities of up to 97% ee were obtained for a range of α,β-unsaturated aldehydes combined with high yields and moderate to high diastereoselectivities by using 2-(fluorodiphenylmethyl)pyrrolidine as the chiral organocatalyst, as shown in Scheme 7.4.

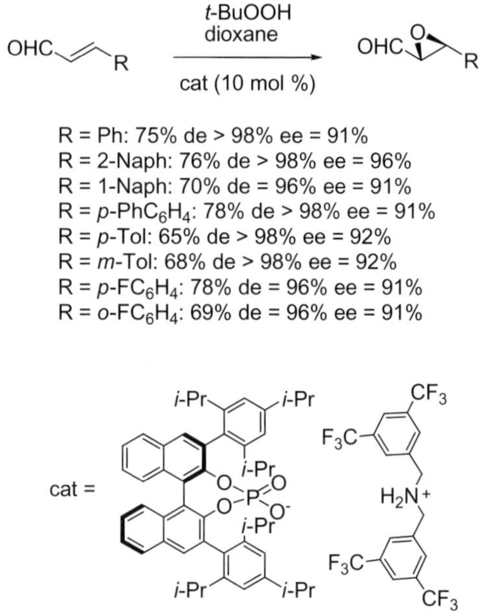
t-BuOOH
dioxane
OHC
R
cat (10 mol %)
R = Ph: 75% de > 98% ee = 91%
R = 2-Naph: 76% de > 98% ee = 96%
R = 1-Naph: 70% de = 96% ee = 91%
R = p-PhC6H4: 78% de > 98% ee = 91%
R = p-Tol: 65% de > 98% ee = 92%
R = m-Tol: 68% de > 98% ee = 92%
R = p-FC6H4: 78% de = 96% ee = 91%
R = o-FC6H4: 69% de = 96% ee = 91%
cat =
i-Pr
CF3
F3C
H2N+

Scheme 7.3 Epoxidations of α,β-unsaturated aldehydes with *t*-BuOOH.

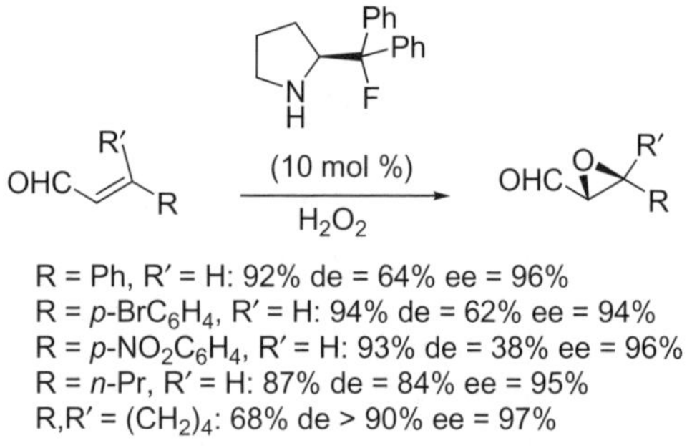
Ph
Ph
F
N
H
(10 mol %)
H2O2
OHC
R'
R
R = Ph, R' = H: 92% de = 64% ee = 96%
R = p-BrC6H4, R' = H: 94% de = 62% ee = 94%
R = p-NO2C6H4, R' = H: 93% de = 38% ee = 96%
R = n-Pr, R' = H: 87% de = 84% ee = 95%
R,R' = (CH2)4: 68% de > 90% ee = 97%

Scheme 7.4 Epoxidations of α,β-unsaturated aldehydes with H_2O_2.

On the other hand, an enantioselective epoxidation of α,β-unsaturated ketones was developed by Zhu *et al.* by using easily accessible and recoverable fluorous α,α-diaryl-L-prolinol as an organocatalyst and *tert*-butyl hydrogenperoxide (TBHP) as an oxidant.[8] As shown in Scheme 7.5, the corresponding epoxides have been obtained for a number of substrates with moderate to good enantioselectivities of up to 84% ee.

A series of cyclic and acyclic β-amino alcohols were examined as organocatalysts by Lattanzi and Russo in the epoxidation of *trans*-chalcones with *tert*-butyl hydrogenperoxide as the oxidant.[9] It was demonstrated that primary, secondary and tertiary β-amino alcohols were able to promote the reaction with

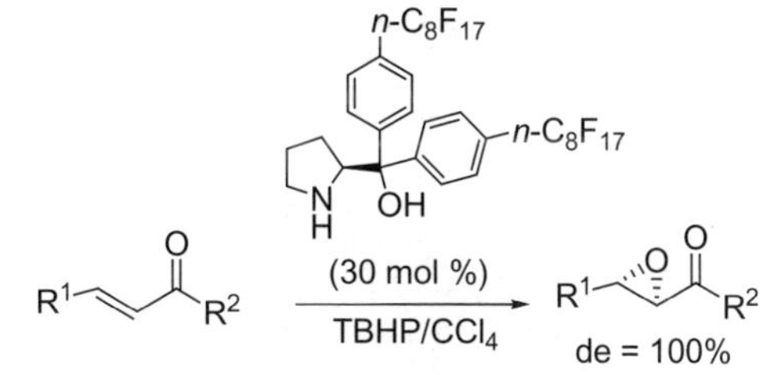

Scheme 7.5 Epoxidations of α,β-unsaturated ketones.

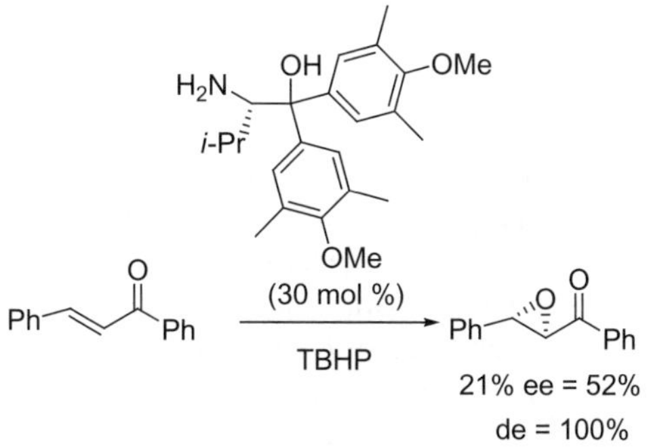

R^1 = R^2 = Ph: 65% ee = 84%
R^1 = p-ClC$_6$H$_4$, R^2 = Ph: 63% ee = 80%
R^1 = p-ClC$_6$H$_4$, R^2 = p-MeOC$_6$H$_4$: 41% ee = 83%
R^1 = Ph, R^2 = 2-Fu: 56% ee = 81%
R^1 = Ph, R^2 = p-ClC$_6$H$_4$: 61% ee = 80%
R^1 = Ph, R^2 = p-BrC$_6$H$_4$: 60% ee = 82%
R^1 = Ph, R^2 = p-MeOC$_6$H$_4$: 31% ee = 00%
R^1 = Ph, R^2 = p-FC$_6$H$_4$: 67% ee = 83%

Scheme 7.6 Epoxidation of *trans*-chalcone.

variable activity and level of asymmetric induction. Subtle modifications to the substitution pattern of the skeleton of primary β-amino alcohols strongly influenced their efficiency in the epoxidation. The best result affording *trans*-chalcone epoxide with enantioselectivity of 52% ee is shown in Scheme 7.6. In addition, the authors have shown that secondary pyrrolidine-based compounds were the best promoters in comparison to four- and six-membered ring analogues.

The first highly enantioselective epoxidation of cyclic enones was developed by List *et al.* by using hydrogen peroxide as the oxidant and a chiral primary amine salt as the organocatalyst.[10] A series of primary amine salts were investigated for this reaction and the best results were obtained with a quinine-derived primary amine salt, which provided excellent enantioselectivities of up to 99% ee for a variety of cyclic enones, as shown in Scheme 7.7.

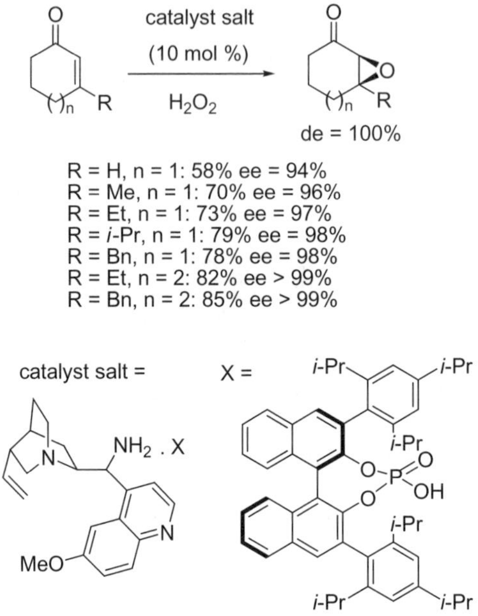
catalyst salt
(10 mol %)
H2O2
de = 100%
R = H, n = 1: 58% ee = 94%
R = Me, n = 1: 70% ee = 96%
R = Et, n = 1: 73% ee = 97%
R = i-Pr, n = 1: 79% ee = 98%
R = Bn, n = 1: 78% ee = 98%
R = Et, n = 2: 82% ee > 99%
R = Bn, n = 2: 85% ee > 99%
catalyst salt =
X =
NH2 . X
MeO
i-Pr
OH

Scheme 7.7 Epoxidations of cyclic enones.

The versatility of asymmetric organocatalysis was demonstrated by the practical synthesis of methyl (2*R*,3*S*)-3-(4-methoxyphenyl)glycidate, a key intermediate in the synthesis of diltiazem.[11] Therefore, the key step of this synthesis was the asymmetric epoxidation of methyl (*E*)-4-methoxycinnamate using a chiral dioxirane generated from Yang's catalyst. This reaction provided the desired epoxide in high chemical and optical yields, as shown in Scheme 7.8.

In addition, Shi *et al.* have reported the organocatalytic epoxidation of *cis*-1-propenylphosphoric acid with hydrogen peroxide as the oxidant by using a D-mannitol-derived chiral amine as the organocatalyst.[12] The reaction performed in a MeCN–H_2O mixture as the solvent afforded the corresponding epoxide in a quantitative yield and a good enantioselectivity, as shown in Scheme 7.9.

7.3 Other Oxidations

The number of biologically interesting natural products possessing structural motifs is substantial and still growing.[13] Many peroxy natural products display antitumour, anticancer and antiparasite activities, which are attributed to the propensity of the peroxide to initiate radical reactions in an iron-rich environment. In this context, the first example of an asymmetric β-peroxidation of nitroalkenes was disclosed by Lattanzi and Russo, in 2008.[14] This very simple reaction was promoted by catalytic loadings of a chiral diarylprolinol and employed *tert*-butyl hydroperoxide as the oxidant, which gave access to chiral peroxides in moderate to good yields and synthetically useful levels of

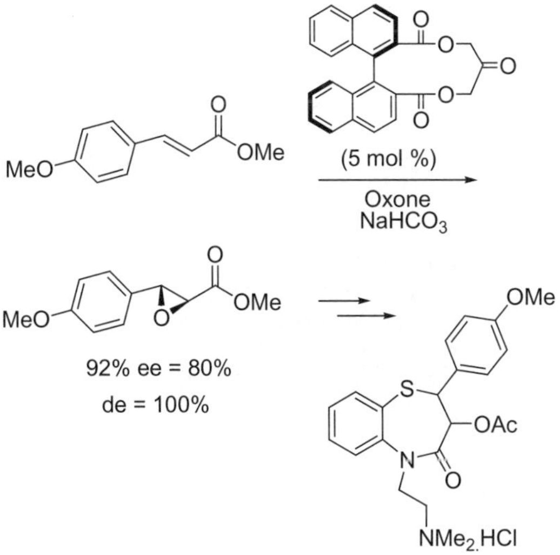

Scheme 7.8 Synthesis of diltiazem.

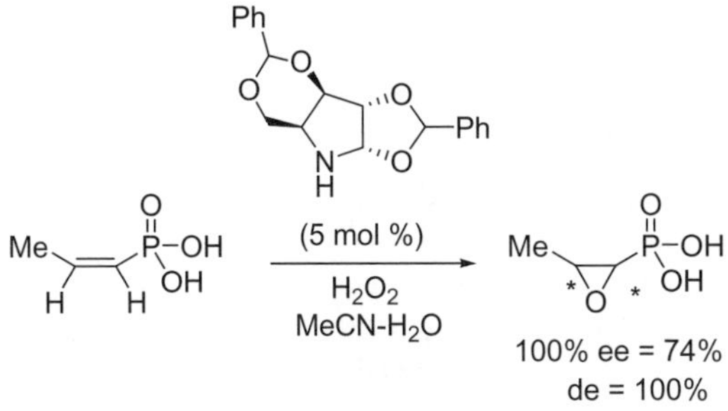

Scheme 7.9 Epoxidation of *cis*-1-propenylphosphoric acid.

enantioselectivity of up to 84% ee (Scheme 7.10). The utility of this asymmetric β-peroxidation was demonstrated by a straightforward and efficient conversion of peroxides into 1,2-amino alcohols.

In the same area, Deng and List independently developed the organocatalytic β-peroxidation of α,β-unsaturated ketones by using a cinchona alkaloid derivative as the organocatalyst.[15,16] According to the nature of the oxidant used, the process led to the formation of either the expected peroxides when employing a hydroperoxide as the oxidant,[15] or to the formation of the corresponding cyclic peroxyhemiketals as almost equimolar diastereomeric mixtures when using hydrogen peroxide as the oxidant.[16] In both cases, the products were obtained in good to high yields combined with excellent

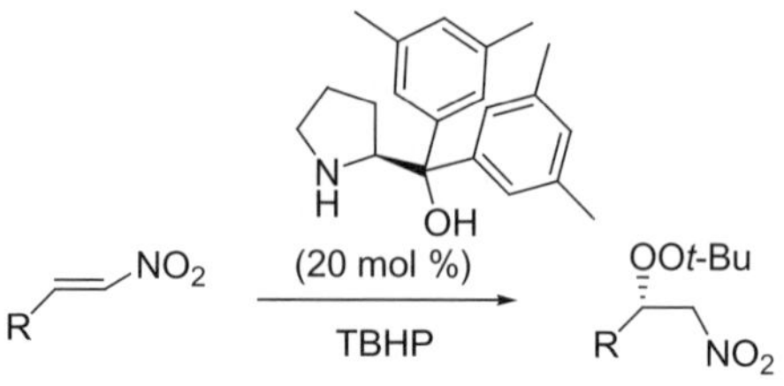

R = Ph: 72% ee = 84%
R = p-Tol: 59% ee = 83%
R = 3,4-Me$_2$C$_6$H$_3$: 70% ee = 83%
R = p-t-BuC$_6$H$_4$: 83% ee = 83%
R = p-ClC$_6$H$_4$: 80% ee = 84%
R = 3-Fu: 45% ee = 83%
R = Cy: 73% ee = 72%

Scheme 7.10   β-Peroxidations of nitroalkenes.

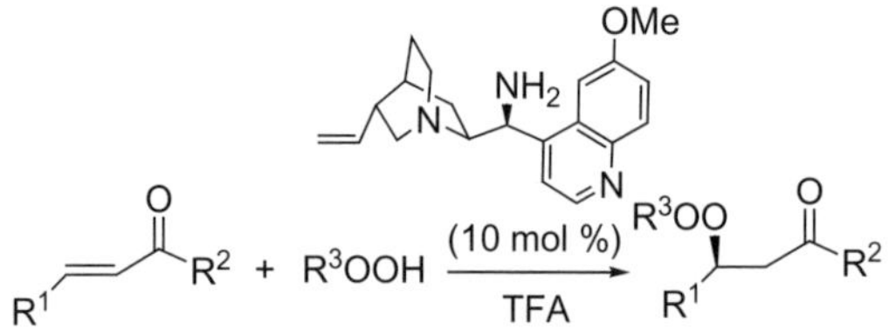

R^1 = R^2 = Me, R^3 = t-Bu: 85% ee = 91%
R^1 = BnCH$_2$, R^2 = Me, R^3 = t-Bu: 86% ee = 93%
R^1 = n-Pr, R^2 = n-Bu, R^3 = t-Bu: 64% ee = 94%
R^1 = R^2 = Me, R^3 = C(Me)$_2$Ph: 74% ee = 94%
R^1 = BnCH$_2$, R^2 = Me, R^3 = C(Me)$_2$Ph: 77% ee = 95%
R^1 = (CH$_2$)$_3$OBn, R^2 = Me, R^3 = C(Me)$_2$Ph: 82% ee = 96%
R^1 = n-Pr, R^2 = Me, R^3 = C(Me)$_2$OMe: 70% ee = 95%
R^1 = BnCH$_2$, R^2 = Me, R^3 = C(Me)$_2$OMe: 62% ee = 95%
R^1 = R^2 = Et, R^3 = C(Me)$_2$OMe: 63% ee = 95%
R^1 = Me, R^2 = n-Bu, R^3 = C(Me)$_2$OMe: 60% ee = 94%

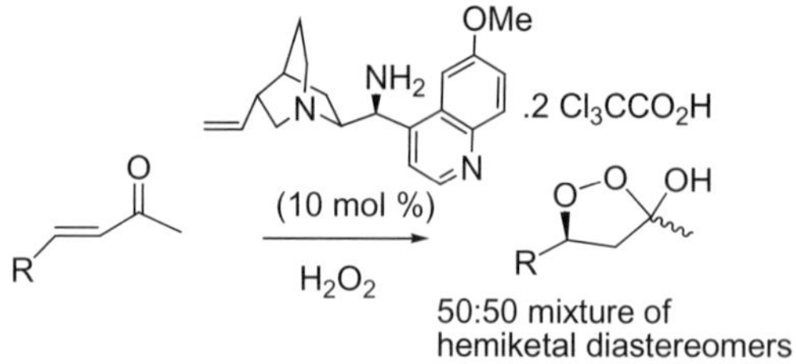

50:50 mixture of
hemiketal diastereomers

R = n-Hex: 65% ee = 95%
R = BnCH$_2$: 68% ee = 94%
R = (CH$_2$)$_2$CH=CH$_2$: 69% ee = 94%
R = n-Bu: 61% ee = 95%
R = Cy: 54% ee = 95%

Scheme 7.11   β-Peroxidations of α,β-unsaturated ketones.

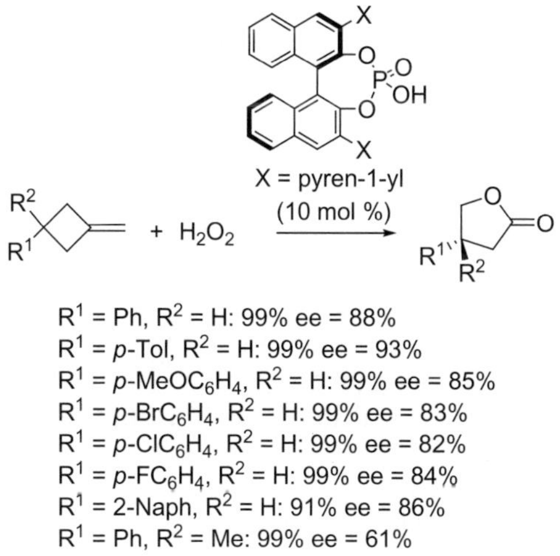

X = pyren-1-yl
(10 mol %)

R¹ = Ph, R² = H: 99% ee = 88%
R¹ = *p*-Tol, R² = H: 99% ee = 93%
R¹ = *p*-MeOC₆H₄, R² = H: 99% ee = 85%
R¹ = *p*-BrC₆H₄, R² = H: 99% ee = 83%
R¹ = *p*-ClC₆H₄, R² = H: 99% ee = 82%
R¹ = *p*-FC₆H₄, R² = H: 99% ee = 84%
R¹ = 2-Naph, R² = H: 91% ee = 86%
R¹ = Ph, R² = Me: 99% ee = 61%

Scheme 7.12   Baeyer–Villiger oxidations of 3-substituted cyclobutanones.

(10 mol %)

Pt(+)-Pt(-), 1.5 *F*/mol, 20 mA
NaBr
NaHCO₃/H₂O/CH₂Cl₂

R = *o*-Tol: 47% ee = 72% + 43% of ketone
R = 2,4,6-Me₃C₆H₂: 47% ee = 64% + 49% of ketone
R = 1-Naph: 60% ee = 39% + 40% of ketone
R = 2-Naph: 45% ee = 76% + 52% of ketone
R = Ph: 53% ee = 65% + 42% of ketone

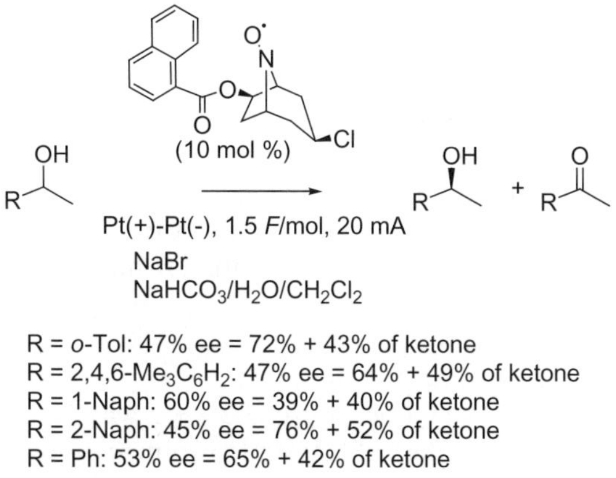

Scheme 7.13   Electrooxidations of racemic *sec*-alcohols.

enantioselectivities of up to 95–97% ee, as shown in Scheme 7.11. The versatility of these intermediates, which for the first time could be prepared directly in an enantiopure form, has been illustrated with the syntheses of the corresponding epoxides.

   Although more than one century has gone by since its discovery in 1899, the Baeyer–Villiger reaction is still far from being fully developed. In particular, there are only a few catalyst systems which afford products from the Baeyer–Villiger oxidation of 3-substituted cyclobutanones in more than 80% ee. The first example of the enantioselective Baeyer–Villiger oxidation of 3-substituted cyclobutanones catalysed by a chiral organocatalyst and 30%

aqueous H_2O_2 as the oxidant was developed by Ding *et al.*[17] This process yielded the corresponding γ-lactones in almost quantitative yields in all cases of substrates and enantioselectivities of up to 93% ee, as shown in Scheme 7.12. The best catalyst was a chiral phosphoric acid derived from BINOL and featuring pyren-1-yl groups at the 3,3′-positions, which was applied to the Baeyer–Villiger oxidation of a variety of 3-aryl- as well as alkyl-substituted cyclobutanones. Furthermore, an excellent reactivity was also observed in the reaction of 3,3-disubstituted cyclobutanones, albeit with moderate enantio-selectivities ($\leq -61\%$ ee).

In addition, Onomura *et al.* demonstrated that enantiomerically pure aza-bicyclo-*N*-oxyls prepared from L-hydroxyproline mediated the enantioselective electrooxidation of racemic *sec*-alcohols to afford optically active *sec*-alcohols with moderate to high enantioselectivities of up to 76% ee (Scheme 7.13).[18] The oxidation was conducted using platinum electrodes in an undivided beaker-type cell, containing a catalytic amount of the organocatalyst, an excess amount of NaBr and a mixture of CH_2Cl_2 and saturated aqueous $NaHCO_3$ as the solvent. After passing through $1.5\,F/mol$ of electricity at constant current (20 mA, terminal voltage: *ca.* 3V) at 0 °C, acetophenone and the optically active (*S*)-*sec*-alcohol were obtained.

7.4   Conclusions

Recently, several combinations of different types of chiral organocatalysts with oxidative reagents have been elaborated and successfully applied to the asymmetric epoxidation of α,β-unsaturated carbonyl compounds. As an example, the combination of a chiral dibenzyl ammonium salt of a chiral phosphoric acid derived from BINOL as the organocatalyst with *tert*-butyl hydroperoxide as the oxidant allowed to convert a wide range of α,β-unsaturated aldehydes into the corresponding *trans*-epoxides in good to excellent yields, high *trans*-diastereo-selectivities and high enantioselectivities of up to 96% ee. In addition, the first highly enantioselective epoxidation of cyclic enones was developed in 2008 by using hydrogen peroxide as the oxidant and a quinine-derived primary amine salt as the organocatalyst, providing excellent enantioselectivities of up to 99% ee for a variety of cyclic enones.

References

1. Q.-H. Xia, H.-Q. Ge, C.-P. Ye, Z.-M. Liu and K.-X. Su, *Chem. Rev.*, 2005, **105**, 1603–1662.
2. M. C. José, T. M. Maria and A. Shazia, *Chem. Rev.*, 2004, **104**, 2857–2900.
3. O. A. Wong and Y. Shi, *Chem. Rev.*, 2008, **108**, 3958–3987.
4. P. C. Bulman Page, F. Marken, C. Williamson, Y. Chan, B. R. Buckley and D. Bethell, *Adv. Synth. Catal.*, 2008, **350**, 1149–1154.
5. P. C. Bulman Page, P. Parker, G. A. Rassias, B. R. Buckley and D. Bethell, *Adv. Synth. Catal.*, 2008, **350**, 1867–1874.

6. X. Wang and B. List, *Ang. Chem., Int. Ed. Engl.*, 2008, **47**, 1119–1122.
7. C. Sparr, W. B. Schweizer, H. M. Senn and R. Gilmour, *Ang. Chem., Int. Ed. Engl.*, 2009, **48**, 3065–3068.
8. H. Cui, Y. Li, C. Zheng, G. Zhao and S. Zhu, *J. Fluorine Chem.*, 2008, **129**, 45–50.
9. A. Russo and A. Lattanzi, *Eur. J. Org. Chem.*, 2008, 2767–2773.
10. X. Wang, C. M. Reisinger and B. List, *J. Am. Chem. Soc.*, 2008, **130**, 6070–6071.
11. M. Seki, *Synlett*, 2008, **2**, 164–176.
12. Z. Zhang, J. Tang, X. Wang and H. Shi, *J. Mol. Catal.*, 2008, **285**, 68–71.
13. F. Rahm, P. Y. Hayes and W. Kitching, *Heterocycles*, 2004, **64**, 523–524.
14. A. Russo and A. Lattanzi, *Adv. Synth. Catal.*, 2008, **350**, 1991–1995.
15. X. Lu, Y. Liu, B. Sun, B. Cindric and L. Deng, *J. Am. Chem. Soc.*, 2008, **130**, 8134–8135.
16. C. M. Reisinger, X. Wang and B. List, *Ang. Chem., Int. Ed. Engl.*, 2008, **47**, 8112–8115.
17. S. Xu, Z. Wang, X. Zhang, X. Zhang and K. Ding, *Ang. Chem., Int. Ed. Engl.*, 2008, **47**, 2840–2843.
18. H. Shiigi, H. Mori, T. Tanaka, Y. Demizu and O. Onomura, *Tetrahedron Lett.*, 2008, **49**, 5247–5251.

CHAPTER 8
Reductions

8.1 Reductions of Imines and Ketones

The asymmetric hydrogenation of unsaturated organic compounds is currently becoming a standard procedure in both academic laboratories and industrial applications.[1] Until recently, all the methods developed for the reduction of organic compounds have been dominated by the use of metal catalysts surrounded by proper stereodiscriminating chiral ligands. A shift of this paradigm was, however, recently made by the discovery that simple small organic molecules were able to catalyse chemoselective reductions.[2] The enantioselective reduction of imines to obtain chiral amines still represents a challenging topic. Although many highly enantioselective hydrogenations of ketones and alkenes are known, only less effective reductions of imines are available. Indeed, the development of efficient catalysts giving high enantioselectivity has proved to be much more difficult in the case of imines, compared with alkenes and ketones. However, several efficient chiral organocatalysts have been very recently developed for the enantioselective reduction of ketimines. As an example, Jones et al. reported, in 2009, the asymmetric reduction of N-aryl ketimines with trichlorosilane employing an imidazole organocatalyst derived from L-proline.[3] A wide variety of N-aryl ketimines could be reduced into the corresponding amines in high yields and enantioselectivities of up to 87% ee at a low catalyst loading of 1 mol % (Scheme 8.1). Surprisingly, the ratio of the ketimine geometric isomer did not seem to have a great influence on the outcome of the reaction. This observation is of particular interest, since the substrates needed not to be geometrically pure, which is of significant advantage when one considers the difficulty in preparing and handling the sensitive ketimine substrates, let alone purifying them as a single geometric isomer.

Higher enantioselectivities were reported by Sun et al. by using a chiral bissulfinamide with a polymethylene tether connecting the two chiral sulfinamide functional groups for similar reactions.[4] At a catalyst loading of 10 mol %, a wide variety of chiral amines were obtained in high yields and enantioselectivities of up to 96% ee (Scheme 8.2). Furthermore, these authors also

RSC Catalysis Series No. 3
Recent Developments in Asymmetric Organocatalysis
By Hélène Pellissier
© Hélène Pellissier 2010
Published by the Royal Society of Chemistry, www.rsc.org

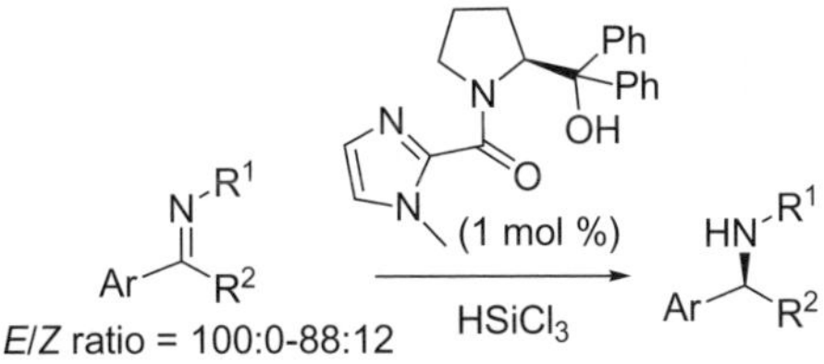

Ar = R^1 = Ph, R^2 = Me: 82% ee = 85%
Ar = *p*-ClC$_6$H$_4$, R^1 = Ph, R^2 = Me: 77% ee = 85%
Ar = Ph, R^1 = PMP, R^2 = Me: 96% ee = 87%
Ar = *p*-CF$_3$C$_6$H$_4$, R^1 = PMP, R^2 = Me: 72% ee = 82%
Ar = *p*-MeOC$_6$H$_4$, R^1 = PMP, R^2 = Me: 81% ee = 85%
Ar = *p*-NO$_2$C$_6$H$_4$, R^1 = PMP, R^2 = Me: 85% ee = 86%
Ar = Ph, R^1 = PMP, R^2 = Et: 95% ee = 83%
Ar = 2-Naph, R^1 = PMP, R^2 = Me: 86% ee = 86%

Scheme 8.1 Reductions of *N*-aryl ketimines.

investigated a series of L-pipecolinic acid-derived *N*-formamides as chiral organocatalysts for these reactions.[5] The substituents on N4 of the piperazinyl backbone and the 2-carboxamide group both proved to have profound effects on the efficacy of the catalyst. The best enantioselectivities of up to 97% ee were obtained with the catalyst bearing a bulky *para-tert*-butyl group on the benzene ring of the benzenesulfonyl group at N4 (Scheme 8.2). Most remarkably, this catalyst promoted the reduction of the relatively bulky non-methyl ketimines in higher yields and enantioselectivities than that of the methyl ketimines as substrates. This so far unprecedented feature rendered this catalyst a good complement to the other existing catalyst systems for the highly enantioselective reduction of imines in terms of the substrate spectrum. Similar reactions were also studied by Benaglia *et al.* in the presence of chiral binaphthyl-diamine-derived organocatalysts, giving almost quantitative yields for the formed chiral amines, albeit with lower enantioselectivities ($\leq$ 83% ee).[6] The best result (98%, 83% ee) was obtained for the reduction of the *N*-2-methoxyphenyl imine derived from methyl phenyl ketone by using the *N*,*N*'-dimethyl amino derivative of a C_2-symmetric bis-picolinamide derived from (*R*) binaphthyl diamine.

In addition, the asymmetric reduction of *N*-aryl ketimines with trichlorosilane could be achieved on polymer-supported organocatalysts by Kocovsky *et al.*[7] Indeed, *N*-methylvaline-derived formamide anchored to a polymeric support, used at a catalyst loading of 15 mol %, allowed good enantioselectivities of up to 82% ee combined with good yields to be obtained for the formed chiral amines (Scheme 8.3). This novel methodology simplified the recovery of the catalyst, which could be reused at least five times without any loss of the activity. The best results were obtained with the catalysts directly attached to the polymer or *via* a suitable spacer. A strong influence of the solvents on the catalytic performance was observed with chloroform giving the

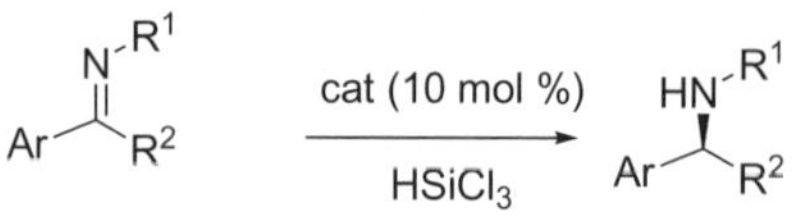

with cat 1:
Ar = R^1 = Ph, R^2 = Me: 91% ee = 96%
Ar = p-$CF_3C_6H_4$, R^1 = Ph, R^2 = Me: 95% ee = 95%
Ar = p-$NO_2C_6H_4$, R^1 = Ph, R^2 = Me: 90% ee = 93%
Ar = p-BrC_6H_4, R^1 = Ph, R^2 = Me: 92% ee = 95%
Ar = PMP, R^1 = Ph, R^2 = Me: 83% ee = 95%
Ar = 2-Naph, R^1 = Ph, R^2 = Me: 77% ee = 94%
Ar = Ph, R^1 = p-ClC_6H_4, R^2 = Me: 93% ee = 95%
Ar = Ph, R^1 = p-Tol, R^2 = Me: 88% ee = 92%
Ar = Ph, R^1 = PMP, R^2 = Me: 92% ee = 92%
Ar = Ph, R^1 = PMP, R^2 = Et: 84% ee = 92%
Ar = R^1 = Ph, R^2 = n-Pr: 87% ee = 91%
Ar = R^1 = Ph, R^2 = i-Bu: 82% ee = 93%
with cat 2:
Ar = R^1 = Ph, R^2 = Et: 92% ee = 94%
Ar = R^1 = Ph, R^2 = Me: 95% ee = 89%
Ar = p-FC_6H_4, R^1 = Ph, R^2 = Et: 87% ee = 95%
Ar = p-ClC_6H_4, R^1 = Ph, R^2 = Me: 83% ee = 94%
Ar = p-BrC_6H_4, R^1 = Ph, R^2 = Me: 89% ee = 95%
Ar = PMP, R^1 = Ph, R^2 = Me: 83% ee = 90%
Ar = R^1 = Ph, R^2 = c-Pr: 85% ee = 97%
Ar = R^1 = Ph, R^2 = i-Bu: 86% ee = 91%
Ar = p-BrC_6H_4, R^1 = Ph, R^2 = Me: 81% ee = 89%

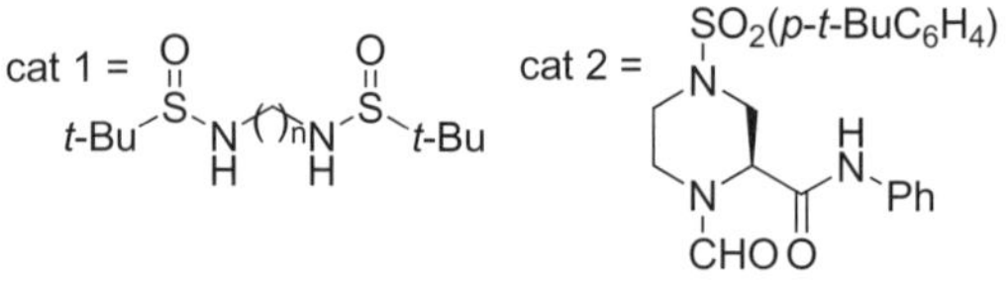

Scheme 8.2 Reductions of *N*-aryl ketimines.

best results. The reduced efficiency of these catalyst systems has hampered their further developments. In this context, the same authors have developed gold nanoparticles functionalised with a valine-derived formamide as homogeneous catalysts for the asymmetric reduction of *N*-aryl ketimines with trichlorosilane in toluene.[8] This methodology simplified both the recovery of the catalyst and its separation from the product, since the nanoparticles could be readily removed and subsequently recycled by precipitation from the reaction mixture. The application of this novel methodology to the reduction of *N*-PMP phenyl methyl ketone afforded the corresponding chiral amine in 90% yield and with an enantioselectivity of 84% ee.

The involvement of *N*-aryl ketimines as substrates undoubtedly limits their applications since the aryl group on the nitrogen atom of the resulting amine in

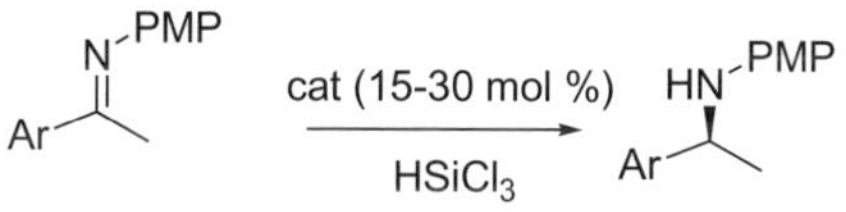

Ar = Ph: 84% ee = 82%
Ar = 2-Naph: 72% ee = 79%
Ar = p-CF$_3$C$_6$H$_4$: 67% ee = 81%
Ar = PMP: 62% ee = 77%
Ar = 2,5-Me$_2$-3-furyl: 67% ee = 78%

cat =

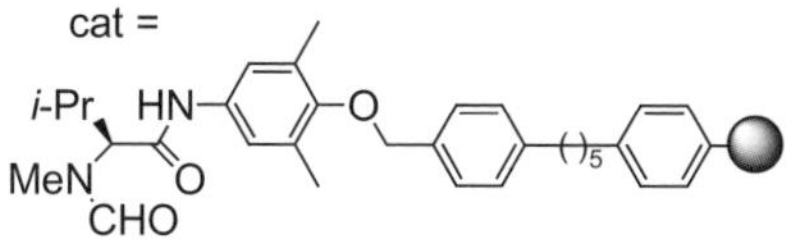

Scheme 8.3 Reductions of *N*-aryl ketimines catalysed by polymer-supported organocatalyst.

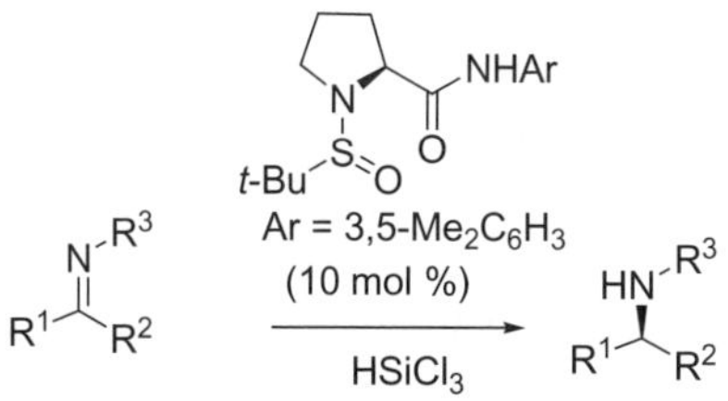

R^1 = Ph, R^2 = Me, R^3 = Bn: 98% ee = 96%
R^1 = p-ClC$_6$H$_4$, R^2 = Me, R^3 = Bn: 92% ee = 95%
R^1 = m-ClC$_6$H$_4$, R^2 = Me, R^3 = Bn: 98% ee = 97%
R^1 = m-BrC$_6$H$_4$, R^2 = Me, R^3 = Bn: 93% ee = 97%
R^1 = p-CF$_3$C$_6$H$_4$, R^2 = Me, R^3 = Bn: 94% ee = 98%
R^1 = p-NO$_2$C$_6$H$_4$, R^2 = Me, R^3 = Bn: 80% ee > 99%
R^1 = 2-Naph, R^2 = Me, R^3 = Bn: 96% ee = 96%
R^1 = o-ClC$_6$H$_4$, R^2 = Me, R^3 = CH$_2$CH=CH$_2$: 75% ee = 97%
R^1 = p-BrC$_6$H$_4$, R^2 = Me, R^3 = CH$_2$CH=CH$_2$: 83% oo = 03%
R^1 = p-CF$_3$C$_6$H$_4$, R^2 = Me, R^3 = CH$_2$CH=CH$_2$: 65% ee = 97%
R^1 = p-NO$_2$C$_6$H$_4$, R^2 = Me, R^3 = CH$_2$CH=CH$_2$: 97% ee = 96%
R^1 = Ph, R^2 = Me, R^3 = CH$_2$PMP: 85% ee = 93%
R^1 = Ph, R^2 = Et, R^3 = Bn: 80% ee = 89%

Scheme 8.4 Reductions of *N*-alkyl ketimines.

many cases needs to be removed or replaced with alkyl groups. In this context, Sun *et al.* have developed the first example of organocatalytic methods which allowed the highly enantioselective reduction of a wide variety of *N*-alkyl keti-mines.[9] Therefore, in the presence of trichlorosilane and a catalytic amount of a chiral sulfinamide derived from L-proline as the organocatalyst in toluene, the reduction of *N*-alkyl ketimines afforded the corresponding amines in high yields and high to excellent enantioselectivities of up to 99% ee, as shown in Scheme 8.4.

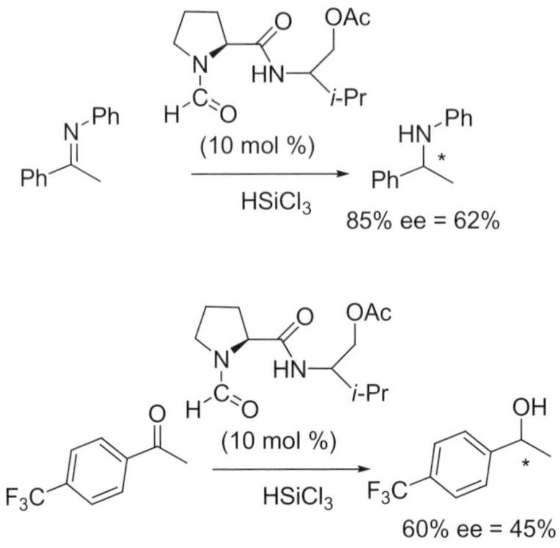

Scheme 8.5   Reductions of ketone and imine.

A series of (*S*)-1-formylpyrrolidine-2-carboxylic acid derivatives has been synthesised and examined as chiral organocatalysts by Zhang *et al.* in the asymmetric reduction of both *para*-trifluoromethylphenyl methyl ketone and *N*-phenyl methyl phenyl imine.[10] These catalysts afforded moderate enantios-electivities of 41% ee and 52% ee, respectively, for the asymmetric reductions of both the ketone and the imine. The best results are shown in Scheme 8.5.

In 2008, Curran *et al.* demonstrated that a fluorous prolinol precatalyst bearing only 34 fluorine atoms could be immobilised in the hydrofluoroether solvent HFE-7500.[11] Performed in the presence of this novel organocatalyst, the reduction of acetophenone proceeded rapidly, in an almost quantitative yield and with an excellent enantioselectivity of up to 95% ee in the absence of solvent (Scheme 8.6). The organic product was stripped from the HFE-7500 phase with a polar solvent, and the HFE-7500 phase was reused with satis-factory results through eight runs.

On the other hand, Toru *et al.* reported the first organocatalytic hydro-phosphonylation of sulfonylimines catalysed by commercially available cinchona alkaloids, such as quinine, hydroquinine or hydroquinidine.[12] In these conditions, the products were obtained in almost quantitative yields and excellent enantioselectivities of up to 97% ee, as shown in Scheme 8.7. In this reaction, it was demonstrated that the heteroarenesulfonyl group worked as a good activating group of the imino group in the hydrophosphonylation. Indeed, the reactions involving *N*-(arenesulfonyl)imines instead of *N*-(hetero-arenesulfonyl)imines did not afford good results ($\leq$68% yields, $\leq$49% ee).

In 2008, Kocovsky *et al.* reported a novel methodology based on the organo-catalytic asymmetric hydrosilylation of enamines that allowed a direct access to a range of β^3- and $\beta^{2,3}$-amino acid derivatives for some of which other methods proved less satisfactory.[13] This method relied on the fast equilibration

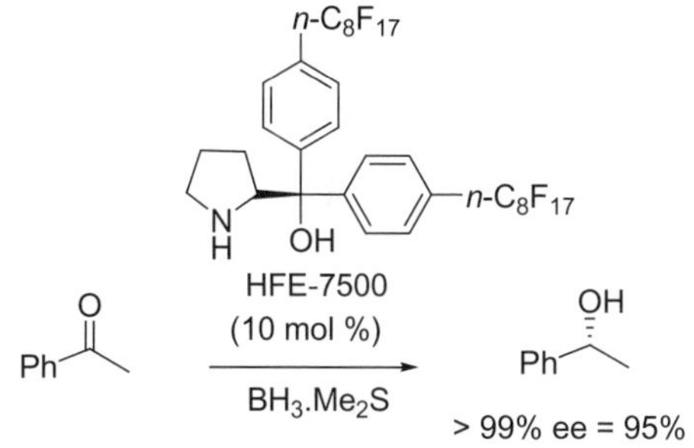

Scheme 8.6 Reduction of acetophenone in neat HFE-7500.

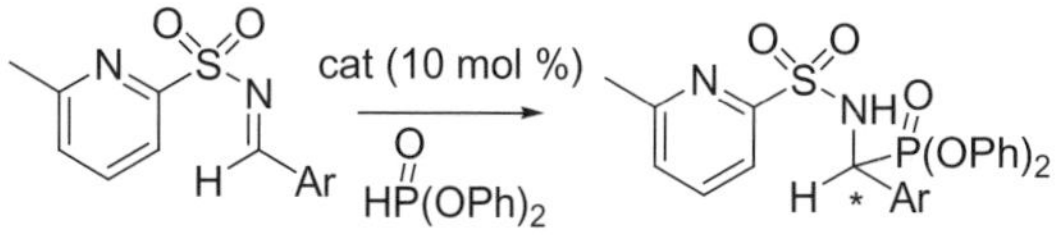

with cat = hydroquinine:
Ar = Ph: 98% ee = 97%
Ar = *p*-Tol: > 99% ee = 92%
Ar = PMP: > 99% ee = 94%
Ar = 1-Naph: > 99% ee = 92%
with cat = quinine:
Ar = 2-Naph: > 99% ee = 92%
Ar = (*E*)-PhCH=CH: > 99% ee = 85%
with cat = hydroquinidine:
Ar = Ph: 95% ee = 98%
Ar = PMP: > 99% ee = 93%
Ar = *o*-MeOC$_6$H$_4$: > 99% ee = 91%
Ar = *m*-MeOC$_6$H$_4$: > 99% ee = 91%
Ar = 1-Naph: > 99% ee = 91%
Ar = 2-Naph: > 99% ee = 92%

Scheme 8.7 Hydrophosphonylations of *N*-(heteroarenesulfonyl)imines.

between the enamine and imine forms. The reduction of the equilibrated mixture with Cl$_3$SiH catalysed by an L-valine-derived formamide afforded the corresponding amino esters and amino nitriles in excellent yields and with good to high enantioselectivities ($\leq$90% ee), as shown in Scheme 8.8. An efficient dynamic kinetic resolution[14] occurred in the case of reaction of α-aryl and α-alkyl derivatives, providing a set of highly diastereomerically enriched β^{2,3}-amino acid derivatives (*syn/anti* $\geq$ 95:5) with a good enantioselectivity (76–86% ee).

8.2 Other Reductions

Optically active β-amino acids are very important chiral building blocks for the synthesis of β-peptides, β-lactams, various natural products and physiologically

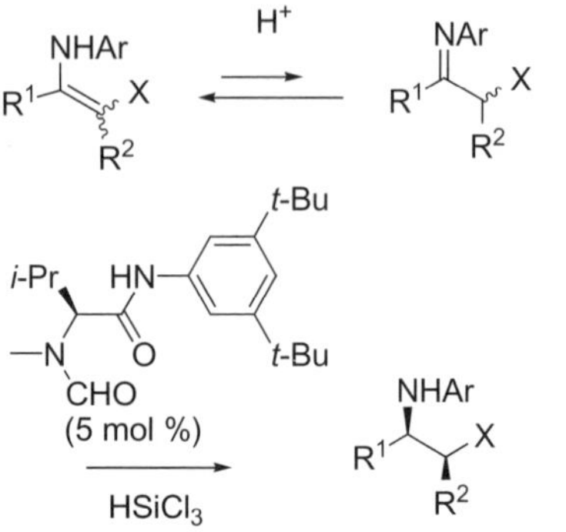

Ar = PMP, X = CO_2Et, R^1 = Ph, R^2 = H:
98% ee = 89%
Ar = PMP, X = CO_2Et, R^1 = PMP, R^2 = H:
80% ee = 88%
Ar = PMP, X = CO_2Et, R^1 = *p*-FC_6H_4, R^2 = H:
95% ee = 90%
Ar = PMP, X = CO_2Et, R^1 = *p*-$CF_3C_6H_4$, R^2 = H:
94% ee = 88%
Ar = PMP, X = CO_2Et, R^1 = *m*-$MeOC_6H_4$, R^2 = H:
95% ee = 88%
Ar = PMP, X = CO_2Et, R^1 = *m*-ClC_6H_4, R^2 = H:
91% ee = 86%
Ar = PMP, X = CN, R^1 = *m*-$MeOC_6H_4$, R^2 = H:
75% ee = 87%
Ar = PMP, X = CO_2Et, R^1 = R^2 = Ph:
84% *syn*/*anti* > 99:1 ee = 76%
Ar = PMP, X = CO_2Et, R^1 = Ph, R^2 = Bn:
87% *syn*/*anti* = 27:73 ee = 86%
Ar = PMP, X = CO_2Et, R^1 = Ph, R^2 = Et:
86% *syn*/*anti* = 95:5 ee = 82%
Ar = PMP, X = CN, R^1 = R^2 = Ph:
46% *syn*/*anti* > 99:1 ee = 83%
Ar = PMP, X = CN, R^1 = Ph, R^2 = PMP:
43% *syn*/*anti* > 99:1 ee = 83%

Scheme 8.8 Reductions of enamines.

active substances.[15] In this context, efficient methods for the synthesis of chiral β-amino acids are of a great value for drug discovery and organic synthesis. One of them was successfully developed by Zhang *et al.* through a general and highly enantioselective organocatalytic hydrosilylation of *N*-aryl β-enamino esters performed in the presence of $HSiCl_3$ in chloroform as the solvent.[16] This process, involving chiral *N*-picolinoylpyrrolidine derivatives as organocatalysts, enabled the straightforward and mild synthesis of a broad range of β-amino acid derivatives in high yields and enantioselectivities of up to 96% ee, as shown in Scheme 8.9.

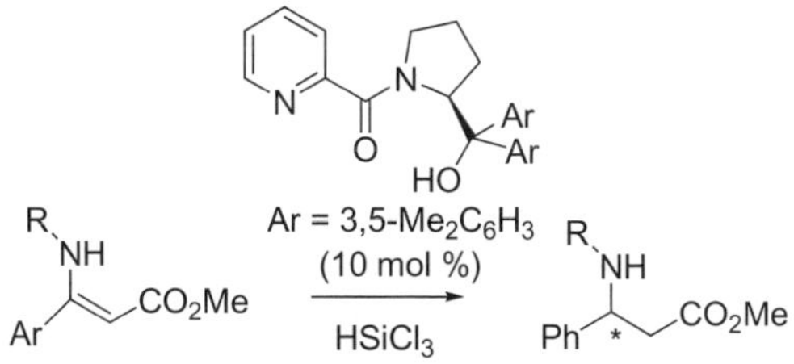

R = Ar = Ph: 94% ee = 95%
R = PMP, Ar = Ph: 82% ee = 95%
R = *m*-MeOC$_6$H$_4$, Ar = Ph: 86% ee = 96%
R = *p*-ClC$_6$H$_4$, Ar = Ph: 97% ee = 93%
R = PMP, Ar = *p*-FC$_6$H$_4$: 86% ee = 96%
R = PMP, Ar = *p*-Tol: 95% ee = 94%
R = Ar = PMP: 95% ee = 95%
R = PMP, Ar = 1-Naph: 96% ee = 95%

Scheme 8.9 Hydrosilylations of *N*-aryl β-enamino esters.

Another organocatalytic approach to chiral β-amino acids was reported by List *et al.* based on a highly enantioselective thiourea-catalysed reduction of β-nitroacrylates to their saturated analogues mediated by Hantzsch ester.[17] This asymmetric transfer hydrogenation process occurred in general high yields (61–97%) and high enantioselectivities of up to 95% ee, as shown in Scheme 8.10. While the yield was high in all cases of substrates, the enantioselectivity increased slightly with size and bulkiness of the ester moiety. Gratifyingly, branched as well as unbranched aliphatic nitroacrylates were equally suitable substrates. Moreover, the enantioselectivity of the reductions was shown to be strongly dependent on the substrate olefin geometry. Accordingly, nitroolefins (*E*) or (*Z*) gave opposite enantiomers of products, each with high enantioselectivity, while a 1:1 (*E*)/(*Z*)-mixture gave essentially racemic products.

Finally, Kudo *et al.* have developed a resin-supported *N*-terminal prolyl peptide having a β-turn motif and a hydrophobic polyleucine tether to be applied to the asymmetric transfer hydrogenation in aqueous media.[18] The polyleucine tether provided a hydrophobic cavity in aqueous media that brought about a remarkable acceleration of the reaction. In addition, the polyleucine chain also turned out to be essential for high enantioselectivity. This novel methodology, imitating the performance of enzymes in the cell, allowed the transfer hydrogenation of α,β-unsaturated aldehydes mediated by Hantzsch ester to be achieved in good yields of up to 76% and excellent enantioselectivities of up to 96% ee, as shown in Scheme 8.11.

8.3 Conclusions

The enantioselective reduction of imines to obtain chiral amines still represents a challenging topic. In the last year, several chiral organocatalysts have been

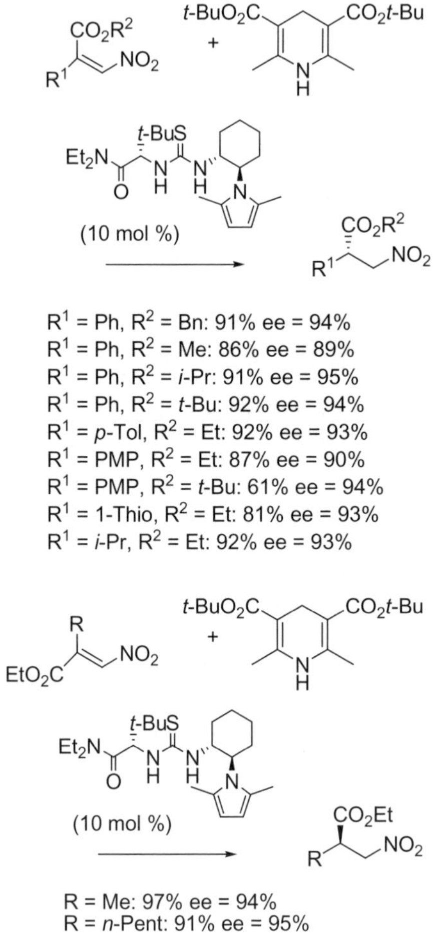

Scheme 8.10   Reductions of β-nitroacrylates.

developed for the enantioselective reduction of ketimines. Among them, the use of a chiral sulfinamide derived from L-proline as organocatalyst allowed the first example of organocatalytic highly enantioselective reduction of a wide variety of *N*-alkyl ketimines to be achieved. Performed in the presence of trichlorosilane the corresponding amines were obtained in high yields and excellent enantioselectivities of up to 99% ee.

Optically active β-amino acids are very important chiral building blocks for the synthesis of biologically active products. In this context, a general and highly enantioselective organocatalytic hydrosilylation of *N*-aryl β-enamino esters was developed by using chiral *N*-picolinoylpyrrolidine derivatives as the organocatalysts, enabling the straightforward and mild synthesis of a broad range of β-amino acid derivatives in high yields and enantioselectivities of up to 96% ee.

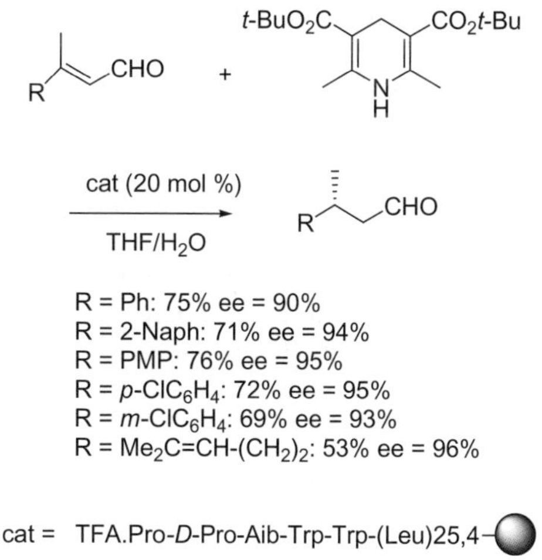

R = Ph: 75% ee = 90%
R = 2-Naph: 71% ee = 94%
R = PMP: 76% ee = 95%
R = p-ClC$_6$H$_4$: 72% ee = 95%
R = m-ClC$_6$H$_4$: 69% ee = 93%
R = Me$_2$C=CH-(CH$_2$)$_2$: 53% ee = 96%

cat = TFA.Pro-*D*-Pro-Aib-Trp-Trp-(Leu)25,4-⬤

Scheme 8.11   Transfer hydrogenations of α,β-unsaturated aldehydes.

References

1. (a) H. U. Blaser, C. Malan, B. Pugin, F. Spindler, H. Steiner and M. Studer, *Adv. Synth. Catal.*, 2003, **345**, 103–151; (b) R. Noyori, M. Kitamura and T. Ohkuma, *Proc. Natl. Acad. Sci. USA*, 2004, **101**, 5356–5362; (c) H. Adolfsson, *Ang. Chem., Int. Ed. Engl.*, 2005, **44**, 3340–3342.
2. J. W. Yang, M. T. Hechavarria Fonseca and B. List, *Ang. Chem., Int. Ed. Engl.*, 2004, **43**, 6660–6662.
3. F.-M. Gautier, S. Jones and S. J. Martin, *Org. Biomol. Chem.*, 2009, **7**, 229–231.
4. D. Pei, Y. Zhang, S. Wei, M. Wang and J. Sun, *Adv. Synth. Catal.*, 2008, **350**, 619–623.
5. P. Wu, Z. Wang, M. Cheng, L. Zhou and J. Sun, *Tetrahedron*, 2008, **64**, 11304–11312.
6. S. Guizzetti, M. Benaglia, F. Cozzi, S. Rossi and G. Celentano, *Chirality*, 2009, **21**, 233–238.
7. A. V. Malkov, M. Figlus and P. Kocovsky, *J. Org. Chem.*, 2008, **73**, 3985–3995.
8. A. V. Malkov, M. Figlus, G. Cooke, S. T. Caldwell, G. Rabani, M. R. Prestly and P. Kocovsky, *Org. Biomol. Chem.*, 2009, **7**, 1878–1883.
9. C. Wang, X. Wu, L. Zhou and J. Sun, *Chem. Eur. J.*, 2008, **14**, 8789–8792.
10. Z. Chen, A. Zhang, L. Zhang, J. Zhang and X. Lei, *J. Chem. Res.*, 2008, 266–269.
11. Q. Chu, M. S. Yu and D. P. Curran, *Org. Lett.*, 2008, **10**, 749–752.
12. S. Nakamura, H. Nakashima, A. Yamamura, N. Shibata and T. Toru, *Adv. Synth. Catal.*, 2008, **350**, 1209–1212.
13. A. V. Malkov, S. Stoncius, K. Vrankova, M. Arndt and P. Kocovsky, *Chem. Eur. J.*, 2008, **14**, 8082–8085.

14. (a) H. Pellissier, *Tetrahedron*, 2003, **59**, 701–730; (b) H. Pellissier, *Tetra-hedron*, 2008, **64**, 1563–1601.
15. I. Ojima, S. N. Lin and T. Wang, *Curr. Med. Chem.*, 1999, **6**, 927–954.
16. H.-J. Zheng, W.-B. Chen, Z.-J. Wu, J.-G. Deng, W.-Q. Lin, W.-C. Yuan and X.-M. Zhang, *Chem. Eur. J.*, 2008, **14**, 9864–9867.
17. N. J. A. Martin, X. Cheng and B. List, *J. Am. Chem. Soc.*, 2008, **130**, 13862–13863.
18. (a) K. Akagawa, H. Akabane, S. Sakamoto and K. Kudo, *Org. Lett.*, 2008, **10**, 2035–2037; (b) K. Akagawa, H. Akabane, S. Sakamoto and K. Kudo, *Tetrahedron: Asymmetry*, 2009, **20**, 461–466.

CHAPTER 9

Kinetic Resolutions and Desymmetrisations

The kinetic resolution of racemic alcohols *via* asymmetric acylation has been widely used to construct various useful chiral building blocks in the synthesis of complex natural products.[1] Most methods reported to date have employed enzymes, such as lipases or esterases. The challenge of developing easily accessible and effective non-enzymatic asymmetric acylation catalysts has attracted many research groups over the last decade.[2] The asymmetric acylation of alcohols using molecular catalysts has emerged as a viable alternative to the well-established enzyme-catalysed acylation. As an example, Chen *et al.* investigated the kinetic resolution of a series of 2,2-difluoro-3-hydroxy-3-aryl-propionates with (R)-benzotetramisole as the catalyst.[3] The results showed that when the aryl group in the substrate was a phenyl (or a phenyl substituted by an electron-donating group) or a naphthyl group, the enantioselectivity factor (s) could reach 20 or higher (Scheme 9.1). On the other hand, the presence of electron-withdrawing substituents on the benzene ring dramatically lowered the enantioselectivity factor. This study represented the first kinetic resolution of fluorinated secondary alcohols. In addition, these authors applied the similar methodology to a series of 2,2,2-trifluoro-1-aryl ethanol in the presence of isobutyric anhydride.[4] As shown in Scheme 9.1, the kinetic resolution led to even better results with an enantioselectivity factor of up to 71.

In 2009, Iwabuchi *et al.* accomplished a highly enantioselective organo-catalytic kinetic resolution of various secondary alcohols by using chirally modified 2-azaadamantane *N*-oxyls as organocatalysts in the presence of trichloro isocyanuric acid (TCCA).[5] The best results were obtained when α-substituted cyclopentanols and cyclohexanols were employed as the substrates with enantioselectivities of up to 99% ee, as shown in Scheme 9.2.

The catalytic asymmetric desymmetrisation of *meso* anhydrides *via* the addition of an alcohol nucleophile represents a simple and elegant method for

RSC Catalysis Series No. 3
Recent Developments in Asymmetric Organocatalysis
By Hélène Pellissier
© Hélène Pellissier 2010
Published by the Royal Society of Chemistry, www.rsc.org

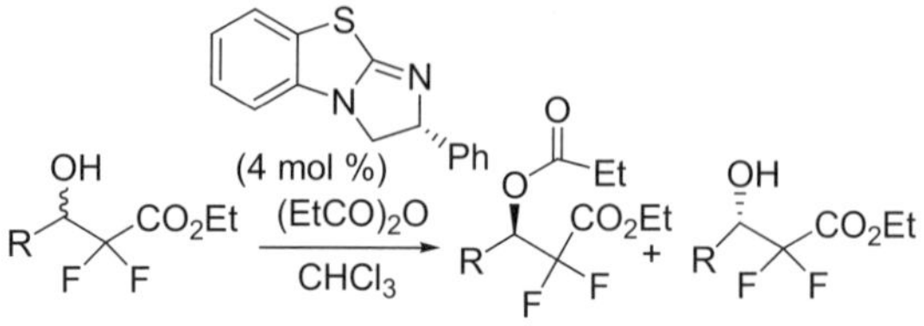

R = Ph: ee (ester) = 85% ee (alcohol) = 98% s = 20
R = p-Tol: ee (ester) = 82% ee (alcohol) = 71% s = 20
R = PMP: ee (ester) = 81% ee (alcohol) = 75% s = 21
R = p-MeSC$_6$H$_4$: ee (ester) = 85% ee (alcohol) = 58% s = 22
R = p-FC$_6$H$_4$: ee (ester) = 72% ee (alcohol) = 52% s = 10
R = 1-Naph: ee (ester) = 88% ee (alcohol) = 72% s = 34
R = 2-Naph: ee (ester) = 89% ee (alcohol) = 85% s = 47
R = 2-Fu: ee (ester) = 67% ee (alcohol) = 76% s = 11

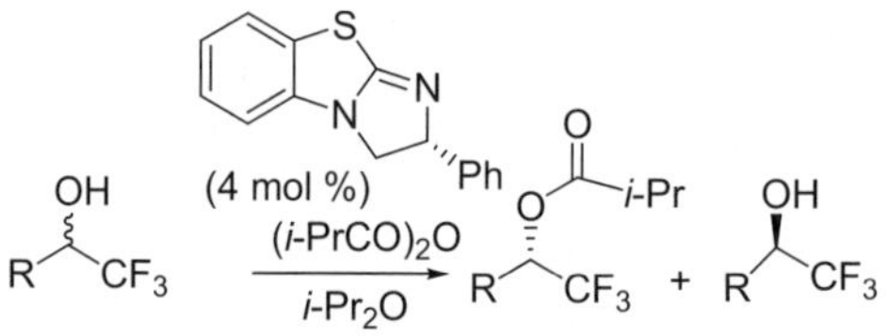

R = Ph: ee (ester) = 83% ee (alcohol) = 68% s = 23
R = o-Tol: ee (ester) = 87% ee (alcohol) = 65% s = 28
R = p-Tol: ee (ester) = 87% ee (alcohol) = 70% s = 31
R = PMP: ee (ester) = 89% ee (alcohol) = 69% s = 35
R = p-MeSC$_6$H$_4$: ee (ester) = 82% ee (alcohol) = 82% s = 25
R = p-FC$_6$H$_4$: ee (ester) = 71% ee (alcohol) = 72% s = 13
R = p-ClC$_6$H$_4$: ee (ester) = 76% ee (alcohol) = 86% s = 20
R = 1-Naph: ee (ester) = 86% ee (alcohol) = 92% s = 44
R = 2-Naph: ee (ester) = 90% ee (alcohol) = 96% s = 71
R = 2-Fu: ee (ester) = 67% ee (alcohol) = 76% s = 11

Scheme 9.1 Kinetic resolutions of 2,2-difluoro-3-hydroxy-3-aryl-propionates and 2,2,2-trifluoro-1-aryl ethanols.

the preparation of synthetically pliable hemiesters with the generation of either single or multiple stereocentres with high levels of enantiocontrol.[6] While chiral Lewis acid-catalysed anhydride desymmetrisations have been reported, the use of chiral amine catalysts has emerged as a more effective strategy for the enantioselective promotion of these reactions.[7] For example, Connon *et al.* have successfully developed the highly enantioselective desymmetrisation of *meso* anhydrides by a bifunctional thiourea-based cinchona alkaloid as the organocatalyst.[8] Both succinic and challenging glutaric anhydride derivatives could be cleanly converted at ambient temperature to the corresponding methyl hemiesters with excellent enantioselectivities of up to 93% ee by using an unprecedented catalyst loading of 1 mol % without requiring a stoichiometric amine additive. As shown in Scheme 9.3, the scope of this methodology could

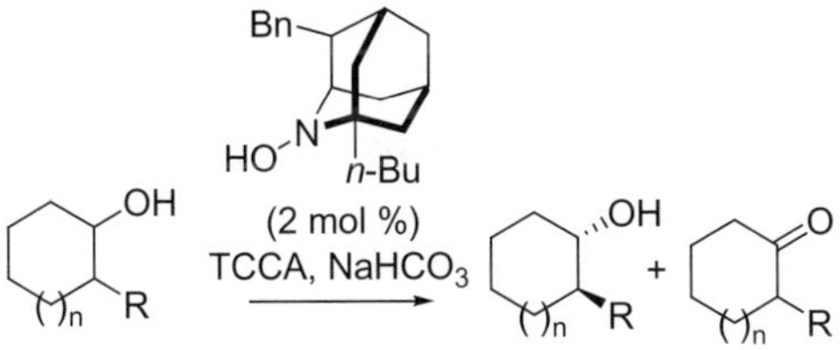

R = Ph, n = 0: 57% ee = 99%
R = Ph, n = 1: 52% ee = 98%
R = p-FC$_6$H$_4$, n = 1: 53% ee = 99%
R = Ph, n = 2: 58% ee = 60%
R = OAc, n = 1: 57% ee = 88%
R = OBz, n = 1: 58% ee = 90%

Scheme 9.2   Kinetic resolutions of cycloalkanols.

be extended to a wide range of anhydrides. It must be noted that the same studies were also independently reported by Chin *et al.*, providing slightly higher enantioselectivities albeit lower yields in some cases of substrates, and using a higher catalyst loading of 10 mol %.[9] Furthermore, allylic alcohol could be used as the nucleophile, which allowed the convenient enantioselective synthesis of either antipodal methyl hemiester product of a given *meso* anhydride with a single catalyst enantiomer (Scheme 9.3).

Moreover, the same authors employed a closely related organocatalyst and the corresponding urea derivative to promote the enantioselective dynamic kinetic resolution of azalactones with allylic alcohol.[10] In this case of substrates, the urea derivatives proved to be superior to their thiourea analogues and, most usefully, these catalysts were insensitive to the steric bulk of the amino acid residue, allowing alanine-, methionine- and phenylalanine-derived azalactones to undergo dynamic kinetic resolution with unprecedented levels of enantioselectivity, as shown in Scheme 9.4. Furthermore, the compatibility of these catalysts with thiol nucleophiles was exploited in the first enantioselective catalytic dynamic kinetic resolution of azalactones by thiolysis to furnish enantioenriched amino acid thioesters of potential use with moderate enantioselectivities ($\leq 64\%$ ee).

Finally, the methanolytic desymmetrisation of a variety of cyclic anhydrides was also achieved by using a thermally robust sulfonamide based bifunctional cinchona alkaloid as the organocatalyst.[11] Under these conditions, an unprecedented catalytic activity combined with an excellent level of enantioselectivity of up to 98% ee were obtained at a catalyst loading of 5 mol %, as shown in Scheme 9.5. No appreciable effects of the concentration and temperature on the reactivity and enantioselectivity were observed with this catalyst.

Conclusions

The challenge of developing easily accessible and effective non-enzymatic asymmetric acylation catalysts has attracted many research groups over the last

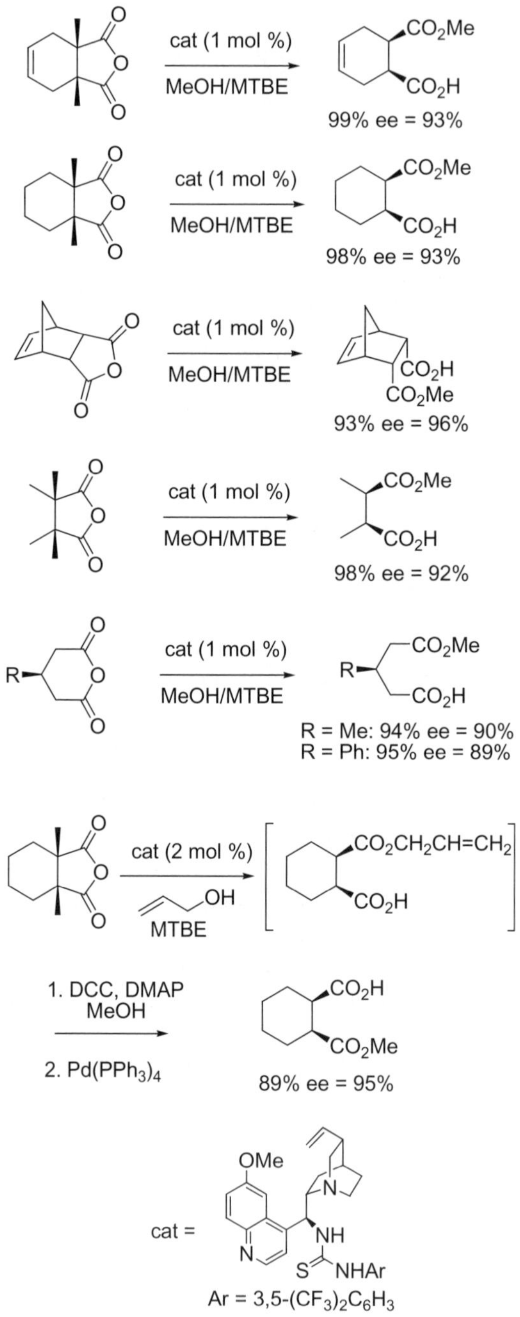

Scheme 9.3   Desymmetrisations of succinic and glutaric anhydrides.

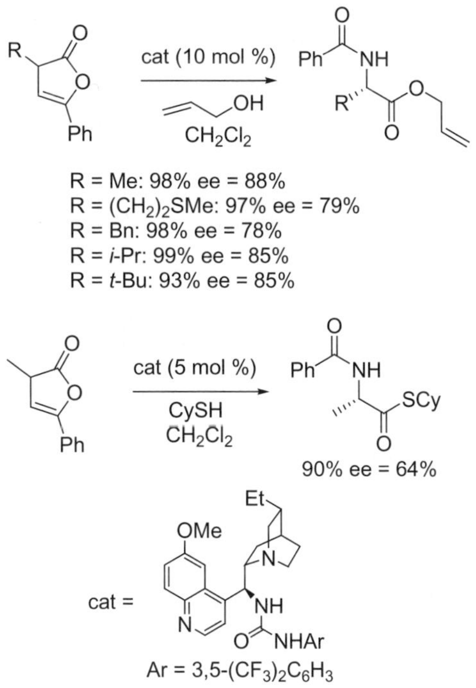
cat (10 mol %)
OH
CH2Cl2
Ph
O
NH
R = Me: 98% ee = 88%
R = (CH2)2SMe: 97% ee = 79%
R = Bn: 98% ee = 78%
R = i-Pr: 99% ee = 85%
R = t-Bu: 93% ee = 85%
cat (5 mol %)
CySH
CH2Cl2
SCy
90% ee = 64%
cat =
OMe
Et
NH
NHAr
Ar = 3,5-(CF3)2C6H3

Scheme 9.4 Dynamic kinetic resolutions of azalactones.

decade. In this context, the asymmetric acylation of alcohols using organocatalysts has emerged as a viable alternative to the well-established enzyme-catalysed acylation. In 2009, a highly enantioselective organocatalytic kinetic resolution of various secondary alcohols by using chirally modified 2-aza-adamantane *N*-oxyls as organocatalysts was accomplished, providing enantioselectivities of up to 99% ee when α-substituted cyclopentanols and cyclohexanols were employed as the substrates.

On the other hand, the catalytic asymmetric desymmetrisation of *meso* anhydrides *via* the addition of an alcohol represents a simple and elegant method for the preparation of chiral hemiesters. In this context, the use of chiral amine catalysts has emerged as a powerful strategy for the enantioselective promotion of these reactions. For example, the use of a bifunctional thiourea-based cinchona alkaloid as a chiral organocatalyst in these reactions allowed the desymmetrisations of a wide range of *meso* anhydrides with excellent enantioselectivities of up to 93% ee by using an unprecedented catalyst loading of 1 mol%. Moreover, higher enantioselectivities of up to 98% ee were obtained for the methanolytic desymmetrisation of a wide variety of cyclic anhydrides by using a thermally robust sulfonamide-based bifunctional cinchona alkaloid as the organocatalyst at a catalyst loading of 5 mol %.

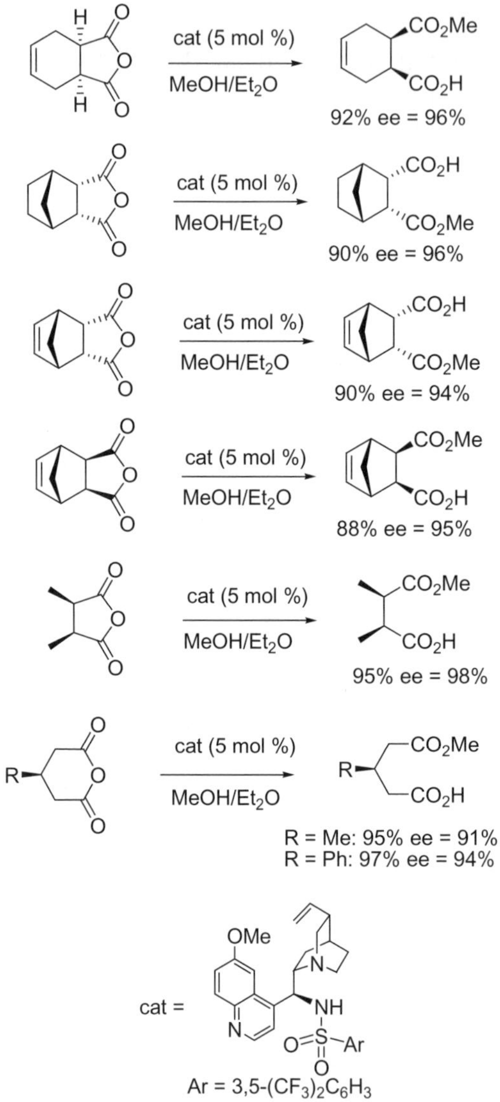

cat (5 mol %)
MeOH/Et$_2$O
92% ee = 96%

CO$_2$Me
CO$_2$H

cat (5 mol %)
MeOH/Et$_2$O
90% ee = 96%

CO$_2$H
CO$_2$Me

cat (5 mol %)
MeOH/Et$_2$O
90% ee = 94%

CO$_2$H
CO$_2$Me

cat (5 mol %)
MeOH/Et$_2$O
88% ee = 95%

CO$_2$Me
CO$_2$H

cat (5 mol %)
MeOH/Et$_2$O
95% ee = 98%

CO$_2$Me
CO$_2$H

cat (5 mol %)
MeOH/Et$_2$O

R
CO$_2$Me
CO$_2$H

R = Me: 95% ee = 91%
R = Ph: 97% ee = 94%

OMe

cat =

NH

O=S—Ar
O

Ar = 3,5-(CF$_3$)$_2$C$_6$H$_3$

Scheme 9.5 Desymmetrisations of cyclic anhydrides.

References

1. F. Theil, *Chem. Rev.*, 1995, **95**, 2203–2227.
2. (a) E. Vedejs and M. Jure, *Ang. Chem., Int. Ed. Engl.*, 2005, **44**, 3974–4001; (b) S. France, D. J. Guerin, S. J. Miller and T. Lectka, *Chem. Rev.*, 2003, **103**, 2985–3012.
3. H. Zhou, Q. Xu and P. Chen, *Tetrahedron*, 2008, **64**, 6494–6499.
4. Q. Xu, H. Zhou, X. Geng and P. Chen, *Tetrahedron*, 2009, **65**, 2232–2238.

5. M. Tomizawa, M. Shibuya and Y. Iwabuchi, *Org. Lett.*, 2009, **11**, 1829–1831.
6. I. Atodiresei, I. Schiffers and C. Bolm, *Chem. Rev.*, 2007, **107**, 5683–5712.
7. Y. Chen, P. McDaid and L. Deng, *Chem. Rev.*, 2003, **103**, 2965–2983.
8. A. Peschiulli, Y. Gun'ko and S. J. Connon, *J. Org. Chem.*, 2008, **73**, 2454–2457.
9. H. Sik Rho, S. Ho Oh, J. W. Lee, J. Y. Lee, J. Chin and C. E. Song, *Chem. Commun.*, 2008, 1208–1210.
10. A. Peschiulli, C. Quigley, S. Tallon, Y. K. Gun'ko and S. J. Connon, *J. Org. Chem.*, 2008, **73**, 6409–6412.
11. S. Ho, H. Oh, J. W. Sik Rho, J. Lee, E. Lee, S. H. Youk, J. Chin and C. E. Song, *Ang. Chem., Int. Ed. Engl.*, 2008, **47**, 7872–7875.

CHAPTER 10
Miscellaneous Reactions

The enantioselective Friedel–Crafts alkylation of indoles is one of the most important carbon–carbon bond-forming reactions for the preparation of biologically active compounds such as indole alkaloids.[1] Indeed, the indole framework represents a privileged structure motif in a large number of natural products and therapeutic agents.[2] The asymmetric Friedel–Crafts alkylation has become a powerful strategy for the construction of indole architectures. A number of highly selective metal-catalysed asymmetric Friedel–Crafts reactions have been developed; however, the corresponding organocatalytic asymmetric process has been much less widely explored. In this area, Hiemstra *et al.* developed the asymmetric Friedel–Crafts reaction between a bulky tritylsulfenyli-mine and indole promoted by an (*R*)-octahydrobinol phosphoric acid as an organocatalyst.[3] This process afforded the corresponding *N*-triphenylmethyl-sulfenyl-substituted (*R*)-indolylglycine in low yield and high enantioselectivity of up to 88% ee (Scheme 10.1). Another organocatalyst, such as an (*R*)-binol-phosphoric acid, was applied to the asymmetric Friedel–Crafts reaction between a 2-nitrophenylsulfenamide-substituted glyoxylimine and indole, providing the corresponding *N*-(2-nitrophenyl)sulfenyl-substituted (*S*)-indolylglycine in a mode-rate yield and a good enantioselectivity of 69% ee (Scheme 10.1).[3] It must be noted that a low catalyst loading of only 2 mol % was used in this process, and the alkylation products of indole were obtained with opposite configuration depending on the sulfur substituent when the phosphoric acid catalyst with the (*R*) configuration was used exclusively.

This phosphoric acid derived from (*R*)-BINOL was also applied by Akiyama *et al.* as an organocatalyst to promote the highly enantioselective Friedel–Crafts alkylation of indoles with nitroalkenes, generating the corresponding Friedel–Crafts adducts with high yields and excellent enantioselectivities of up to 94% ee for a broad range of substrates in the presence of 3-Å molecular sieves (Scheme 10.2).[4]

In addition, the asymmetric organocatalytic Friedel–Crafts alkylation of indoles with simple α,β-unsaturated aromatic ketones was studied by Zhou *et al.*, in 2008.[5] This Michael-type Friedel–Crafts alkylation of indoles was

RSC Catalysis Series No. 3
Recent Developments in Asymmetric Organocatalysis
By Hélène Pellissier
© Hélène Pellissier 2010
Published by the Royal Society of Chemistry, www.rsc.org

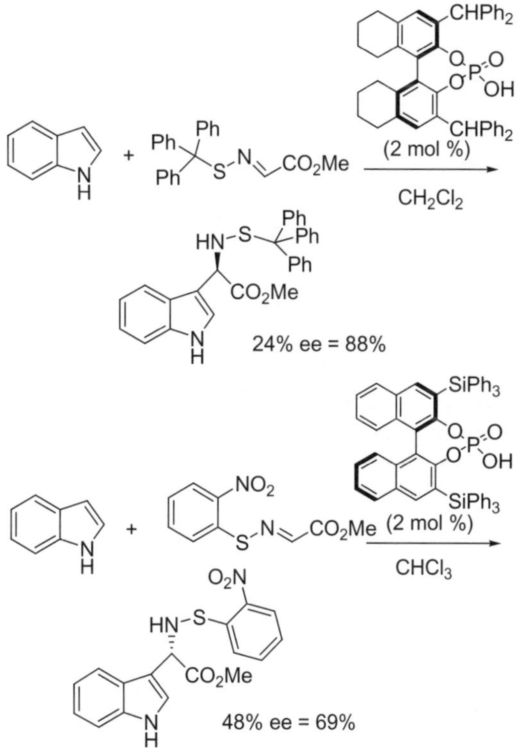

Scheme 10.1 Friedel–Crafts alkylations of indole.

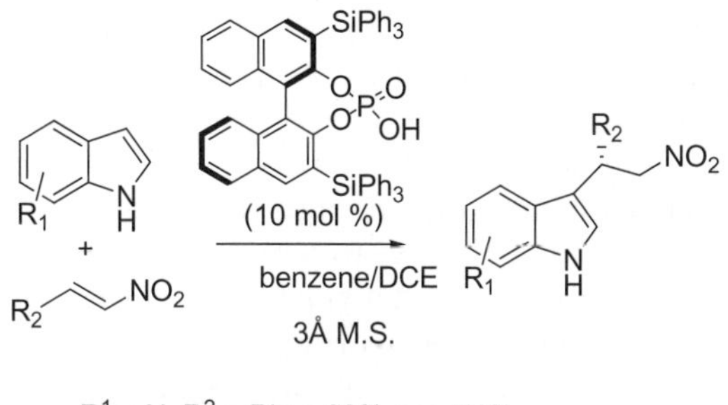

R^1 = H, R^2 = Ph: > 99% ee = 91%
R^1 = H, R^2 = *p*-Tol: 64% ee = 90%
R^1 = H, R^2 = *p*-MeOC$_6$H$_4$: 74% ee = 91%
R^1 = H, R^2 = *p*-ClC$_6$H$_4$: 73% ee = 91%
R^1 = H, R^2 = *p*-CF$_3$C$_6$H$_4$: 84% ee = 91%
R^1 = H, R^2 = 2-thienyl: 71% ee = 90%
R^1 = H, R^2 = *n*-Pent: 77% ee = 90%
R^1 = H, R^2 = *n*-Non: 62% ee = 91%
R^1 = 5-Cl, R^2 = Ph: 63% ee = 90%
R^1 = 7-Me, R^2 = Ph: 70% ee = 94%

Scheme 10.2 Friedel–Crafts alkylations of indoles.

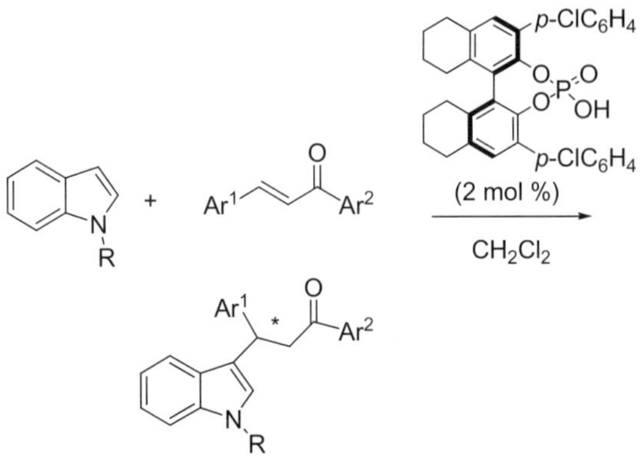

R = H, Ar^1 = Ar^2 = Ph: 73% ee = 56%
R = Me, Ar^1 = Ar^2 = Ph: 63% ee = 47%
R = H, Ar^1 = p-AcOC$_6$H$_4$, Ar^2 = p-NO$_2$C$_6$H$_4$: 88% ee = 51%
R = H, Ar^1 = Ph, Ar^2 = p-BrC$_6$H$_4$: 89% ee = 39%
R = H, Ar^1 = p-AcOC$_6$H$_4$, Ar^2 = o-ClC$_6$H$_4$: 80% ee = 36%
R = H, Ar^1 = o-AcOC$_6$H$_4$, Ar^2 = o-ClC$_6$H$_4$: 69% ee = 38%

Scheme 10.3   Friedel–Crafts alkylations of indoles.

catalysed by an (*R*)-octahydrobinol phosphoric acid used at only 2 mol % of catalyst loading, affording the expected Friedel–Crafts adducts in good yields combined with moderate enantioselectivities ($\leq$56% ee), as shown in Scheme 10.3.

In 2009, Nicolaou *et al.* reported the asymmetric total synthesis of the antitumour natural product demethyl calamenene based on an enantioselective organocatalytic intramolecular Friedel–Crafts type α-arylation of aldehydes bearing electron-donating groups on their aromatic nucleus.[6] This reaction, catalysed by chiral *tert*-butyl-3-methyl-5-benzyl-4-imidazolidinone, afforded the corresponding polycyclic products in good to high yields and with excellent enantioselectivities for a broad variety of aldehydes, as shown in Scheme 10.4.

C_2-Symmetric guanidinium ions derived from *trans*-1-pyrrolo-2-amino-cyclohexane have proved to be effective to promote the enantioselective Claisen rearrangement of ester-substituted allyl vinyl ethers.[7] High enantioselectivities of up to 96% ee were obtained in the rearrangement of a range of substrates in reactions carried out in hexanes as the solvent, despite the fact that the catalyst was virtually insoluble in this solvent. Disubstituted compounds rearranged to form adjacent tertiary stereogenic centres in high enantio- and diastereoselectivities with the *anti* stereochemical relationship predicted by a six-membered chair-like transition state. Furthermore, quaternary stereogenic centres could also be generated with a good stereocontrol. The best results are collected in Scheme 10.5.

In 2008, Jorgensen *et al.* reported the first organocatalysed enantioselective [1,3]-sigmatropic *O*- to *N*-rearrangement reactions.[8] These reactions took place under regio- and enantioselective control, and were catalysed by cinchona alkaloids, such as [DHQD]$_2$PHAL derived from dihydroquinidine. A first

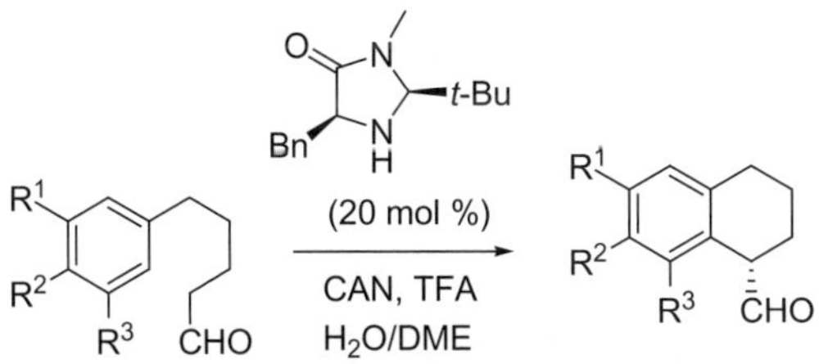

$R^1 = R^3 = MeO, R^2 = H: 80\%$ ee = 94%
$R^1 = MeO, R^2 = R^3 = H: 77\%$ ee = 97%
$R^1 = R^2 = MeO, R^3 = H: 58\%$ ee = 94%
$R^1, R^2 = (CH=CH)_2, R^3 = H: 52\%$ ee = 92%

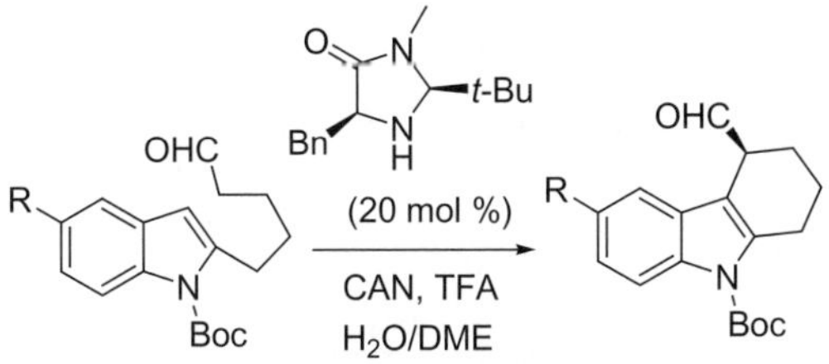

R = H: 64% ee = 98%
R = OMe: 55% ee = 94%

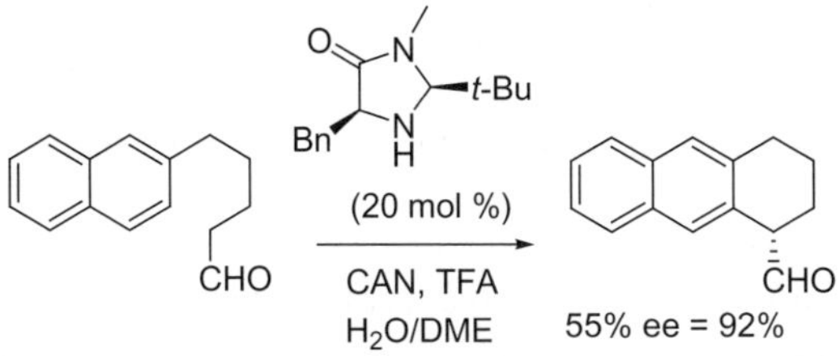

55% ee = 92%

Scheme 10.4 Intramolecular Friedel–Crafts α-arylations of aldehydes.

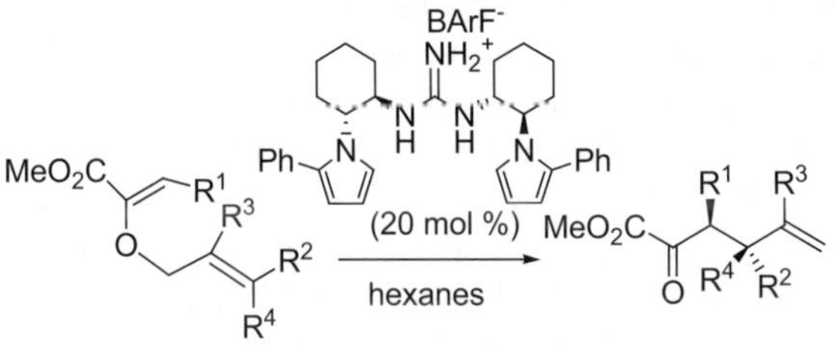

$R^1 = Me, R^2 = R^3 = R^4 = H: 80\%$ ee = 92%
$R^1 = Et, R^2 = R^3 = R^4 = H: 86\%$ ee = 92%
$R^1 = Et, R^2 = n\text{-Pr}, R^3 = R^4 = H: 92\%$ de > 90% ee = 85%
$R^1 = Et, R^2 = Ph, R^3 = R^4 = H: 91\%$ de = 90% ee = 81%
$R^1 = R^3 = Me, R^2 = R^4 = H: 73\%$ ee = 96%
$R^1 = Et, R^2 = Ph, R^3 = H, R^4 = Me: 89\%$ de > 90% ee = 82%

Scheme 10.5 Claisen rearrangements of ester-substituted allyl vinyl ethers.

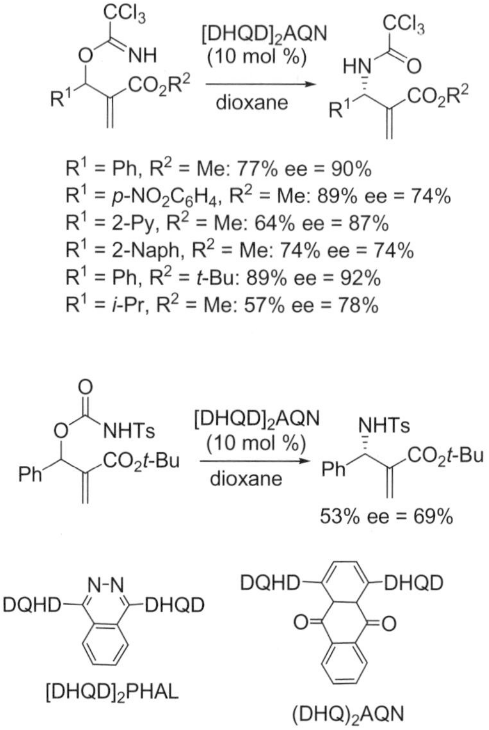

R^1 = Ph, R^2 = Me: 77% ee = 90%
R^1 = *p*-NO$_2$C$_6$H$_4$, R^2 = Me: 89% ee = 74%
R^1 = 2-Py, R^2 = Me: 64% ee = 87%
R^1 = 2-Naph, R^2 = Me: 74% ee = 74%
R^1 = Ph, R^2 = *t*-Bu: 89% ee = 92%
R^1 = *i*-Pr, R^2 = Me: 57% ee = 78%

53% ee = 69%

[DHQD]$_2$PHAL

(DHQ)$_2$AQN

Scheme 10.6 [1,3]-Sigmatropic *O*- to *N*-rearrangement reactions.

reaction concerned the rearrangement of allylic trichloroacetimidates into chiral trichloroacetamides, assuring a complete regioselectivity as well as high enantioselectivities of up to 92% ee (Scheme 10.6). Moreover, a second enantioselective [1,3]-sigmatropic *O*- to *N*-rearrangement reaction was developed, occurring from carbamates to amines *via* a decarboxylation with moderate to good enantioselectivities of up to 69% ee (Scheme 10.6).

The Pictet–Spengler reaction is an important acid-catalysed transformation frequently used in organic synthesis, as well as by various organisms for the synthesis of tetrahydro-β-carbolines or tetrahydroisoquinolines from carbonyl compounds and phenyl ethylamines or tryptamines, respectively.[9] In this area, Jacobsen *et al.* developed organocatalytic regio- and enantioselective cyclisations of pyrroles onto *N*-acyliminium ions generated *in situ* from hydroxylactams.[10] Modest to excellent enantioselectivities of up to 97% ee were obtained in these Pictet–Spengler-type reactions which were promoted by a chiral thiourea-pyrrole catalyst. When the pyrrole nitrogen of the substrate was not protected, a regioselective C2-cyclisation occurred, while a regioselective C4-cyclisation occurred with a TIPS-protected pyrrole, as shown in Scheme 10.7, affording the corresponding versatile pyrroloindolizidinone and pyrroloquinolizidinone products, respectively.

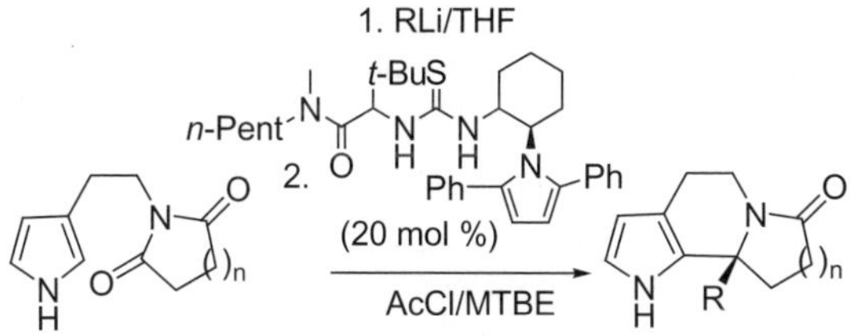

R = Me, n = 1: 77% ee = 90%
R = n-Bu, n = 1: 84% ee = 91%
R = i-Bu, n = 1: 57% ee = 88%
R = i-Pr, n = 1: 86% ee = 93%
R = n-Bu, n = 2: 54% ee = 61%

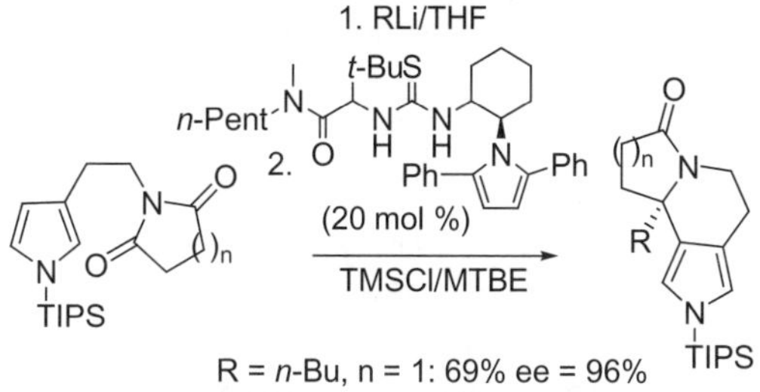

R = n-Bu, n = 1: 69% ee = 96%
R = Me, n = 1: 77% ee = 92%
R = i-Pr, n = 1: 49% ee = 93%
R = i-Bu, n = 1: 68% ee = 96%
R = Ph, n = 1: 63% ee = 92%
R = Me, n = 2: 75% ee = 93%
R = n-Bu, n = 2: 70% ee = 97%

Scheme 10.7 Cyclisations of pyrroles onto *N*-acyliminium ions.

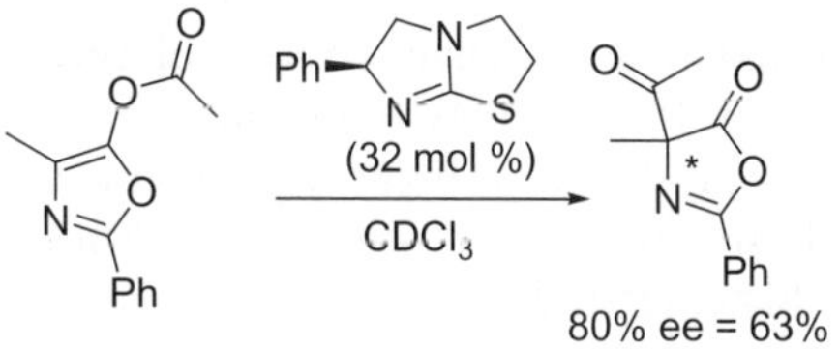

80% ee = 63%

Scheme 10.8 Acetyl migration in a Steglich rearrangement reaction.

The first enantioselective acetyl migration in Steglich rearrangement reactions was developed by Gröger and Dietz as a key step in a novel short synthetic route to protected α-methyl threonine.[11] It was demonstrated that this reaction could be promoted by a chiral heterocyclic organocatalyst, providing the rearranged adduct in a good yield and an enantioselectivity of 63% ee, as shown in Scheme 10.8.

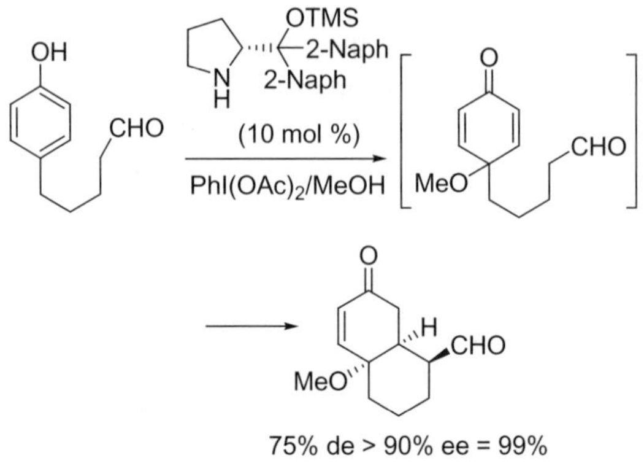

75% de > 90% ee = 99%

Scheme 10.9   Domino oxidative dearomatisation-Michael reaction.

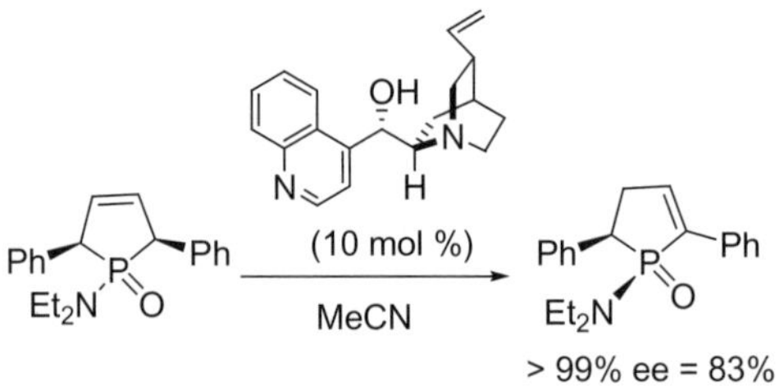

> 99% ee = 83%

Scheme 10.10   Alkene isomerisation.

In another context, Gaunt *et al.* reported a process that directly converted a *para*-substituted phenol into a highly functionalised chiral molecule *via* an oxidative dearomatisation and an organocatalysed enantioselective desymmetrising Michael reaction.[12] This one-pot transformation revealed a complex bicyclic structure formed with exquisite control of three stereogenic centres. This oxidative dearomatisation strategy was catalysed by a chiral diarylprolinol silyl ether as an organocatalyst and provided the corresponding decalin in a good yield, a high diastereoselectivity and an excellent enantioselectivity of 99% ee, as shown in Scheme 10.9.

Another novel extension of asymmetric organocatalysis was reported by Hintermann and Schmitz towards the successful development of an organocatalytic enantioselective double-bond isomerisation, which has been previously associated with the field of metal catalysis.[13] Therefore, an asymmetric synthesis of the 2,5-diphenylphosphol-2-ene fragment was achieved *via* the enantioselective cinchonine-catalysed double-bond isomerisation of a *meso*-2,5-diphenylphosphol-3-ene amide into a 2,5-diphenylphosphol-2-ene amide with an enantioselectivity of up to 83% ee (Scheme 10.10). This new asymmetric concept opened the way to a catalytic enantioselective synthesis of 2,5-diarylphospholane building blocks for many applications in transition metal catalysis.

The enantioselective protonation of prochiral enol derivatives is an attractive route for the preparation of optically active α-carbonyl compounds.[14] However, it is difficult to control the enantioselectivity in an acidic condition because of the bonding flexibility between the proton and its chiral counterion and the orientational flexibility of the proton. An excellent solution for the asymmetric protonation reaction of silyl enol ethers was proposed by Yamamoto and Cheon by using a chiral *N*-triflyl thiophosphoramide derived from BINOL as an organocatalyst.[15] Therefore, in the presence of this catalyst and phenol as the proton source, a range of silyl enol ethers of cyclic ketones could be converted into the corresponding chiral cyclic ketones in quantitative yields and with high enantioselectivities of up to 90% ee, as shown in Scheme 10.11. It must be noted that the catalyst loading for this reaction could be decreased up to 0.05 mol % without any significant loss of the enantioselectivity.

In the same context, the asymmetric protonation of 2-methyl-1-tetralone-derived trimethylsilyl enol ether was achieved by Levacher *et al.* in the presence of a cinchona alkaloid [DHQ]₂AQN derived from hydroquinine as the organocatalyst and citric acid as the proton source in DMF.[16] This reaction occurred with a good yield and an enantioselectivity of 73% ee, as shown in Scheme 10.12.

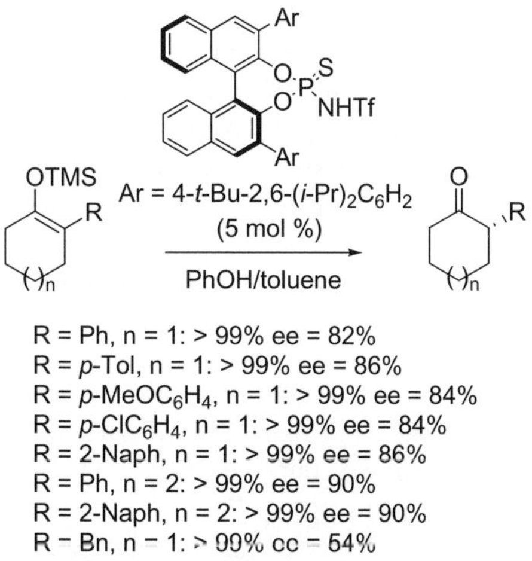

R = Ph, n = 1: > 99% ee = 82%
R = *p*-Tol, n = 1: > 99% ee = 86%
R = *p*-MeOC$_6$H$_4$, n = 1: > 99% ee = 84%
R = *p*-ClC$_6$H$_4$, n = 1: > 99% ee = 84%
R = 2-Naph, n = 1: > 99% ee = 86%
R = Ph, n = 2: > 99% ee = 90%
R = 2-Naph, n = 2: > 99% ee = 90%
R = Bn, n = 1: > 99% ee = 54%
R = Cy, n = 1: > 99% ee = 64%

Scheme 10.11 Protonations of silyl enol ethers of cyclic ketones.

72% ee = 73%

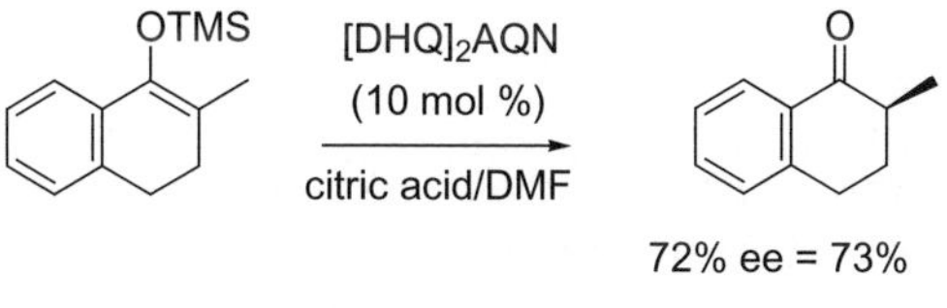

Scheme 10.12 Protonation of 2-methyl-1-tetralone-derived trimethylsilyl enol ether.

Examples dealing with asymmetric organocatalytic radical reactions remain rare in the literature.[17] In 2007, the groups of MacMillan and Sibi almost simultaneously introduced a new mode of organocatalytic activation, termed SOMO (singly occupied molecular orbital) catalysis,[18] that was founded upon the transient production of a 3π-electron radical cation species that could function as a generic platform of induction and reactivity. In this context, MacMillan applied his catalyst to the first asymmetric α-vinylation of aldehydes using vinyl trifluoroborate salts, which allowed the production of α-formyl, α-vinyl methine stereogenic centres without olefin transposition or subsequent erosion in the enantiopurity.[19] This new mode of organocatalytic activation, termed SOMO catalysis, was founded upon the mechanistic hypothesis that one-electron oxidation of a transient enamine intermediate, derived from the aldehyde and the chiral amine catalyst, rendered a 3π-electron SOMO-activated species, which could readily participate in asymmetric bond construction. As shown in Scheme 10.13, the reaction of various boron salts as coupling reagents in these radical-based processes with a range of aldehydes led to the corresponding α-vinyl aldehydes in good to high yields and high enantioselectivities of up to 95% ee.

Another application of this new mode of organocatalytic activation was developed by the same authors with the first asymmetric SOMO-catalysed carbo-oxidation of styrenes to provide chiral γ-nitrate-α-alkyl aldehydes.[20] This new organo-SOMO reaction allowed simple styrenes to function as α-alkylation partners for aldehydes. As highlighted in Scheme 10.14, a wide array of styrenes readily participated as SOMOphiles in this new catalytic carbo-oxidation, providing the corresponding γ-nitrate-α-alkyl aldehydes in excellent yields and enantioselectivities of up to 97% ee. Indeed, the SOMO activation constitutes a new highly promising strategy for organocatalysis which has a great potential given the numerous radical-based C–X (X $=$ C, O, N, S,

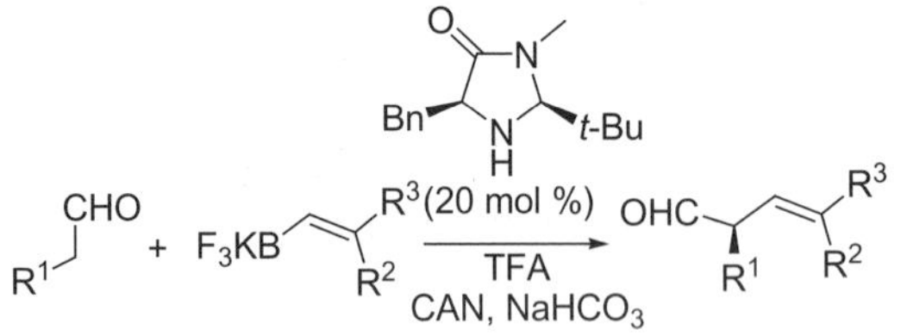

$R^1 = n$-Hex, $R^2 = $ H, $R^3 = $ Ph: 81% ee $= $ 94%
$R^1 = n$-Hex, $R^2 = $ H, $R^3 = p$-FC$_6$H$_4$: 63% ee $= $ 93%
$R^1 = n$-Hex, $R^2 = $ H, $R^3 = p$-FC$_6$H$_4$: 63% ee $= $ 93%
$R^1 = n$-Hex, $R^2 = $ H, $R^3 = p$-MeOC$_6$H$_4$: 61% ee $= $ 95%
$R^1 = n$-Hex, $R^2 = $ Me, $R^3 = $ Ph: 93% ee $= $ 94%
$R^1 = $ Me, $R^2 = $ H, $R^3 = $ Ph: 72% ee $= $ 94%
$R^1 = $ Cy, $R^2 = $ H, $R^3 = $ Ph: 82% ee $= $ 96%
$R^1 = $ BnO(CH$_2$)$_3$, $R^2 = $ H, $R^3 = $ Ph: 78% ee $= $ 93%

Scheme 10.13   Vinylations of aldehydes with boron salts.

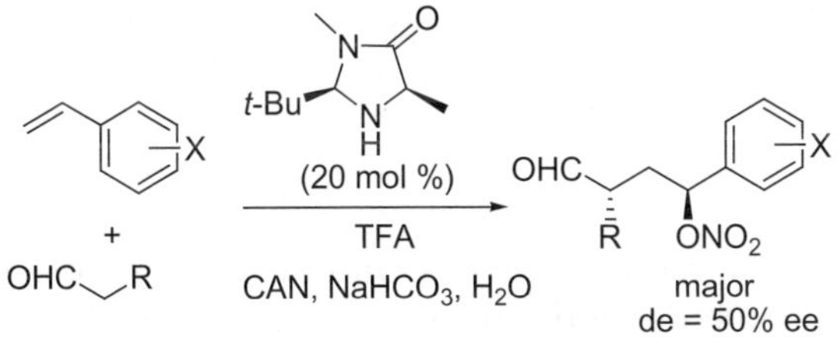

R = *n*-Hex, X = 4-F: 93% ee = 96%
R = *n*-Hex, X = 4-Cl: 92% ee = 96%
R = *n*-Hex, X = 4-Br: 92% ee = 97%
R = *n*-Hex, X = 4-NO_2: 95% ee = 92%
R = *n*-Hex, X = 4-CF_3: 90% ee = 96%
R = *n*-Hex, X = 3-MeO: 94% ee = 96%
R = *n*-Hex, X = 3,4-$(CH{=}CH)_2$: 90% ee = 97%
R = *n*-Hex, X = H: 91% ee = 96%
R = Cy, X = H: 88% ee = 96%
R = Bn, X = H: 81% ee = 96%

Scheme 10.14 Carbo-oxidations of styrenes.

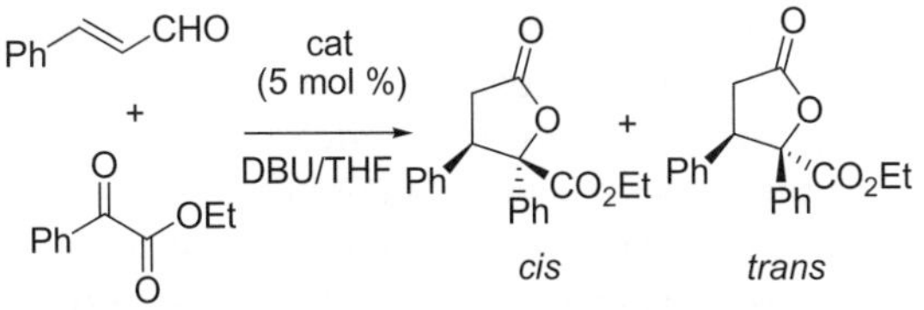

with cat 1:
85% *cis:trans* = 81:19 ee (*cis*) = ee (*trans*) = 14%
with cat 2:
74% *cis:trans* = 40:60 ee (*cis*) = 55% ee (*trans*) = 78%

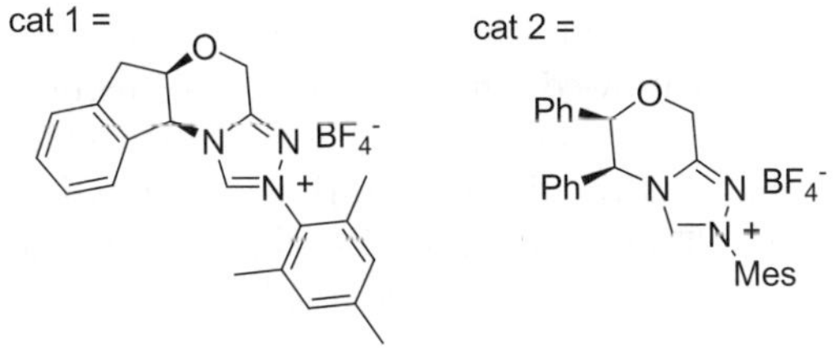

Scheme 10.15 Annulation of enal and keto ester.

halogen) bond-forming reactions that can be carried out in a catalytic and asymmetric manner.

In addition, chiral 4,5,5-trisubstituted γ-butyrolactones bearing two stereocentres including one quaternary carbon centre were synthesised by You *et al.* on the basis of chiral *N*-heterocyclic carbene-catalysed annulations of enals and keto esters.[21] The use of this type of chiral organocatalysts allowed tuning the

diastereoselectivity up to an 81:19 *cis/trans* ratio and the enantioselectivity up
to 78% ee, as shown in Scheme 10.15.

Conclusions

This chapter is very representative of the extensive efforts that have been
recently made for the development of novel organocatalysts applicable in a
broad variety of reaction types. Indeed, a great number of very different chiral
organocatalysts have been investigated in a wide range of reactions, such as
Friedel–Crafts reactions, Claisen rearrangements, [1,3]-sigmatropic rearrangements, Pictet–Spengler reactions, Steglich rearrangements, domino oxidative
dearomatisation-Michael reactions, alkene isomerisations, protonations of silyl
enol ethers, vinylations of aldehydes, carbo-oxidation of styrene and annulations of enals with keto esters.

Among the most successful examples is a Friedel–Crafts alkylation of
indoles with nitroalkenes promoted by a chiral phosphoric acid derived from
R-BINOL, which generated the corresponding Friedel–Crafts adducts with
high yields and excellent enantioselectivities of up to 94% ee for a broad range
of substrates.

In 2007, the groups of MacMillan and Sibi almost simultaneously introduced
a new mode of organocatalytic activation, termed SOMO (singly occupied
molecular orbital) catalysis, which was founded upon the transient production
of a 3π-electron radical cation species that could function as a generic platform
of induction and reactivity. This new mode of organocatalytic activation, was
founded upon the mechanistic hypothesis that one-electron oxidation of a
transient enamine intermediate, derived from the aldehyde and the chiral amine
catalyst, rendered a 3π-electron SOMO-activated species, which could readily
participate in asymmetric bond construction.

An application of this new mode of organocatalytic activation was developed by the same authors with the first asymmetric SOMO-catalysed carbo-
oxidation of styrenes to provide chiral γ-nitrate-α-alkyl aldehydes. This new
organo-SOMO reaction allowed simple styrenes to function as α-alkylation
partners for aldehydes. As highlighted in Scheme 10.14, a wide array of styrenes
readily participated as SOMOphiles in this new catalytic carbo-oxidation,
providing the corresponding γ-nitrate-α-alkyl aldehydes in excellent yields and
enantioselectivities of up to 97% ee. Indeed, the SOMO activation constitutes a
new highly promising strategy for organocatalysis which has great potential
given the numerous radical-based C–X (X = C, O, N, S, halogen) bond-forming
reactions that can be carried out in a catalytic and asymmetric manner.

References

1. S. B. Tsogoeva, *Eur. J. Org. Chem.*, 2007, 1701–1716.
2. D. J. Faulkner, *Nat. Prod. Rep.*, 2002, **19**, 1–49.

3. M. J. Wanner, P. Hauwert, H. E. Schoemaker, R. de Gelder, J. H. van Maarseveen and H. Hiemstra, *Eur. J. Org. Chem.*, 2008, **1**, 180–185.
4. J. Itoh, K. Fuchibe and T. Akiyama, *Ang. Chem., Int. Ed. Engl.*, 2008, **47**, 4016–4018.
5. H.-Y. Tang, A.-D. Lu, Z.-H. Zhou, G.-F. Zhao, L.-N. He and C.-C. Tang, *Eur. J. Org. Chem.*, 2008, 1406–1410.
6. K. C. Nicolaou, R. Reingruber, D. Sarlah and S. Bräse, *J. Am. Chem. Soc.*, 2009, **131**, 2086–2087.
7. C. Uyeda and E. N. Jacobsen, *J. Am. Chem. Soc.*, 2008, **130**, 9228–9229.
8. S. Kobbelgaard, S. Brandes and K. A. Jorgensen, *Chem. Eur. J.*, 2008, **14**, 1464–1471.
9. M. Chrzanowska and M. D. Rozwadowska, *Chem. Rev.*, 2004, **104**, 3341–3370.
10. I. T. Raheem, P. S. Thiara and E. N. Jacobsen, *Org. Lett.*, 2008, **10**, 1577–1580.
11. F. R. Dietz and H. Gröger, *Synlett*, 2008, **5**, 663–666.
12. N. T. Vo, R. D. M. Pace, F. O'Hara and M. J. Gaunt, *J. Am. Chem. Soc.*, 2008, **130**, 404–405.
13. L. Hintermann and M. Schmitz, *Adv. Synth. Catal.*, 2008, **350**, 1469–1473.
14. L. Duhamel, P. Duhamel and J. C. Plaquevent, *Tetrahedron: Asymmetry*, 2004, **15**, 3653–3691.
15. C. H. Cheon and H. Yamamoto, *J. Am. Chem. Soc.*, 2008, **130**, 9246–9247.
16. T. Poisson, S. Oudeyer, V. Dalla, F. Marsais and V. Levacher, *Synlett*, 2008, **16**, 2447–2450.
17. (a) G. J. Rowlands, *Chem. New Zealand*, 2008, **72**, 92–96; (b) S. Bertelsen, M. Nielsen and K. A. Jorgensen, *Ang. Chem., Int. Ed. Engl.*, 2007, **46**, 7356–7359.
18. (a) T. D. Beeson, A. Mastracchio, J. B. Hong, K. Ashton and D. W. C. MacMillan, *Science*, 2007, **316**, 582–585; (b) H.-Y. Jang, J.-B. Hong and D. W. C. MacMillan, *J. Am. Chem. Soc.*, 2007, **129**, 7004–7005; (c) M. P. Sibi and M. Hasegawa, *J. Am. Chem. Soc.*, 2007, **129**, 4124–4125.
19. H. Kim and D. W. C. MacMillan, *J. Am. Chem. Soc.*, 2008, **130**, 398–399.
20. T. H. Graham, C. M. Jones, N. T. Jui and D. W. C. MacMillan, *J. Am. Chem. Soc.*, 2008, **130**, 16494–16495.
21. Y. Li, Z.-A. Zhao, H. He and S. L. You, *Adv. Synth. Catal.*, 2008, **350**, 1885–1890.

General Conclusion

Asymmetric organocatalysis is one of the most important focal areas in organic synthesis. Extensive efforts have been made for the development of new advantageous chiral organocatalysts applicable in a broad variety of reaction types and their preparation continues to be an important area of synthetic organic research. The impressive number of reports dealing with the use of chiral organocatalysts for asymmetric synthesis is well representative of the success of such catalysts to promote numerous catalytic transformations.

This book clearly demonstrates the diversity and power of asymmetric organocatalysed reactions in the field of synthetic organic chemistry. Enantioselective organocatalytic processes have reached maturity in recent years with an impressive and steadily increasing number of publications, regarding the applications of these types of reactions, which paint a comprehensive picture for their real possibilities in organic synthesis. Even though transition-metal-catalysed enantioselective reactions will certainly continue to play a central role in synthetic organic chemistry in the future, the last years have, however, seen an increasing trend to the use of metal-free catalysts. The reasons are the often high costs of transition metals, the high effort for their preparations, the use of noxious metals, which, although present in trace amounts, contaminate the final organic product in particular for pharmaceutical products, the lack of orthogonality with a wide range of functional groups, and in some cases the need to operate under rigorously anhydrous or anaerobic conditions. In contrast, organocatalysts, some of which are natural products, appear to provide an answer to these problems. After an initial period of investigation on the scope of organocatalysis by using model systems, the time has now been reached where this approach, in combination with other modern reaction concepts and synthetic tools, can be applied to the construction of more sophisticated targets and can be used to address specific issues and solve pending problems of stereochemical relevance. The search for new organocatalysts is particularly important for the advancement of one of the central themes of modern organic synthesis, such as the creation of new structure

RSC Catalysis Series No. 3
Recent Developments in Asymmetric Organocatalysis
By Hélène Pellissier
© Hélène Pellissier 2010
Published by the Royal Society of Chemistry, www.rsc.org

classes with a wide range of chemical and stereochemical properties. In this context, organocatalysis is ideally suited since it allows the structure of basic molecular fragments to be modified efficiently. A number of organocatalytic reactions promoted by customised organocatalysts show great resemblance to enzymatic reactions. The application of chiral organocatalysts has permitted the preparation of a number of very valuable chiral products with the exclusion of any trace of hazardous metals and with several advantages from an economical and environmental point of view. The ultimate validation of any synthetic method is its successful application to the synthesis of structurally complex molecular targets, especially those of biological or pharmaceutical relevance. Organocatalysis appeared to have all the credentials for use in drug and natural product synthesis, and the first successes were achieved recently.[1] Today, the scope of organocatalysis spans from the generation of complex molecular systems to the consideration of technical synthesis processes, particularly in regard to environmentally friendly techniques. Mechanistic schemes and basic operational procedures have been established, thus giving great confidence in the success of many challenging endeavours that rely on organocatalysed key steps.

References

1. R. M. De Figueiredo and M. Christmann, *Eur. J. Org. Chem.*, 2007, 2575–2600.

Subject Index

Abbreviations used in the text can be found on pages ix–xi.

acetaldehyde
  aldolisation 92–4
  Mannich reactions 123–4, 127–9
acetone 47–50, 53
  aldol reactions with trifluoromethyl-
    ketones 79
  aldolisation 83–8, 91–3, 100–1
  dihydroxyacetones, aldolisations of
    94–6
  Mannich reaction with cyclic imine
    125–6
acetophenone, reduction of 206–7
acetyl migration, Steglich
  rearrangements 225
acetylene dicarboxylates, oxa-Diels–
  Alder reactions with 3-formyl-
  chromones 179–80
adamantane oxyls 217
aldehydes 3–13
  aldolisations of cyclic ketones 79–81,
    84–91
  alkylation with alcohols 160, 162
  allylations 107–10
  amination 152, 167–8
  α-aminoxylation of 152–5
  aza-Diels–Alder reactions with
    butadienes 178–9
  cycloadditions with
    enecarbamates 186–7
    Nazarov reagents 186–7
  Diels–Alder reactions with
    butadienes 176–7

diphenylprolinol-catalysed 3, 4
domino cyclopropanation-Wittig
  reactions 186–8
epoxidation 192–4
fluorination of α-alkyl α-
  chlorinated 159–60
Friedel–Crafts alkylations 222–3
α-hydroxyamination of 155–6
iodination 160, 162
oxa-Diels–Alder reactions with α,β-
  unsaturated trifluoromethyl
  ketones 179–80
synthesis of allenamides 168–9
transfer hydrogenations 209, 211
transition-state model 3, 6
vinylations with boron salts 228
aldol reactions, CO double bonds
  molecular sieve controlled 84–5
  nitroaldol reactions 105–8
  non-proline derivative catalysed
    94–105
  proline derivative catalysed 77–94
alkenes
  epoxidations of 192
  isomerisation 226
N-alkyl ketimines, reductions of 204–5
alkylation
  aldehydes with alcohols 160, 162
  N-diphenylmethylene glycine tert-
    butyl ester 160–4
  Morita–Baylis–Hillman carbonates
    164–5

allenamides, synthesis of 168–9
allenes, cycloadditions with imines
183, 185
allyl silanes, fluorination 159–60, 161
allyl vinyl ethers, Claisen
rearrangements 222–3
allylations
aldehydes 107–10
Morita–Baylis–Hillman acetates
164–5
amaminol B, synthesis of 176–7
amide, phosphine derived 140, 142
amination
carbonyl compounds 167–8
citronellal-derived aldehyde 152
α-cyanoketones 150–1
ethyl α-phenyl-α-cyano acetate
150–1
β-keto esters 152
amine catalysts, primary 15–16, 117
bispidine-derived 99, 100
quinine-derived 102
amine catalysts, secondary
amides 106
aza-Henry reactions 135–7
binapthyl-based 102–3
amine catalysts, tertiary 13–14
aza-Henry reactions 137–8
amine-thioureas 13–17, 48–51, 115–16
amino acids
as catalysts 99–100
derivatives, synthesis of 208–9
synthesis of 141–3
amino alcohols
epoxidation 194–5
peroxidation production of 197–8
amino carbonyl compounds 123–4
amino phosphonate 12–13, 44–5
amino sulfonamide 87, 129–30
aminosulfenylation 62–4
aminoxylation of aldehydes 152–5
annulation, enal and keto esters 229–30
anthrones, Diels–Alder reactions of
174–5
arsonium ylides, cyclopropanation-
Wittig reactions with α,β-
unsaturated aldehydes 186–8

aryl enamino esters, hydrosilylations
of 208–9
aryl ethanols, kinetic resolutions of
213–14
aryl ketimines, reductions of 202–5
aryl-propionates
kinetic resolutions of 213–14
2-aza-adamantane *N*-oxyls 217
aza-Henry reactions 133–9, 146
aza-Michael additions 58–9, 68–9
aza-Michael-Michael reaction 62–3
aza-Morita–Baylis–Hillman reactions
139–41, 146
aziridine, ring-opening of 165–6
1,2,4-triazolium salts 110–11
azomethine ylide, cycloadditions to
nitroalkenes 181–3

Baeyer–Villiger oxidation 199–200
benzoin condensations 110–11
benzotetramisole 213
biarylprolinol ethers 27–8
Biginelli reactions 116–17
BINAM 48, 137
BINOL-derived catalysts 110, 116–17,
138, 144, 164–5, 181, 227
biphenylprolinol catalysts 3–6, 8, 27–32
bisindole, synthesis of 166–7
boronic acids 62, 63
butadienes, Diels–Alder reactions
with aldehyde 176–7, 178–9
butenolide derivatives, synthesis of
134–6
butyldiphenylsilyl ether 42–3
butyrolactones, syntheses of 78

carbonyl compounds and derivatives
α-alkylation 160–7
α-amination 167–8
epoxidations 192–6
α-halogenation 158–60
carbo-oxidations of styrenes 228–9
carboxylic acid derived catalysts
126–8, 206
CC double bonds, electron-deficient
1–70

chalcone, epoxidation of 194–5
cholic acids 43–4
chromene synthesis, domino Michael-
    aldol reactions 35–6
cinchona alkaloids and derivatives
    aldolisations 89–90
    alkene isomerisation 226
    asymmetric desymmetrisation 217
    aza-Henry reactions 137–9
    Diels–Alder reactions 172–4
    hydrophosphonylation of
        sulfonylimines 206–7
    Mannich reactions 134, 136
    Michael addition, C-nucleophiles
        15–28, 48, 53, 54–6
    Michael addition, non C-
        nucleophile 60–1, 64–6
    nitroaldolisation 106–8
    O- to *N*-rearrangements 222–4
    β-peroxidation 197–9
    vinylic substitutions 164
Claisen rearrangements, allyl vinyl
    ethers 222–3
CN double bonds, nucleophilic
    additions to 123–47
    aza-Henry reactions 133–9, 146
    aza-Morita–Baylis–Hillman
        reactions 139–41, 146
    Mannich reactions 123–35, 145–6
    Strecker reaction 141–5, 146–7
C-nucleophiles
    domino Michael additions 26–37
    domino nitro-Michael additions
        54–8
    nitro-Michael additions 37–54
    non-proline derived catalysed
        Michael additions 13–26
    proline catalysed Michael additions
        3–13
CO double bonds, nucleophilic
    additions to 77–118
    aldol reactions
        non-proline derivative catalysed
            94–105
        proline derivative catalysed 77–94
    allylations of aldehydes 107–10
    benzoin condensations 110–11

Biginelli reactions 116–17
    hydrophosphonylation of α-keto
        esters 110–11
    Morita–Baylis–Hillman reactions
        32, 33, 110–14
    nitroaldol reaction 105–8
    Petasis reactions 114–16
α-cyanoketones, amination of 150–1
cyclic anhydrides, desymmetrisations
    of 215, 218
cyclisation, intramolecular 125–6
cycloaddition reactions 172–89
    Diels–Alder reactions 172–9
    1,3-dipolar additions 179–86
    domino cyclopropanation-Wittig
        reactions 186–8
    double reactions 181–7
    oxa-Diels–Alder reactions 179–80
cycloalkanols, kinetic resolutions of
    213, 215
cyclobutanones, Baeyer–Villiger
    oxidation 199–200
cyclohexane derived catalysts 15, 35,
    42, 48, 50
    diaminocyclohexanes 94–6, 98,
        117–18
cyclohexanone 41–7, 50, 117
    aldolisation 81–3, 85–6, 90–1, 95–9,
        102–3
    cross-aldol reactions with keto
        esters 78–9
2-cyclohexen-1-one, Diels–Alder
    reaction 78–9
cyclohexyldiamine 94–8, 117
cyclopentanone 47

DABCO 113–14
desymmetrisations 213–18
diaminocyclohexanes 94–6, 98
diarylprolinol and derivatives
    aldehyde additions 3, 10, 59–63, 92
    aldehyde fluorination 159–60
    asymmetric Mannich reaction 128–30
    domino Michael reactions 33
    intramolecular aza-Michael
        additions 68–9, 70
    oxidative dearomatisation 226

dicarboxylates, oxa-Diels–Alder
    reactions with 3-formyl-
    chromones 179–80
dicyanoolefin additions 9, 10
Diels–Alder reactions
    anthrones 174–5, 177–8
    aza-Diels–Alder reactions 178–9
    butadienes with aldehydes 176–7,
        178–9
    cinchona alkaloid-catalysed 172–4
    guanidine-catalysed 174–5
    hydroquinine-derived thiourea
        catalysed 172–3
    intramolecular 176
    maleimides 177–8
    oxa-Diels–Alder reactions 179–80
    2-pyrones 172–4
    sulfonylhydrazine-catalysed 172–3
    synthesis of amaminol B 176–7
    in water 175–6
difluoro-hydroxy-aryl-propionate,
    kinetic resolutions of 213–14
dihydroxyacetones, aldolisations of
    94–6
diltiazem, synthesis of 196, 197
4-(dimethylamino)pyridine (DMAP)
    112–13
N-diphenylmethylene glycine tert-
    butyl ester, alkylation of 160–4
diphosphine oxides 108–9
DMAP 112–13
DMF 81–3, 102
domino aldol-oxa-Michael reactions 94
domino α-aminoxylation-aza-Michael
    reactions 152–3
domino Michael-aldol reactions
    addition-α-alkylation reaction 29, 32
    addition-cyclisation reactions 29, 32
    chromene synthesis 35–6
    Knoevenagel reactions 33, 34
    Morita–Baylis–Hillman mechanism
        32, 33
    1,4-dihydropyridine synthesis 33–5
    quinone synthesis 35–6, 37
    silylated biphenylprolinol-catalysis
        27–32
    triple cascade aldol reactions 33, 34

domino nitro-Michael additions
    asymmetric double Michael
        reaction 56–8
    C-nucleophiles 54–8
    Michael–Henry reactions 55, 57–8
    Michael–Michael-aldol reactions
        55–6
    three-component additions 55–6
domino oxidative dearomatisation-
    Michael reaction. 226
domino thia-Michael-aldol reactions
    64–5

eicosanoid, synthesis of 187 8
enals
    annulation of esters 229–30
    intramolecular hydroarylation 68–9
enamines
    catalysis 1–2
    hydrosilylations of esters 208–9
    reductions of 207–8
enecarbamates, cycloadditions with
    α,β-unsaturated aldehydes 186–7
enol ethers, protonation 227
enones
    α,β-unsaturated 15–17
    cyclic, epoxidations of 195–6
epiquinine 15–16
epoxidations
    alkenes 192–3
    α,β-unsaturated carbonyl compounds
        192–6
ethyl α-phenyl-α-cyano acetate,
    amination of 150–1
ethyl malonate 132
ethyl vinyl ether, 1,3-dipolar
    cycloadditions to nitrones 181–2

FBSM 21–2, 27
fluorination
    α-alkyl α-chlorinated aldehyde
        159–60
    allyl silanes 159–60, 161
    oxindoles 159–60, 161
    silyl enols 159–60, 161
fluoroamines, synthesis of 158–9

formylchromones, oxa-Diels–Alder
    reactions with acetylene
    dicarboxylates 179–80
Friedel–Crafts reactions 220–3, 230

GABA analogues 32
glutaric anhydride, desymmetrisations
    of 213–15, 216
guanidine catalysis
    Diels–Alder reactions 174–5
    Michael reactions 24–5, 29
    sulfa-Michael reactions 59, 61
guanidine-thiourea catalysis 106–7

halogenation, carbonyl compounds
    158–60
3-halo-3-pyrrolidin-2-ones, synthesis
    of 164
Henry reactions
    aza-Henry reactions 133–9, 146
    Michael–Henry reactions 55, 57–8
    thia-Michael–Henry reactions 67–8
Horner–Wadsworth–Emmons
    olefination 4, 7
hydrazonoesters 132–3
hydrophosphonylation, keto esters
    110–11
hydroxyamination of aldehydes 155–6
3-hydroxyl-2-alkanones, synthesis of
    153–4
hydroxylations of oxindoles 169–70

imides, α,β-unsaturated 13–14
imines
    aza-Henry reactions 135–8
    aza-Morita–Baylis–Hillman
        reaction 139–41
    cycloadditions
        with allenes 183, 185
        with ketenes 186
        with ketones 184–5
    hydrophosphonylation of
        sulfonylimines 206–7
    Mannich reactions 123–35
    reductions 202–7
iminium catalysis 1–2

indoles 26, 30, 53
    Friedel–Crafts alkylations 220–2
    oxindoles 134, 159–60, 161, 169–70
    vinyl indoles 172–3
iodinations of aldehyde 160, 162
isatins 89
isobutyric anhydride 213
isochromans, synthesis of 166–7
isocupreidine derivatives 139–41
isocyanoesters, cycloadditions to
    nitroolefins 183–4

Janus kinase inhibitor 59
jaspine B, synthesis of 77–8

ketenes, cycloadditions with imines 186
ketimines, reductions of 202–5
keto esters
    annulation of 229–30
    cross-aldol reactions with cyclo-
        hexanone 78–9
    domino Michael additions 30, 32, 35
    hydrophosphonylation 110–11
    β-keto esters, amination of 152
    Michael additions 18, 20, 22–6, 28,
        48, 50, 53–4
    as Michael donors 13
    nitroaldolisation 106–7
ketones
    acetone 47–50, 53, 79, 83–8, 91–3,
        100–1, 125–6
    aldolisations of 99–101, 103–5
        cyclic 79–81, 84–90
    aza-Michael reactions 21–4, 27, 58–9
    α-cyanoketones, amination of 150–1
    cyclic imines, reactions with 125–6
    cycloadditions with imines 184–5
    cyclohexanone 41–7, 50
    cyclopentanone 47
    diketones 33–5
    epoxidation 194–5
    fluoromethyl ketones 106–8
    Michael additions 12–15, 17–24
    oxa-Diels–Alder reactions with
        aldehydes 179–80
    β-peroxidation 197–9

reductions 202–7
vinyl ketones 23–4
see also specific ketones
α-ketophosphonates 105–6
kinetic resolution 213–18
Knoevenagel reactions, domino
    Michael additions 33, 34, 66–7

lactams, syntheses of 4, 7
lactones
    kinetic resolutions of azalactones
        215, 217
    syntheses of 4, 7, 78
Lycopodium alkaloids, cernuane-type,
    syntheses of 152

maleimides, Diels–Alder reactions
    177–8
malonates
    β-formyl-substituted 3, 4
    diphenylprolinol-catalysed
        additions 5–6, 8
    pyrrolidinyl tetrazole-catalysed
        additions 10–11, 12
malononitriles 14–15, 17, 20–1
Mannich reactions 123–35, 145–6
    nitro-Mannich (*see* aza-Henry
        reactions)
Merrifield resin-supported dipeptide
    catalysts 86–7
methyl *tert*-butyl ether (MTBE)
    catalysis 216
Michael reactions
    aza-Michael additions 58–9, 68–9
    aza-Michael-Michael reaction 62–3
    C-nucleophiles
        domino nitro-Michael additions
            54–8
        intermolecular domino additions
            26–37
        intermolecular nitro-Michael
            additions 37–54
        non-proline derived catalysis 13–26
        proline catalysis 3–13
    domino aldol-oxa-Michael reactions
        94

domino α-aminoxylation-aza-
    Michael reactions 152–3
domino oxidative dearomatisation-
    Michael reaction. 226
enamine catalysis 1–2
guanidine-catalysed 24–5, 29
iminium catalysis 1–2
intramolecular additions 68–9
organoboronic acids 62, 63
oxa-Michael additions 60–1
phase-transfer catalysis 1–2
sulfa-Michael reactions 59, 61
thia-Michael-aldol reactions 64–5
thia-Michael–Henry reactions 67–8
thia-Michael–Knoevenagel reaction
    66–7
thia-Michael–Michael reactions 65–6
thioester 6–7
N-triflylphosphoramide-catalysed
    25–6, 30
Morita–Baylis–Hillman reactions
    acetates, allylic substitutions of 164–5
    aliphatic aldehydes 110–14
    aza-Morita–Baylis–Hillman reactions
        139–41, 146
    carbonates, allylic-allylic alkylations
        of 164–5
    domino Michael additions 32, 33
MTBE catalysis 216

Nazarov reagents, cycloadditions
    with aldehydes 186–7
nitroacrylates, reductions of 209, 210
nitroaldolisation 105–8
nitroalkanes 12–13, 15, 17–19
nitroalkenes
    cycloadditions to azomethine ylide
        181–3
    intermolecular nitro-Michael
        additions 37–54
    β-peroxidation 197–8
nitromethane 105–6
nitro-Michael intermolecular
    additions, C-nucleophiles 37–54
nitrones, 1,3-dipolar cycloadditions
    ethyl vinyl ether 181–2
    nitroolefins 179–81

nitroolefins
  cycloadditions to isocyanoesters
    183–4
  1,3-dipolar cycloadditions to nitrones
    179–81
(E)-5-iodo-1-nitropent-1-ene 55
nitrostyrene 38–9
NN double bonds, nucleophilic
  additions to 150–2
NO double bonds, nucleophilic
  additions to 152–6
nucleophilic additions to double
  bonds
  CC electron-deficient bonds 1–70
  CN bonds 123–47
  CO bonds 77–118
  NN bonds 150–2
  NO bonds 152–6
nucleophilic substitutions, aliphatic
  carbon 158–70

O- to *N*-rearrangements 222–4
octahydrobinol phosphoric acid 220–2
organoboronic acids 62, 63
oxa-Michael additions 60–1
oxazolone additions 9–10, 11
oxidations
  Baeyer–Villiger oxidation 199–200
  carbo-oxidations of styrenes 228–9
  electrooxidation 199–200
  epoxidations of α,β-unsaturated
    carbonyl compounds 192–6
  epoxidations of alkenes 192–3
  β-peroxidation 196–9
oxindoles 134
  fluorination 159–60, 161
  α-hydroxylations of 169–70

β-peroxidation 196–9
Petasis reactions 114–16
phase transfer catalysis 1–2, 21–3, 27,
  68–9, 162–6
phenylcyclohexene, epoxidation of
  192–3
phosphine amide 140, 142
diphosphine oxides 108–9

phosphinothiourea catalysis 113–14
phosphoric acid-catalysis
  aza-Henry reactions 138–40
  vinylogous Mannich reactions 134,
    136
Pictet–Spengler reaction 224–5
PMP 126–32
polysiloxane derived catalysts 97–9
prolinamide derivatives 43–4, 83–6, 144
proline catalysis 117
  aldol reactions in CO double bonds
    77–94
  α-amination of carbonyl
    compounds 167–8
  α-aminoxylation of aldehydes 152–3
  C-nucleophile Michael reactions 3–13
  Mannich reactions 123–6
  Morita–Baylis–Hillman reactions
    110–13
  Strecker reactions 141–3
prolinethioamide 90–1
cis-1-propenylphosphoric acid,
  epoxidation of 196, 197
propionates, kinetic resolutions of
  213–14
protonation, silyl enol ethers 227
4-(dimethylamino)pyridine (DMAP)
  112–13
pyridine *N*-oxide and *N,N'*-dioxide
  catalysts 107–9
dihydropyridine synthesis 33–5, 38
2-pyrones, Diels–Alder reactions of
  172–4
pyrroles, cyclisations of 224–5
pyrrolidine catalysts 3–4, 6, 9, 12, 14,
  34, 40–7
  chiral 40–5, 62–4
pyrrolidine-carboxylic acid catalysis
  126–8
pyrrolidinones, synthesis of 164
pyrrolidinyl-camphor thiourea-amine
  90–1

quinolines 115
quinones 20–1, 26
  synthesis, domino Michael-aldol
    reactions 35–6, 37

reductions
β-nitroacrylates 209, 210
enamines 207–8
imines and ketones 202–7
transfer hydrogenations of
aldehydes 209, 211
ring-opening of aziridine 165–6

serine derivative 131–2
silyl enols
ether protonation 227
fluorination 159–60, 161
SNAP-7941 synthesis 116–17, 133–4
SOMO catalysts 228, 230
Steglich rearrangement reaction 225
Strecker reaction 141–5, 146–7
styrenes, carbo-oxidations of 228–9
succinic anhydride, desymmetrisations of 213–15, 216
sulfa-Michael reactions 59–61
sulfonamide catalysts 87–9
N-sulfonyl-1-aza-1,3-butadienes,
Diels–Alder reactions with
aldehyde 176–7, 178–9
sulfonylhydrazine-catalysed Diels–
Alder reactions 172–3
sulfonylimines,
hydrophosphonylation of 206–7
sulfoxide catalysts 109–10

TBDPS 42–3
TEA 25, 29
tetrahydro-1,2-oxazines, synthesis of
154–5
tetraMe-BITIOPO 108–9
tetrazole catalysts 116–17
thia-Michael-aldol reactions 64–5
thia-Michael–Henry reactions 67–8
thia-Michael–Knoevenagel reaction
66–7

thia-Michael–Michael reactions 65–6
thioesters, Michael additions 6–7
thiourea catalysis 13–19, 48–51
amine-thioureas 115–16
aza-Henry reactions 135–8, 144–5,
146
Diels–Alder reactions 172–3
Mannich reactions 132–4
Morita–Baylis–Hillman reactions
112–14
phosphinothiourea 113–14
primary amine catalysis 15–16
Strecker reactions 144–5
tertiary amine catalysis 13–14
threonine derivatives 131
thysanone, synthesis of 153–4
trichloro isocyanuric acid (TCCA)
213, 215
trifluoro-aryl ethanols, kinetic
resolutions of 213–14
N-triflylphosphoramide-catalysed
Michael reactions 25–6, 30
trifunctional catalyst, Liu and Garnier
141, 143
triple cascade domino Michael-aldol
reactions 33, 34

(S)-VAPOL-catalysis 114–15
vinyl
aldehydes 228
bisphosphonates 12–13, 14, 17–18, 22
indoles 172–3
ketones 23 4
sulfones 3–6, 15, 18–19
vinylic substitutions 164–5
vinylogous Mannich reactions 134, 136

water, Diels–Alder reactions in 175–6

xyaldehydes 124